PRELIMINARY EDITION

Organic Chemistry

A TWO-SEMESTER COURSE OF ESSENTIAL
ORGANIC CHEMISTRY

BY VIKTOR ZHDANKIN AND PETER GRUNDT

UNIVERSITY OF MINNESOTA - DULUTH

Bassim Hamadeh, CEO and Publisher
Kassie Graves, Director of Acquisitions
Jamie Giganti, Senior Managing Editor
Miguel Macias, Senior/ Graphic Designer
Amy Stone, Acquisitions Editor
Sean Adams, Project Editor
Luiz Ferreira, Licensing Coordinator

Copyright © 2018 by Cognella, Inc. All rights reserved. No part of this publication may be reprinted, reproduced, transmitted, or utilized in any form or by any electronic, mechanical, or other means, now known or hereafter invented, including photocopying, microfilming, and recording, or in any information retrieval system without the written permission of Cognella, Inc.

Trademark Notice: Product or corporate names may be trademarks or registered trademarks, and are used only for identification and explanation without intent to infringe.

Cover image copyright © 2013 iStockphoto LP/Pixx.

Printed in the United States of America

ISBN: 978-1-63487-897-5 (pbk) / 978-1-63487-898-2 (br)

CONTENTS

Preface — xi

CHAPTER 1
Covalent Bonding and Structure of Molecules — 1

1.1 Electronic Structure of Atoms and Lewis Model of Bonding .. 2
  Problems 7
1.2. Lewis Structures of Molecules and Molecular Ions 8
  Problems 13
1.3 Functional Groups and Introduction to the
IUPAC Nomenclature 14
  Problems 18
1.4. Shape of Molecules 19
  Problems 22
1.5. Polarity of Molecules and Physical Properties of Organic
Compounds 23
  Problems 25
1.6. Resonance 25
  Problems 28
1.7. Introduction to Molecular Orbitals and Valence
Bond Theory 29
  Problems 35

CHAPTER 2
Proton Transfer Reactions in Organic Chemistry — 37

2.1. Definition of Acids and Bases 37
2.2. Acid-Base Equilibrium and Relative Strength of Acids 39
  Problems 42
2.3. Effects of Molecular Structure on Acidity 43
  Problems 46

CHAPTER 3
Alkanes and Cycloalkanes 49

- 3.1. Structure of Alkanes and Constitutional Isomerism 49
  - *Problems* *52*
- 3.2. Nomenclature of Alkanes 52
  - *Problems* *53*
- 3.3. Conformations of Alkanes 54
  - *Problems* *56*
- 3.4. Cycloalkanes and Ring Conformations 57
  - *Problems* *60*
- 3.5. *Cis, Trans* Isomerism in Substituted Cycloalkanes 61
  - *Problems* *63*
- 3.6. Radical Reactions: Oxidation and Halogenation of Alkanes .. 63
  - *Problems* *67*

CHAPTER 4
Stereochemistry 69

- 4.1. Constitutional Isomers and Stereoisomers 69
- 4.2. Chiral Molecules and R,S-Nomenclature of Enantiomers 71
  - *Problems* *77*
- 4.3. Molecules with More than One Chiral Center 78
  - *Problems* *81*
- 4.4. Properties of Enantiomers and Diastereomers 82

CHAPTER 5
Nucleophilic Substitution and β-Elimination Reactions 85

- 5.1. Overview of Nucleophilic Substitution and β-Elimination Reactions 86
  - *Problems* *89*
- 5.2. S_N2 and S_N1 Mechanisms of Nucleophilic Substitution in Haloalkanes 91
  - 5.2.1. S_N2 mechanism 91
  - 5.2.2. S_N1 mechanism 93
    - *Problems* *96*
- 5.3. Stereochemistry of S_N2 and S_N1 Reactions 98
  - *Problems* *99*
- 5.4. E2 and E1 Mechanisms of β-Elimination Reactions 100
  - *Problems* *105*

CHAPTER 6
Alkenes 107

- 6.1. Structure and Nomenclature of Alkenes 107
  - *Problems* *111*

6.2. Electrophilic Addition: Reactions of Alkenes with
Strong Acids 112
    Problems.. *114*
    6.3. Electrophilic Addition of Cl$_2$ or Br$_2$ *116*
    Problems.. *118*
6.4. Hydroboration of Alkenes 119
    Problems.. *121*
6.5. Oxidation and Reduction of Alkenes 122
    6.5.1. Dihydroxylation of alkenes 122
    6.5.2. Oxidation of alkenes with peroxycarboxylic acids:
    Preparation of epoxides 123
    6.5.3. Oxidative cleavage of double bond: Ozonolysis
    of alkenes 125
    6.5.4. Reduction of alkenes: Catalytic hydrogenation......... 127
    Problems.. *127*
6.6. Radical Reactions of Alkenes 129
    6.6.1. Allylic halogenation of alkenes 130
    6.6.2. Radical addition to double bond 131
    6.6.3. Radical polymerization of alkenes: Polymers 132
        Problems *135*

CHAPTER 7
Alkynes   137

7.1. Structure and Nomenclature of Alkynes........................... 137
    Problems.. *139*
7.2. Preparation of Alkynes by β-Elimination Reactions 139
    Problems.. *141*
7.3. Electrophilic Addition Reactions of Alkynes 142
    7.3.1. Electrophilic addition of halogens......................... 142
    7.3.2. Addition of hydrogen halides 143
    7.3.3. Acid-catalyzed hydration of alkynes 144
    7.3.4. Hydroboration-oxidation of alkynes...................... 145
    Problems.. *146*
7.4. Reduction of Alkynes 147
    Problems.. *149*
7.5. Introduction to Organic Synthesis...................................... 149
    Problems.. *150*

CHAPTER 8
Alcohols   151

8.1. Classification, Nomenclature, and Physical Properties of
Alcohols... 151
    Problems.. *153*

8.2. Acidity and Basicity of Alcohols *154*
    Problems *154*
8.3. Reactions of Alkoxides: Preparation of Ethers *155*
    Problems *156*
8.4. Conversion of Alcohols to Alkyl Halides *157*
    Problems *158*
8.5. Acid-Catalyzed Dehydration *159*
    Problems *160*
8.6. Preparation and Reactions of Alkyl Sulfonates *161*
    8.6.1. Thiols and sulfonic acids *161*
    8.6.2. Alkyl tosylates *162*
        Problems *164*
8.7. Oxidation of alcohols *164*
    Problems *167*

CHAPTER 9
Spectroscopy of Organic Compounds   169
9.1. Infrared Spectroscopy *169*
9.2. NMR Spectroscopy *171*
    9.2.1. Overview of the underlying concepts of
    NMR spectroscopy *171*
    9.2.2. Basic information derived from NMR spectroscopy ... *172*
        A. Number of signals *173*
            Problems *176*
        B. Integration *176*
            Problems *177*
        C. Signal splitting *177*
            Problems *180*
        D. Chemical shift *181*
            Problems *184*
9.3. Mass Spectrometry *186*
    9.3.1. The Mass Spectrometer *186*
    9.3.2. Mass spectra *187*
        Problems *188*

CHAPTER 10
Organometallic Compounds and Transition Metal Catalysis   193
10.1. Organolithium and Organomagnesium Compounds *193*
    Problems *195*
10.2. Compounds of Transition Metals *195*
    10.2.1. Organocopper compounds *195*
        Problems *197*
    10.2.2. Palladium-catalyzed coupling reactions *198*
        Problems *200*

    10.2.3. Carbene intermediates and the Simmons-Smith reaction 200
        Problems *203*
    10.2.4. Alkene metathesis 203
        Problems *205*

CHAPTER 11
Aldehydes and Ketones   207

   11.1. Nomenclature of Aldehydes and Ketones 208
       Problems *209*
   11.2. General Characteristics of Aldehydes and Ketones 209
       Problems *210*
   11.3. Reactions of Aldehydes and Ketones with Carbon Nucleophiles 211
       Problems *212*
   11.4. Reaction with Hydride Anion 213
       Problems *214*
   11.5. Reaction with Phosphonium Ylides (the Wittig Reaction) ... 214
       Problems *216*
   11.6. Reaction with Amines and other Nitrogen Nucleophiles 217
       Problems *220*
   11.7. Reaction with Water and Alcohols 221
       Problems *224*
   11.8. Oxidation of Carbonyl Compounds 225
       Problems *226*
   11.9. Enols and Enolate Anions 227
       Problems *229*
   11.10. Reaction of Enolate Anion as Nucleophile: Aldol Condensation 229
       Problems *230*

CHAPTER 12
Carboxylic Acids and their Derivatives   233

   12.1. Nomenclature of Carboxylic Acids and their Derivatives 233
       Problems *236*
   12.2. Acidity and Physical Properties of Carboxylic Acids and their Derivatives 236
       Problems *238*
   12.3. General Principles of Reactivity of Acyl Derivatives: Nucleophilic Acyl Substitution Reactions 239
       Problems *240*
   12.4. Preparation and Interconversion of Acid Derivatives 241
      12.4.1. Acid chlorides 241
        Problems *241*

   12.4.2. Anhydrides 241
   12.4.3. Esters 244
     Problems *244*
   12.4.4. Amides 247
     Problems *249*
 12.5. Reaction with Hydride Anion 249
   Problems *253*
 12.6. Reaction with Carbanions 254
   Problems *254*
 12.7. Reaction of Enolate Anions Derived from Esters: Claisen Condensation and Related Reactions 256
   Problems *258*
 12.8. Polyesters and Polyamides 260
   Problems *261*

CHAPTER 13
Non-cyclic Conjugated Systems   263

 13.1. Structure and Properties of Conjugated Systems 263
   Problems *266*
 13.2. Conjugate Electrophilic Addition Reactions of Dienes 266
   Problems *267*
 13.3. Conjugate Nucleophilic Addition Reactions of α,β-Unsaturated carbonyl compounds 268
   Problems *269*
 13.4. Diels-Alder Cycloaddition Reaction 270
   Problems *273*

CHAPTER 14
Benzene and Aromatic Compounds   275

 14.1. Structure of Benzene and the Concept of Aromaticity 276
   Problems *280*
 14.2. Nomenclature of Benzene Derivatives 281
   Problems *283*
 14.3. Electrophilic Aromatic Substitution Reactions of Benzene . 284
   14.3.1. Halogenation 286
   14.3.2. Nitration 286
   14.3.3. Sulfonation 287
   14.3.4. Friedel-Crafts alkylation and acylation 288
 14.4. Electrophilic Aromatic Substitution Reactions of Substituted Benzenes 290
   Problems *294*
 14.5. Nucleophilic Aromatic Substitution Reactions 296
   14.5.1. Addition/elimination mechanism 296

14.5.2. Elimination/addition mechanism via benzyne
intermediate 297
Problems *299*
14.6. Reactions of Alkylbenzenes at the Benzylic Position 300
Problems *302*
14.7. Phenols 303
Problems *305*

CHAPTER 15
Amines   307

15.1. Classification and Nomenclature 307
Problems *310*
15.2. Physical Properties and Basicity of Amines 311
  15.2.1. Basicity of alkylamines 311
  15.2.2. Basicity of arylamines 313
  15.2.3. Basicity of heterocyclic amines 313
    Problems *314*
15.3. Synthesis of Amines 316
Problems *320*
15.4. Hofmann Elimination 321
Problems *322*
15.5. Aryldiazonium Salts 323
  15.5.1. Synthesis of aryl chlorides, aryl bromides and aryl
  cyanides from diazonium salts (the Sandmeyer reaction).... 325
  15.5.2. Synthesis of aryl iodides from diazonium salts....... 326
  15.5.3. Synthesis of aryl fluorides from diazonium salts (the
  Balz-Schiemann reaction) 327
  15.5.4. Synthesis of phenols from diazonium salts 327
  15.5.5. Replacement of the diazo group with hydrogen 328
  15.5.6. Diazonium salts as electrophiles in electrophilic
  aromatic substitution reactions 328
Problems *329*

CHAPTER 16
Introduction to Biomolecules   331

16.1. Lipids 331
Problems *333*
16.2. Carbohydrates 333
Problems *334*
  16.2.1. General classification and nomenclature
  of monosaccharides 334
    Problems *336*
  16.2.2. Cyclic structures of monosaccharides 336
    Problems *338*

  16.2.3. Chemical reactions of monosaccharides *339*
    Problems 340
  16.2.4. Disaccharides and polysaccharides *340*
    Problems 343
 16.3. Amino acids, Peptides, and Proteins *343*
    Problems... 347
 16.4. Nucleosides, Nucleotides, and Nucleic acids *347*
    Problems... 350

Glossary   351

Preface

This textbook is intended as a basic text for a full year organic chemistry lecture course for science majors at a sophomore/junior level. The book is based on lectures taught by the authors at the University of Minnesota, Duluth for many years, and it covers all essential material within the requirements outlined by the American Chemical Society (ACS). The key organic chemistry concepts are delivered in a systematic and condensed manner, and the book provides excellent preparation for standardized ACS exams, MCAT, PCAT, Chemistry GRE, and other professional proficiency exams. Multidisciplinary researchers can also use it as a basic reference book covering essential concepts, definitions, and nomenclature of organic chemistry.

The authors thank to Dr. Sangeeta Mereddy for her valuable contribution to the preparation of this book.

CHAPTER 1

Covalent Bonding and Molecular Structure

Organic chemistry is the chemistry of the element carbon. Carbon is one of the four most common elements in the universe, along with hydrogen, helium, and oxygen. In the human body, carbon is the second most abundant element after oxygen. Carbon atoms are present in all organic molecules; the other most common elements in organic compounds include hydrogen, oxygen, and nitrogen. Atoms of these elements are the main building units of biological molecules, along with the less-abundant atoms of phosphorus, sulfur, and numerous trace elements. Carbon occupies a central place in the periodic table, in the middle of the second period, and it is the only light element that can form four strong bonds to other elements in electrically neutral molecules. The bonding pattern of carbon makes possible the existence of a virtually unlimited array of structurally diverse organic molecules. Owing to these unique features, carbon became the basic element of life, and knowing the principles of carbon chemistry is essential for understanding biology at the molecular level. Carbon compounds play an irreplaceable role in our everyday lives, as a component of food, pharmaceuticals, fuels, wood- and plastic-based construction materials, and many other commercial products. This chapter explains how organic molecules form from the atoms of carbon and other elements, and how understanding the fundamental principles of chemical bonding is essential for further study of organic chemistry.

1.1. ELECTRONIC STRUCTURE OF ATOMS AND THE LEWIS MODEL OF BONDING

All of the matter surrounding us is built from atoms. An **atom** is the basic unit of a chemical element and the smallest particle of a substance that can exist by itself or can be combined with other atoms to form a molecule. The atom consists of a positively charged small **nucleus** surrounded by negatively charged electrons. The nucleus is composed of positively charged **protons** and uncharged **neutrons**. The number of protons is equal to the number of electrons in an electrically neutral atom. This is the **atomic number**, and is one of the basic principles of organization for the periodic table. The combined number of protons and neutrons in a nucleus corresponds to the **atomic mass** of an element; **isotopes** of an element have different masses because they differ in the number of neutrons. Electrons in atoms occupy **electronic shells**, and the number of electronic shells in atom determines the period (represented as a row) the element belongs to in the periodic table. Electrons of the inner shell are not involved in chemical reactions, while the outer (**valence**) electrons can participate in bonding with other atoms. The overall size of an atom is defined by the size of its outer electronic shell and can vary from 0.6 Å to several Å (Å is angstrom, 1Å = 10^{-10} meters or 100 picometers, pm). The size of the nucleus is about 10,000 times smaller than atom itself. The outer electron shells of atoms shape the matter that we see around us. For example, if all electronic shells in our body were forced to collapse to the size of a nucleus, an average person would be reduced to a single dust particle with about 0.1 mm in diameter. This would be a very heavy dust, however, keeping the weight of the original body.

Carbon has an atomic number of 6 and it belongs to group 14 of the second period, which means that it has 6 protons in the nucleus and 6 electrons in two electronic shells. The first (inner) shell is filled with a pair of electrons, which corresponds to a very stable electronic configuration of the noble gas helium, and the second (outer or valence) shell has four electrons. Natural carbon is a mixture of two stable, non-radioactive isotopes: the major isotope, ^{12}C, has 6 protons and 6 neutrons, and the minor isotope, ^{13}C, has 6 protons and 7 neutrons in the nucleus. Both carbon isotopes have the same electronic configuration and are almost identical in chemical properties.

Molecules represent the smallest fundamental units of a chemical compound responsible for its physical properties and chemical reactions. Molecules are formed from atoms by sharing valence electrons. In order to understand how atoms combine together in a molecule, we need to review the electronic configuration of elements in the first two periods of the periodic table. Electrons are filling available shells according to the basic principles of quantum mechanics. The first electronic shell can hold maximum 2 electrons; the second shell can have no more than 8 electrons, the third and the fourth shells - 18 and 32 electrons. Electronic shells are further divided into subshells or **atomic orbitals**, which are defined as areas of space where electrons can be localized with the highest probability according to the laws of quantum mechanics. The first shell contains only one s orbital (1s orbital) and the second shell has one s orbital (2s orbital) and three p orbitals (2p$_x$, 2p$_y$, and 2p$_z$) arranged perpendicular (orthogonal) to each other along imaginary axis x, y, and z (Figure 1.1). The shape of atomic orbitals is predetermined by the wave-like behavior of electrons, which is mathematically described by quantum mechanics. The s orbitals are spherical and the p orbitals consist of two lobes (which have opposite signs of the wave function) with a node in the center. In addition to s and p orbitals, the electronic shells of elements in the lower periods of the periodic table have d and f orbitals, which are not involved in common organic molecules.

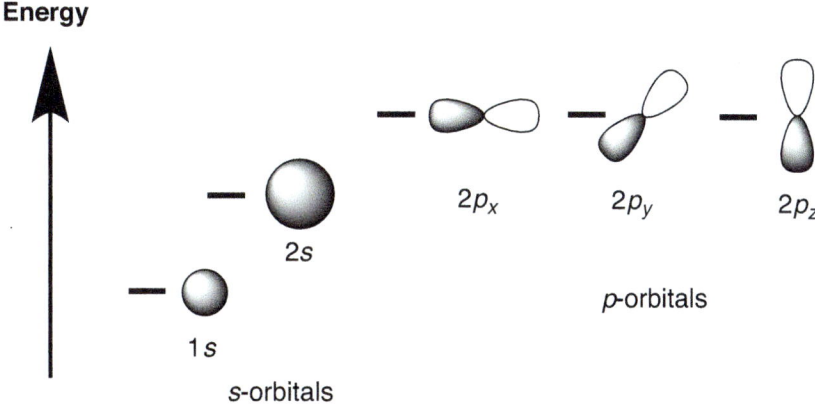

FIGURE 1.1. Shape and relative energy of atomic orbitals within the first two shells.

Any orbital can be occupied either by one electron, or by a pair of electrons with opposite spins (**Pauli principle**). Orbitals are filled starting from the lowest in energy 1s orbital and continuing to the 2s orbital and then the 2p orbitals (**Aufbau principle**). The 2p orbitals (2p$_x$, 2p$_y$, and 2p$_z$) are of equal energy, and initially one electron fills each p-orbital before the second electron can be added (**Hund's rule**). Electronic configurations of elements filled according to these principles are listed in Table 1; superscripts indicate the number of electrons occupying each orbital. The element symbols in Table 1.1 are shown as **Lewis structures,** with dots (•) indicating the number of valence electrons. Main-group elements of the third period in the periodic table (Na, Mg, Al, Si, P, S, Cl, Ar) utilize 3s and 3p orbitals (3p$_x$, 3p$_y$, and 3p$_z$) to achieve valence shell electronic configuration analogous to the related elements of the second period. Likewise, main-group elements of the fourth, fifth, and sixth periods use the appropriate s and p outer orbitals in their chemical compounds.

COVALENT BONDING AND MOLECULAR STRUCTURE   3

TABLE 1.1. Ground-state electronic configurations of the elements in the first two periods of the periodic table.

ELEMENT NUMBER	ELEMENT WITH VALENCE ELECTRONS SHOWN AS DOTS (LEWIS STRUCTURES)	ELECTRONIC CONFIGURATION
1	H·	$1s^1$
2	He:	$1s^2$
3	Li·	$1s^2 2s^1$
4	Be:	$1s^2 2s^2$
5	·B:	$1s^2 2s^2 2p_x^1$
6	·C̈:	$1s^2 2s^2 2p_x^1 2p_y^1$
7	·N̈:	$1s^2 2s^2 2p_x^1 2p_y^1 2p_z^1$
8	:Ö:	$1s^2 2s^2 2p_x^2 2p_y^1 2p_z^1$
9	·F̈:	$1s^2 2s^2 2p_x^2 2p_y^2 2p_z^1$
10	:N̈e:	$1s^2 2s^2 2p_x^2 2p_y^2 2p_z^2$

Noble gases He and Ne have especially stable electronic configurations, because pairs of electrons completely fill all available orbitals within valence shells of these elements. The other elements of the first two periods have only partially filled valence shells. Atoms of these elements can exchange electrons between each other such that each atom acquires the most stable, completely filled outer electronic configuration, like that of a noble gas. This simple idea serves as the foundation of the **Lewis model of bonding**, which was originally proposed in 1916 by Gilbert N. Lewis. When forming a molecule, the atoms of elements combine together either by sharing valence electrons, or by a full transfer of electrons from one to another, in order to reach the most stable electronic configuration, similar to that of a noble gas. For elements of the second period, a completely filled electronic configuration has eight electrons i.e., an octet of electrons) occupying the four available orbitals (2s, $2p_x$, $2p_y$, and $2p_z$) by pairs. No more than the octet of electrons can be placed in the valence shell of these elements.

Three typical examples of electron exchange between atoms that make **covalent** or **ionic** compounds are shown in Figure 1.2. In the first case, two atoms of hydrogen share their electrons in order to achieve a very stable electronic configuration of He. This reaction releases a large amount of energy (about 104 kcal/mol) resulting in the formation of molecule H_2, in which the bonding pair of electrons is located exactly between the two nuclei of H, forming a **nonpolar covalent bond**. In the second case (Figure 1.2), a hydrogen atom combines its electron with the electron of a fluorine atom, forming a **polar covalent bond** in which the bonding pair of electrons is shifted closer to the fluorine atom, owing to the high ability of fluorine to attract electrons (**electronegativity**). Due to this shift in the position of bonding electrons, the molecule of H–F is a **dipole**, characterized by a higher electronic density, shown as δ–, on fluorine and a lower electronic density, shown as δ+, on hydrogen. Covalent bonds in the molecules of H–H and H–F are characterized by a specific bond length, which is defined as a distance between the nuclei of connected atoms and measured in angstroms (Å) or picometers (pm). In addition to bond length, polar covalent bonds are also characterized by a **bond dipole moment**, which is a physical measure of bond polarity, calculated as charge multiplied by bond length.

The polarity of covalent bonds has a strong effect on the physical properties of compounds, such as boiling points and melting points. Nonpolar compounds usually have low boiling and melting points because of the absence of electrostatic attraction between molecules. In contrast, polar molecules have higher boiling and melting points due to intermolecular attractive forces between opposite charges of the dipole, known as **dipole-dipole attraction**.

The third case of electron exchange in Figure 1.2 represents the formation of an ionic compound by the full transfer of one electron from the atom of lithium to the atom of fluorine. The resulting lithium cation, Li^+, has the electronic configuration of He ($1s^2$) and a full positive charge, due the removal of one negatively charged electron. The fluoride anion, F^-, has the electronic configuration of Ne ($1s^2 2s^2 2p_x^2 2p_y^2 2p_z^2$) and a full negative charge. Lithium fluoride, LiF, is an example of ionic compound composed from positively charged cations and negatively charged anions attracted to each other by electrostatic forces. In contrast to the covalent molecules, ionic compounds are not characterized by bond lengths and bond dipole moments and usually exist as crystalline solids with very high melting points. Ionic compounds are formed from elements with a significant difference in electronegativity.

1. Formation of a nonpolar molecule of hydrogen, H₂

H· ⌒⌒ ·H ⟶   H : H   or   H–H
                nonpolar covalent bond
                (bond length 0.74 Å, dipole moment 0 D)

Molecular hydrogen, H₂, is a gas, which becomes liquid at –253 ºC and solid at –259 ºC

2. Formation of a polar covalent molecule of hydrogen fluoride, HF

polar covalent bond
(bond length 0.92 Å, dipole moment 1.86 D)

Hydrogen fluoride, HF, is a liquid, which becomes gas at +19.5 ºC and solid at –84 ºC

3. Formation of ionic compound, lithium fluoride, LiF

Li· ⌒ ·F̈: ⟶   Li⁺   :F̈:⁻
                ionic compound
                (in solid state and in solution exists as separate ions)

Lithium fluoride, LiF, is a crystalline solid, with melting point at +845 ºC
and boiling point at +1,676 ºC

NOTES: • the "fishhook arrow" ⌒ is used to show the movement of a single electron to a new position

• δ+ and δ– indicate partial positive on negative charges due to the shift of electron density to the most electronegative element in the molecule

• dipole moment (in D, Debye units) is a physical measure of bond polarity; the dipole arrow ┼⟶ is used to show the direction of dipole moment from δ+ to δ–

FIGURE 1.2. Formation of covalent or ionic compounds from atoms by exchanging valence electrons.

The concept of **electronegativity** is essential for understanding the different types of bonding shown in Figure 1.2. Electronegativity is defined as a measure of the tendency of an atom to attract a bonding pair of electrons. In the commonly used Pauling scale of electronegativity, fluorine, the most electronegative element, is assigned a value of 4.0, and values range down to the alkali metal caesium, which is the least electronegative at 0.7. In the second period of the periodic table the electronegativity of elements increases from lithium (1.0) to fluorine (4.0) in increments of 0.5. The increase in electronegativity of the elements within a period, from left to right, is due to the increasing positive charge of the nuclei, providing greater attraction to the valence electrons. Within a group of the periodic table, the electronegativity of the elements generally decreases from top to bottom, due to an increasing distance between the nuclei and the valence electrons. The noble gases have no electronegativity because electrons already fill their outer shells; therefore they cannot have bonding pairs of electrons. The electronegativity values of common, main-group elements are listed in Table 1.2.

TABLE 1.2. Electronegativity values of main-group elements of the periodic table.

H 2.1							
Li 1.0	Be 1.5	B 2.0	C 2.5	N 3.0	O 3.5		F 4.0
Na 0.9	Mg 1.2	Al 1.5	Si 1.8	P 2.1	S 2.5		Cl 3.0
K 0.8	Ca 1.0	Ga 1.6	Ge 1.8	As 2.0	Se 2.4		Br 2.8
Rb 0.8	Sr 1.0	In 1.7	Sn 1.8	Sb 1.9	Te 2.1		I 2.5
Cs 0.7	Ba 0.9	Tl 1.8	Pb 1.8	Bi 1.9	Po 2.0		At 2.2

Electronegativity values provide useful guidelines for determining the type of bonding between atoms. As a generally accepted rule, the nonpolar covalent bond is formed between identical elements. Typical examples of nonpolar covalent bonds include the bonds between atoms of F in the molecule F_2, or the C–C bonds in various organic molecules. The polar covalent bond is generally formed by two elements that differ in their electronegativity values by 1.9 or less. For example, the H–F bond and C–O bonds are polar covalent bonds with electronegativity differences of 1.9 and 1.0, respectively. The ionic compounds are generally formed between elements with a difference in electronegativity greater than 1.9. Some exceptions from these general guidelines are possible. For example, the F–B bond in a boron fluoride molecule, BF_3, is characterized by a large difference in the elements' electronegativities (4.0 for F minus 2.0 for B = 2.0), which is typical of the ionic compounds. However, BF_3 has the properties of a typical covalent compound (i.e., a gas, which becomes liquid at –100 °C), and the F–B bond is covalent with a bond length of 1.31 Å. This exception is explained by the special electronic configuration of BF_3, in which boron has an incomplete outer shell with only 6 valence electrons. The salts of alkali metals, such as, LiBr, LiI, NaI, KI, RbI and CsI, represent another exception. These salts have typical properties of ionic compounds despite less than a 1.9 difference in the elements' electronegativities; this exception is explained by the exceptional stabilities of the alkali metal cations and the halide anions.

Problems

1.1 Which elements have the following electronic configurations?

a) $1s^2\, 2s^2\, 2p_x^2\, 2p_y^2\, 2p_z^2$
b) $1s^2\, 2s^2\, 2p_x^1\, 2p_y^0\, 2p_z^0$
c) $1s^2\, 2s^2\, 2p_x^1\, 2p_y^1\, 2p_z^1$
d) $[Ne]\, 3s^2\, 3p_x^1\, 3p_y^1\, 3p_z^0$

1.2 How many valence electrons do the following elements have?

a) Phosphorus   b) Magnesium   c) Sodium   d) Silicon   e) Nitrogen

1.3. Use δ+ and δ- to indicate the polarity of the following bonds:

   a) C–F   b) S–O   c) Li–C

1.4. Which of the following pairs of elements will form an ionic compound? Which pair will form a nonpolar covalent bond?

   a) H and C   b) Br and Cl   c) Cl and Cl   d) Mg and F

1.2. LEWIS STRUCTURES OF MOLECULES AND MOLECULAR IONS

Lewis structures of molecules are drawings that show the bonds between the atoms of a molecule, the lone pairs of electrons, and positive or negative charges that may exist in the molecule. Lewis structures represent an essential component of the language of chemistry, and it is important to be proficient in composing these structures. A Lewis structure of any chemical compound can be put together starting from the Lewis structures of elements (see Table 1.2), and combining electrons as illustrated in Figure 1.2. Usually we draw a Lewis structure by beginning with a known **molecular formula** of a chemical compound that shows the number and kinds of atoms in a molecule. Molecular formulas of chemical compounds are experimentally determined by mass spectrometry (see Chapter 9) or combustion elemental analysis. In combustion elemental analysis, a sample of organic compound is burned in an excess of oxygen and the combustion products, such as carbon dioxide and water, are collected in special traps. The masses of these combustion products are used to calculate the composition of the unknown sample. Knowing the how many atoms of each element are in the molecule, we can connect them together using basic principles of chemical bonding.

Several Lewis structures for common chemical compounds are shown in Figure 1.3. For example, oxygen and hydrogen atoms sharing single electrons form a water molecule (molecular formula H_2O). The Lewis structure of H_2O has two single bonds, each formed by a pair of bonding electrons, and two lone pairs of nonbonding electrons located on the oxygen atom. Likewise, a molecule of ammonia, NH_3, is formed from an atom of nitrogen and three hydrogens, and has three single bonds and one lone pair. A molecule of methane (molecular formula CH_4) is formed similarly from a carbon atom and four hydrogens. The molecule of ethane, C_2H_6, has single bonds between two atoms of carbon forming six C–H bonds. A bigger molecule of butane (molecular formula C_4H_{10}) can be assembled from atoms by two different ways, giving two different compounds: normal butane, in which four carbon atoms are linked in one chain, and the branched molecule of isobutane. Such compounds that have the same molecular formula but differ in how atoms are connected are called **structural** (or **constitutional**) **isomers**. A compound with a **double bond**, ethylene (molecular

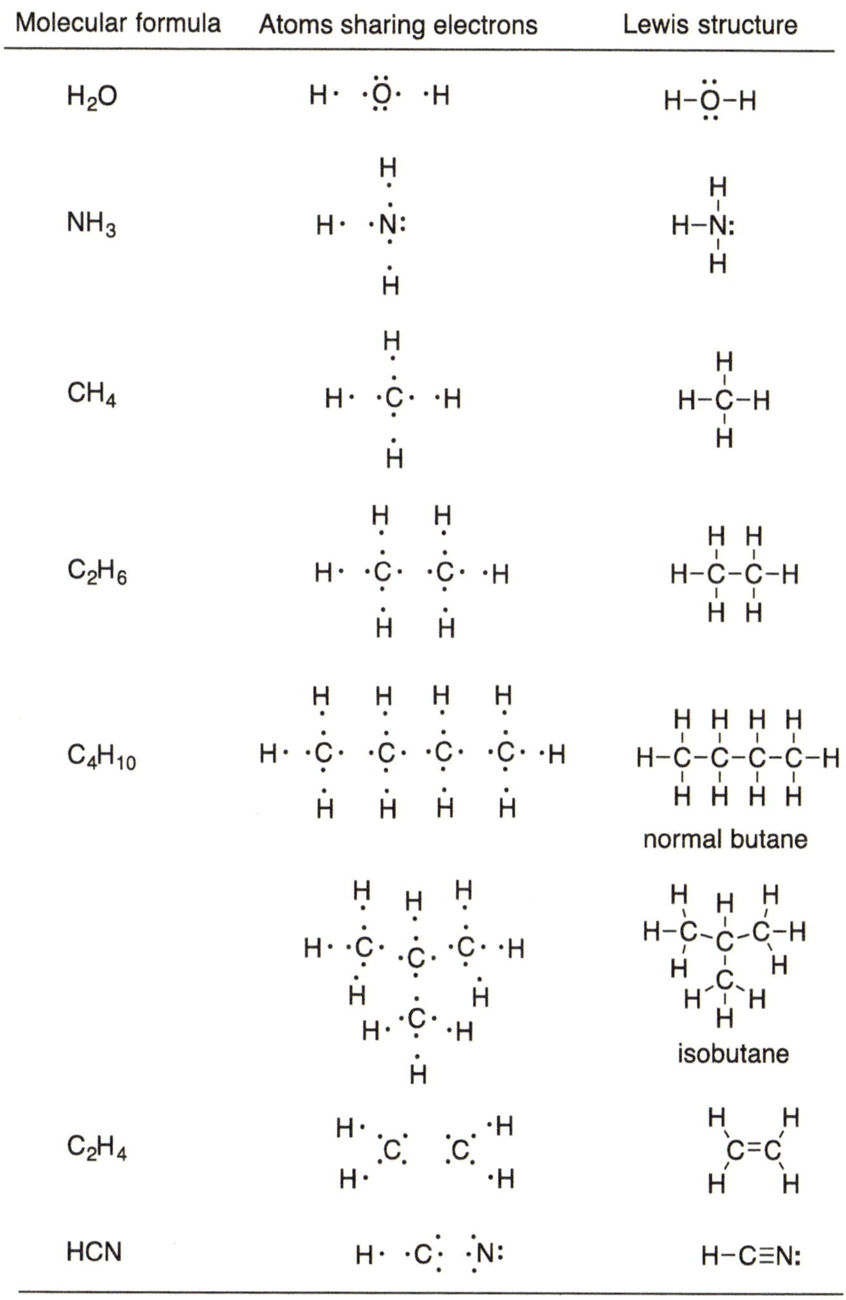

Molecular formula	Atoms sharing electrons	Lewis structure
H_2O	H· ·Ö· ·H	H–Ö–H
NH_3	H· ·N: with H above and H below	H–N: with H above and H below
CH_4	H· ·C· ·H with H above and H below	H–C–H with H above and H below
C_2H_6	H· ·C· ·C· ·H with H's above and below	H–C–C–H with H's above and below
C_4H_{10}	H· ·C· ·C· ·C· ·C· ·H with H's above and below	H–C–C–C–C–H with H's above and below (normal butane)
	(branched arrangement with central C and fourth C below)	(branched structure — isobutane)
C_2H_4	H· :C: :C: ·H with H· and ·H	H₂C=CH₂
HCN	H· ·C: ·N:	H–C≡N:

FIGURE 1.3. Assembling Lewis structures of molecules from atoms.

formula C_2H_4) has four single C–H bonds and two bonds between carbon atoms formed by pairing the other available valence electrons. The last example shows formation of a molecule of hydrogen cyanide, HCN, which has one C–H bond, a **triple bond** between carbon and nitrogen, and a lone pair of electrons on nitrogen. In principle, any complex molecule can be assembled from atoms of elements in a similar manner. It is important to keep in mind that the overall number of valence electrons in the molecule is exactly the same as in the initial atoms, and that the combined number of bonding and nonbonding electrons on any atom of a second period element must be no more than 8.

Assembling molecules directly from atoms is getting complicated for large molecules, especially, when several ways of combining the same group of atoms are possible. It is much easier to build a complex molecule by connecting atoms together while keeping in mind that hydrogen can have only one single bond to other atoms, carbon forms four bonds, oxygen forms two bonds and has two lone pairs, and nitrogen always has three bonds and one lone pair. The important typical bonding patterns of elements in organic chemistry are shown in Figure 1.4.

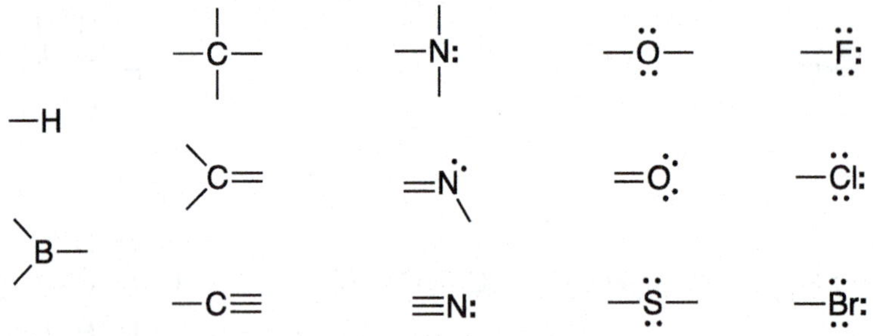

FIGURE 1.4. Typical bonding patterns within Lewis structures of organic molecules.

The Lewis structures of the large organic molecules can become very crowded, if each lone pair and each single bond between elements are shown. Usually organic chemists skip showing lone pairs in structural formulas, and use more condensed ways of drawing. Several examples of organic molecules as **condensed structures**, in which single bonds between elements are not shown, are given in Figure 1.5. The order of groups of atoms listed in condensed structures follow the connectivity of atoms in Lewis structures, which makes possible easy conversion of one type of drawing to the other. The **line-angle structures** (Figure 1.5) represent another common way of simplifying Lewis structures; in this case only the chain of bonds between carbon atoms is shown and the actual symbols of C with attached H atoms are not shown. The line-angle structures can be converted to the Lewis structures by adding symbols of carbon with the attached hydrogens to the ends of each individual line of the chain.

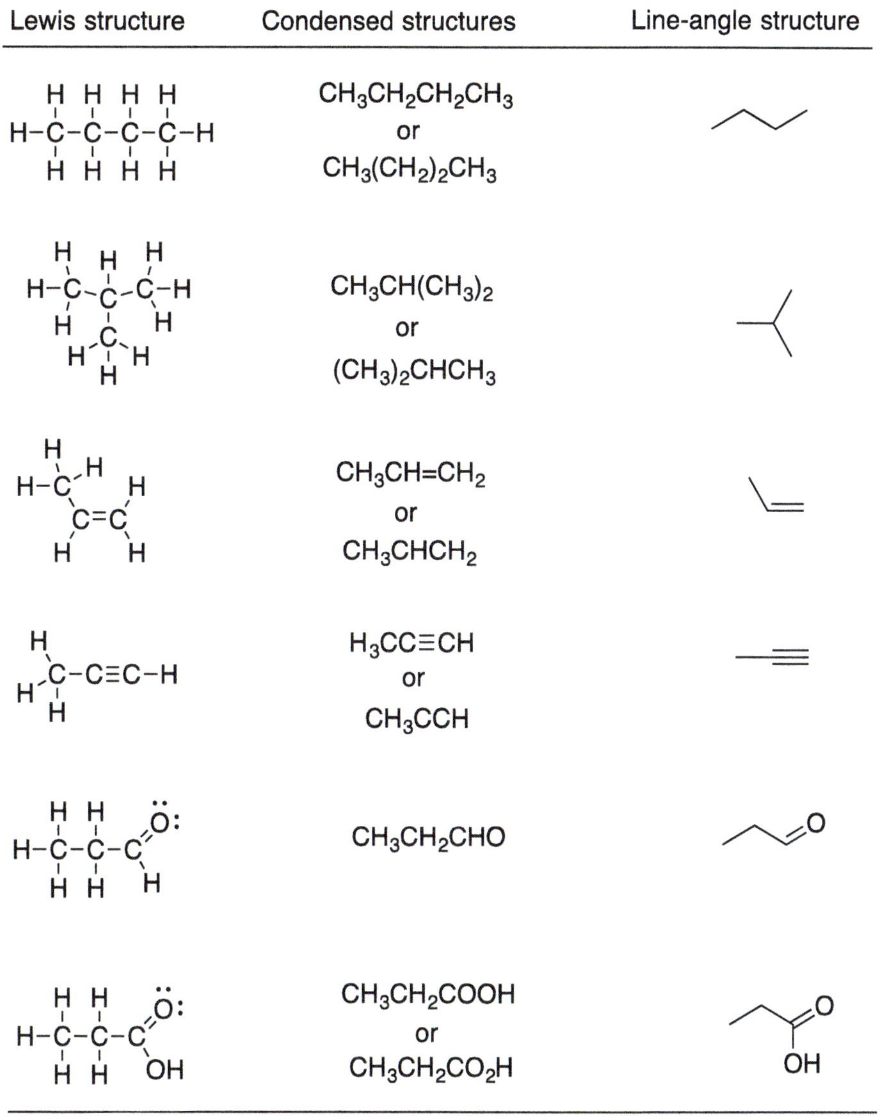

Lewis structure	Condensed structures	Line-angle structure
H-C-C-C-C-H (butane, all H)	CH₃CH₂CH₂CH₃ or CH₃(CH₂)₂CH₃	(line-angle of butane)
(isobutane Lewis)	CH₃CH(CH₃)₂ or (CH₃)₂CHCH₃	(line-angle of isobutane)
(propene Lewis)	CH₃CH=CH₂ or CH₃CHCH₂	(line-angle of propene)
(propyne Lewis)	H₃CC≡CH or CH₃CCH	(line-angle of propyne)
(propanal Lewis)	CH₃CH₂CHO	(line-angle of propanal)
(propanoic acid Lewis)	CH₃CH₂COOH or CH₃CH₂CO₂H	(line-angle with OH)

FIGURE 1.5. Different ways of drawing organic molecule structures.

Some organic compounds have ionic structure with the positively or negatively charged molecular ions incorporated in their structure. **Molecular ions** are defined as charged chemical species composed of two or more covalently bonded atoms. For example, a well-known organic compound, acetic acid, reacts with bases to form acetate salts, such as sodium acetate, with molecular formula CH_3CO_2Na. Sodium acetate has the typical properties of an ionic compound, with a positively charged sodium cation and a negatively charged acetate anion (Figure 1.6). In the Lewis structure of the acetate anion, the negative charge is assigned to one specific oxygen atom as a formal charge. A **formal charge** is defined as the charge assigned to an atom in a molecule, assuming that electrons in a chemical bond are shared equally between atoms, regardless of the relative electronegativity. The presence of a formal negative charge means that the atom has one extra electron, compared to the number of valence electrons for this element in the periodic table. Another example of an ionic organic compound is represented by methylammonium chloride, CH_3NH_3Cl, with a positively charged methylammonium cation and a negatively charged chloride anion (Figure 1.6). The presence of a formal positive charge on nitrogen indicates that the atom has one electron less, compared to the number of valence electrons for nitrogen in the periodic table. To confirm the location of a formal charge on a specific atom of a molecular ion, we need to calculate the balance of electrons on this atom by using the following simple equation:

Formal charge on atom = (number of valence electrons from the periodic table) - (number of nonbonding electrons in lone pairs) - 1/2(number of bonding electrons)

According to this equation, the formal charge on the single-bonded oxygen atom of the acetate anion is equal to $(6 - 6 - 1) = -1$, and the charge on the nitrogen atom of methylammonium chloride is $(5 - 0 - 4) = +1$. All other atoms in these two organic molecular ions have zero formal charges.

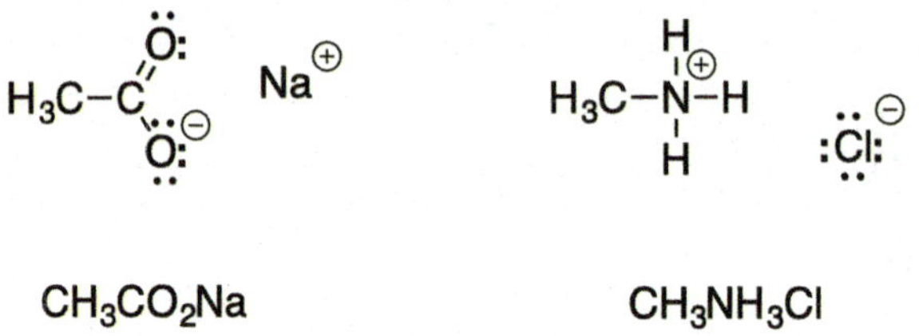

FIGURE 1.6. Lewis structures of molecular ions with assigned formal charges.

Formal charges are an essential part of a Lewis structure, but assigning charges for all atoms in a complex molecule is time-consuming work. It is much easier to keep in mind several common charged molecular fragments, and use these fragments to assign formal charges to atoms in complex molecules or molecular ions. Several important charged molecular fragments in organic chemistry are shown in Figure 1.7.

The positively charged carbon with three covalent bonds (see Figure 1.7) has only 6 valence electrons, which is less than the octet required for the stable electronic configuration of a noble gas. Such positively charged molecular ions can exist only for a very short time as extremely unstable intermediates in organic reactions (i.e. **carbocations**, see Chapter 5). Stable uncharged molecules with 6 valence electrons on the central atoms are known only for the elements of Group 3 of the periodic table, for example, boron fluoride, BF_3 (see also Section 1.1), and aluminum chloride, $AlCl_3$. These compounds are known under the general

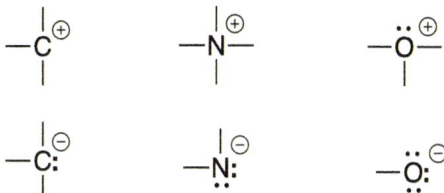

FIGURE 1.7. Important molecular fragments with formal charges.

name of **Lewis acids**, and they readily react with any molecule that can donate a pair of unshared electrons.

While molecules or ions with less than the octet of electrons can exist, molecules with more than eight electrons in their valence shells are impossible for the second period elements, since these elements have only four available atomic orbitals. In principle, main-group elements of the third period and below can have more than the octet, such as molecules of phosphorus pentachloride, PCl_5 (10 valence electrons on P) or iodine heptafluoride, IF_7 (14 valence electrons on I).

Problems

1.5. Draw Lewis structures of two compounds with the molecular formula C_2H_7N.

1.6. The molecules shown below are wrong Lewis structures violating the octet rule and formal charges. Correct these structures by moving electron pairs from bonded to nonbonded positions, showing formal charges if needed.

1.7. For each condensed structure shown, draw the corresponding line-angle structure.

a) $(CH_3)_2CH(CH_2)_3CH_3$
b) $CH_3CONHCH_2CH_3$
c) $CH_3CH_2CH(CH_3)CH(CH_3)_2$
d) $(CH_3)_3COCH_3$

1.8. All structures shown below depict the correct number of nonbonding electrons and bonds. What is the formal charge on the indicated atoms?

a) 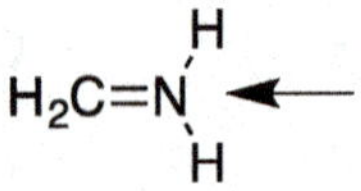 H₂C=N(H)(H) ←

b) 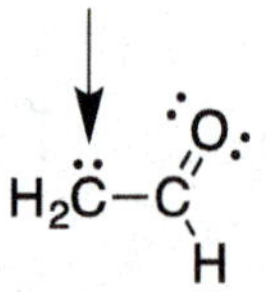 H₂C̈–C(=Ö:)(H) ↓

c) H₂C=Ö(H) ←

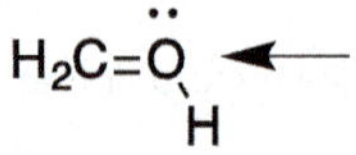

d) 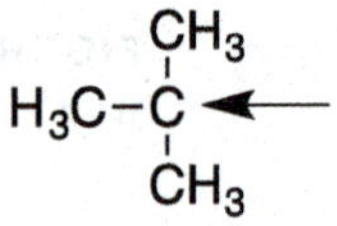 H₃C–C(CH₃)(CH₃) ←

1.3. FUNCTIONAL GROUPS AND INTRODUCTION TO THE IUPAC NOMENCLATURE

Functional groups are important structural fragments within molecules that are responsible for the physical properties and chemical reactions of organic compounds. In combination with the name of the parent carbon chain, functional groups are used in their general classification and systematic nomenclature. Table 1.3 provides a list of important classes of organic compounds, the related functional groups, and specific examples of compounds.

TABLE 1.3. Classes of organic compounds and common functional groups.

CLASS OF COMPOUND	CHARACTERISTIC STRUCTURAL UNIT	NAME FUNCTIONAL GROUP	EXAMPLE OF COMPOUND	SYSTEMATIC NAME
alkane	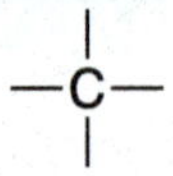 –C–	alkyl	$CH_3CH_2CH_3$	propane
alkene	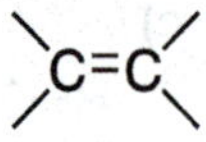 C=C	alkenyl	$CH_3CH=CH_2$	propene
alkyne	–C≡C–	alkynyl	$CH_3C≡CH$	propyne
arene	 ⬡	aryl or phenyl (Ph = C_6H_5)	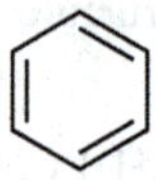 ⬡	benzene
haloalkane or alkyl halide	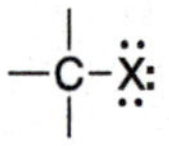 –C–Ẍ: X = F, Cl, Br, or I	fluoro chloro bromo iodo	CH_3CH_2F	fluoroethane

14   ORGANIC CHEMISTRY

Functional group	Structure	Name	Example	Example name
alcohol	\C-Ö-H /	alcohol or hydroxyl	CH_3CH_2OH	ethanol
ether	\C-Ö-C/	ether	CH_3OCH_3	dimethyl ether
aldehyde	—C(=Ö)H	aldehyde or carbonyl	CH_3CH_2CHO	propanal
ketone	\C(=Ö)/	ketone or carbonyl	$CH_3CH_2COCH_3$	butanone
carboxylic acid	—C(=Ö)Ö-H	carboxyl	$CH_3CH_2CO_2H$	propanoic acid
carboxylic ester	—C(=Ö)Ö-C	ester	$CH_3CH_2CO_2CH_3$	methyl propanoate
carboxylic amide	—C(=Ö)N—	amide	$CH_3CH_2CONH_2$	propanamide
amine	—N:	amine	$(CH_3)_3N$	trimethylamine
imine	\C=N: /	imine	$CH_3CH=NH$	ethanimine
nitrile	—C≡N:	nitrile	CH_3CH_2CN	propanonitrile
nitro compound	—N($^+$)(=Ö)Ö$^-$	nitro	CH_3NO_2	nitromethane
thiol	—S̈-H	thiol	CH_3SH	methanethiol

COVALENT BONDING AND MOLECULAR STRUCTURE

Table 1.3 lists the most general classes of organic compounds and further classification is also possible. Alkanes, alkenes, alkynes, and arenes, also known as aromatic compounds, combine to form a general class of **hydrocarbons**, since molecular formulas of these compounds include only atoms of hydrogen and carbon. Alkanes are called **saturated hydrocarbons**, while hydrocarbons with double or triple bonds are known as **unsaturated hydrocarbons**. Cyclic hydrocarbons with a ring of carbon atoms in their structure are called **cycloalkanes**. Each double bond or a cycle in a molecule of hydrocarbon is considered as a unit of unsaturation. A triple bond has two units of unsaturation. The number of **units of unsaturation** in a molecule of a hydrocarbon, also called the **degree of unsaturation**, is defined as the number of hydrogen molecules (H_2) that should be added to the molecule in order to produce a saturated hydrocarbon.

Alcohols are classified as **primary** (1°), **secondary** (2°), and **tertiary** (3°) **alcohols** depending on the number of carbon atoms attached to the carbon bearing the hydroxyl group, –OH. According to this classification, ethanol CH_3CH_2OH is a 1° alcohol, isopropanol $(CH_3)_2CHOH$ is a 2° alcohol, and *tert*-butanol $(CH_3)_3COH$ is a 3° alcohol.

Aldehydes and ketones are known under the combined name of **carbonyl compounds**, since both have the carbonyl functional group, C=O. Carboxylic esters and amides are categorized as **carboxylic acid derivatives**, because both are derived from carboxylic acids.

Like alcohols, amines are classified as **primary** (1°), **secondary** (2°), and **tertiary** (3°) **amines** according to the number of carbon atoms bonded to the nitrogen. For example, CH_3NH_2 is a 1° amine, $(CH_3)_2NH$ is a 2° amine, and $(CH_3)_3N$ is a 3° amine.

The names of functional groups in combination with the name of the parent carbon chain are used for the systematic naming of chemical compounds (the International Union of Pure and Applied Chemistry (**IUPAC**) **nomenclature**). The rules of systematic nomenclature have been established and are periodically updated by IUPAC. According to the basic rules of IUPAC nomenclature, the **systematic name** of any organic compound includes three main parts: 1) a **root** name that indicates a) the number of carbon atoms in the longest carbon chain (or carbon cycle); and b) the presence of single, double, or triple bonds in this chain, 2) **prefixes** at the beginning of the name, identifying **substituents** (or functional groups) attached to specific carbons of the main chain; and 3) a **suffix** at the end of the name specifying the functional group with the highest priority. The **order of priority of the functional groups** has also been established by IUPAC as the following: carboxylic acid and derivatives (the highest priority) > aldehyde > ketone > alcohols > amine (the lowest priority).

In order to indicate the longest carbon chain in the molecule, the names of parent alkanes are used: methane (CH_4), ethane (CH_3CH_3), propane ($CH_3CH_2CH_3$), butane ($CH_3CH_2CH_2CH_3$), followed by the systematic names *pent*ane, *hex*ane, *hept*ane, *oct*ane, *non*ane, *dec*ane, *undec*ane, *dodec*ane, *tridec*ane, etc., in which the initial letters are derived from the Greek or the Latin names for five, six, seven, eight, nine, ten, eleven, twelve, thirteen, etc. Changing the infix -an- to -en- (for alkenes) or -yn- (for alkynes) indicates the presence of double or triple bonds in the carbon chain. The ending -e is present in the name of any hydrocarbon molecule. For example, the name "propyne" means that this is a molecule of hydrocarbon with three carbon atoms and a triple bond in the chain.

In other classes of organic molecules, the ending -e is replaced by the suffix of the functional group with the highest priority, and the presence of other groups can be indicated by the appropriate prefixes. The suffixes and prefixes indicating important functional groups in order of their priority are listed in Table 1.4. When numerous functional groups are present in the molecule, only one, the highest priority group, is shown in the suffix; all other are listed in alphabetical order as prefixes, with numbers corresponding to the specific carbons of the carbon chain to which they are attached. The numbering of the parent carbon chain usually starts from the carbon nearest to the highest priority group, and the carbon atoms of carboxylic acid, aldehyde, and nitrile are counted as part of the parent chain. Several representative examples of organic molecules with their corresponding IUPAC names are shown in Figure 1.8.

TABLE 1.4. Prefixes and suffixes assigned to important functional groups.

NAME OF FUNCTIONAL GROUP OR A CLASS OF COMPOUNDS	SUBSTITUENT ATTACHED TO THE PARENT CARBON CHAIN	PREFIX	SUFFIX
carboxylic acid	$-CO_2H$	none	-oic acid
nitrile	$-CN$	cyano-	-nitrile
aldehyde	$-CHO$	oxo-	-al
ketone	$=O$	oxo-	-one
alcohol	$-OH$	hydroxy-	-ol
thiol	$-SH$	mercapto-	-thiol
amine	$-NH_2$	amino-	-amine
imine	$=NH$	imino-	-imine
ether	$-OCH_3$ $-OCH_2CH_3$ $-OR$ (R = any alkyl)	methoxy- ethoxy- alkoxy	none
haloalkane	$-F$ $-Cl$ $-Br$ $-I$	fluoro- chloro- bromo- iodo-	none
nitro compound	$-NO_2$	nitro-	none
branched alkane	$-CH_3$ $-CH_2CH_3$ $-R$	methyl- ethyl- alkyl-	none

Carboxylic acid is the highest priority group, which is always shown in the suffix. The names of carboxylic acid derivatives (e.g., esters, amides, halides) are based on the names of acids; for specific examples, see Chapter 13. The names of the lowest priority groups, such as alkoxy, alkyl, fluoro, chloro, bromo, iodo, and nitro, are used exclusively in prefixes. The names of the alkyl groups, commonly abbreviated as R, are derived from the names of corresponding alkanes by changing suffix -ane to the suffix -yl (e.g., methyl, ethyl, propyl).

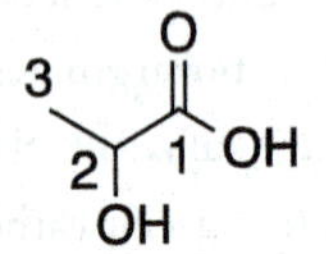

2-hydroxypropanoic acid

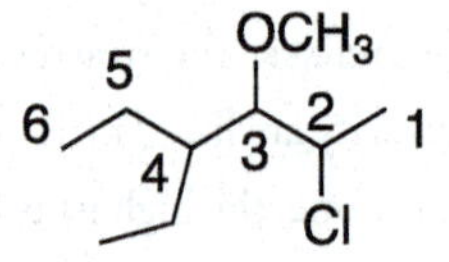

2-chloro-4-ethyl-3-methoxyhexane

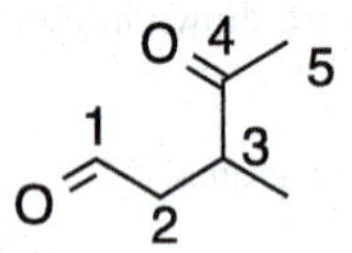

3-methyl-4-oxopentanal

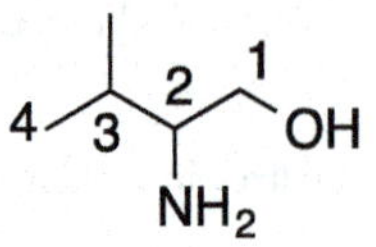

2-amino-3-methyl-1-butanol

FIGURE 1.8. Examples of organic molecules with different functional groups (the numbering of the parent carbon chains is shown)

Problems

1.9 For each of the molecules below, convert the condensed structure into a skeletal structure and name the major functional group present in each molecule.

a) CH_3CH_2CHO b) $(CH_3)_3CCN$
c) $(CH_3)_3COCH_3$ d) $(CH_3)_3CCOOCH_2CH_3$

1.10. Below are ingredients of some personal care products that may be found in your home. Based on the suffix of the name, what functional groups are present in these molecules?

a) Menthol   b) Acetone
c) Benzyl benzoate   d) Cyclohexene

1.11. Identify the functional groups indicated by numbers in the molecule shown below.

18   ORGANIC CHEMISTRY

1.4. SHAPE OF MOLECULES

Lewis structures, as well as the condensed and line-angle structures, are two-dimensional representations showing connectivity of atoms, but not the actual shape of molecules. Knowing the three-dimensional shape of molecules is essential for understanding the physical, chemical, and biological properties of organic compounds. The simple, but very useful concept of valence shell electron pair repulsion (**VSEPR**) makes prediction of the real, three-dimensional geometry of molecules possible. Covalent bonds between atoms in a molecule are formed by pairs of negatively charged electrons. There is repulsion between negatively charged bonds, and therefore, they try to spread out in space as far away from each other as possible. For a molecule of methane (CH_4), arrangement of the four bonds will result in an overall **tetrahedral** geometry, with four hydrogen atoms located at the corners of a tetrahedron and each H–C–H bond angle equal to 109.5° (Figure 1.9). The shape of any saturated hydrocarbon, which has four single bonds at all carbon atoms, can be predicted in a similar way, with the understanding that each carbon atom in alkanes has tetrahedral electronic geometry (the so-called tetrahedral carbon atom). The same tetrahedral geometry is typical of any carbon atom connected by four single bonds to other atoms. Several examples of molecules with tetrahedral carbon atoms are shown in Figure 1.9. Note that in the three-dimensional drawings, bonds in the front of the atom are shown as the dark wedges, bonds in the back are indicated as hashed wedges, and bonds in plane are shown as solid lines. This bond representation is also used for the three-dimensional line-angle structures.

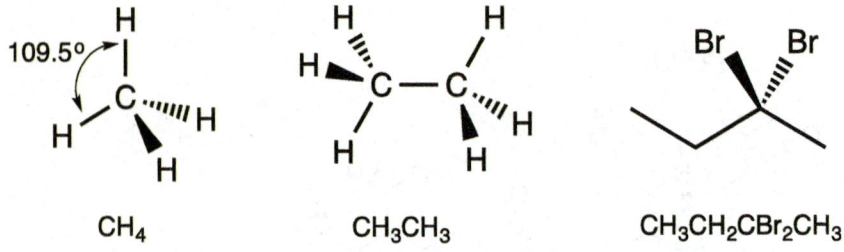

FIGURE 1.9. Three-dimensional drawings of molecules with tetrahedral carbon.

The shapes of the molecules of ammonia (NH$_3$) and water (H$_2$O) can be predicted in a similar way based on repulsion between the lone pairs of unshared electrons and the bonds, in addition to the repulsion just between the bonds. In the molecule of NH$_3$, the three bonds and the lone pair form the overall tetrahedral geometry, the same as in the molecule of H$_2$O with two bonds and two lone pairs. The repulsion between a lone pair and a bond is stronger compared to the repulsion just between two bonds, which results in the smaller bonds angles H–N–H (107.8°) and H–O–H (104.5°), compared to the H–C–H bond angle. Figure 1.10 shows three-dimensional presentations of ammonia, water, an amine, and an alcohol molecule.

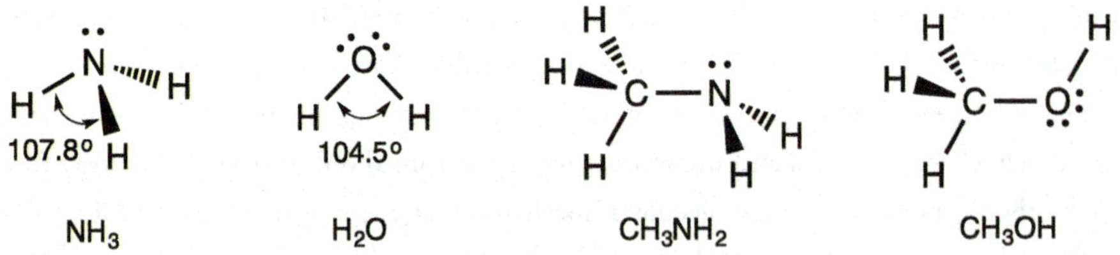

FIGURE 1.10. Three-dimensional drawings of an ammonia, a water, an amine, and an alcohol molecule.

The geometry of an alkene is predetermined by the presence of three bonds to the carbon atom (two single bonds and one double bond). The three bonds are farthest apart when they are in the same plane, with angles of 120° forming the **trigonal planar geometry**. The repulsion between a double bond and a single bond is stronger than the repulsion between two single bonds, and therefore, in the molecule of ethylene (CH$_2$=CH$_2$), the H–C–H angle is smaller than the H–C=C bond angle (117.4° and 121.3° respectively; see Figure 1.11). Any other molecules with a double bond between atoms of carbon, oxygen, or nitrogen have a similar trigonal planar geometry, with lone pairs instead of the covalent bonds for N and

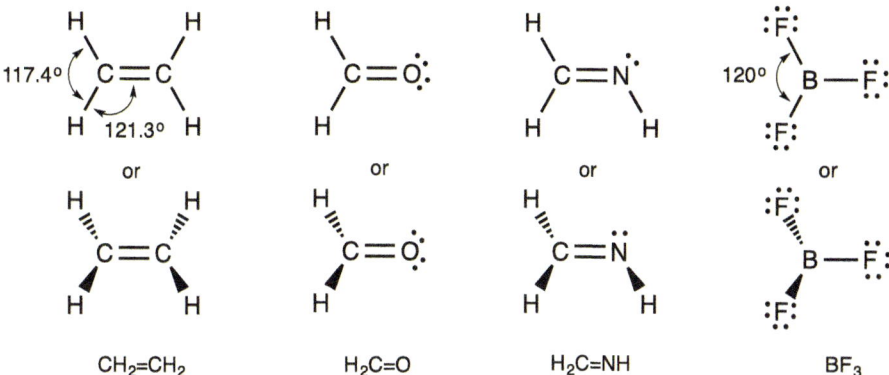

FIGURE 1.11. Examples of molecules with trigonal planar geometry.

O atoms. A molecule of BF_3, which has only three covalent bonds and no lone pairs on the boron atom, represents an example of a trigonal planar molecule. Figure 1.11 shows three-dimensional presentations of several molecules with trigonal planar geometry.

Atoms that have just two covalent bonds (either single, double, or triple bonds) to its neighboring atoms have a linear geometry with 180° angle between the two bonds. Several examples of linear molecules are shown in Figure 1.12.

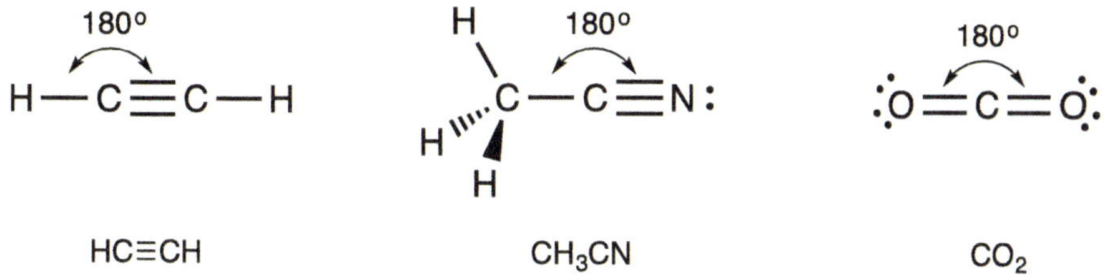

FIGURE 1.12. Examples of linear molecules.

The overall geometry and a three-dimensional shape of a covalent molecule (or a molecular ion) are described using bond angles and bond lengths. As we have just learned, the bonds angles can be estimated by using the VSEPR concept. The bond length (or bond distance) in a molecule depends on the size and the electronegativity of atoms connected and the type of covalent bond (i.e., single, double, or triple bond). Figure 1.13 shows several typical bond lengths for different classes of organic compounds. In general, for a pair of specific atoms, a single bond is longer than a double bond between these atoms, and a triple bond is the shortest. For example, the C–H bond in alkanes is the longest (1.11 Å), in alkenes a bit shorter (1.10 Å), and in alkynes the shortest (1.09 Å). Likewise, the C–H bond in alcohols is longer than in aldehydes.

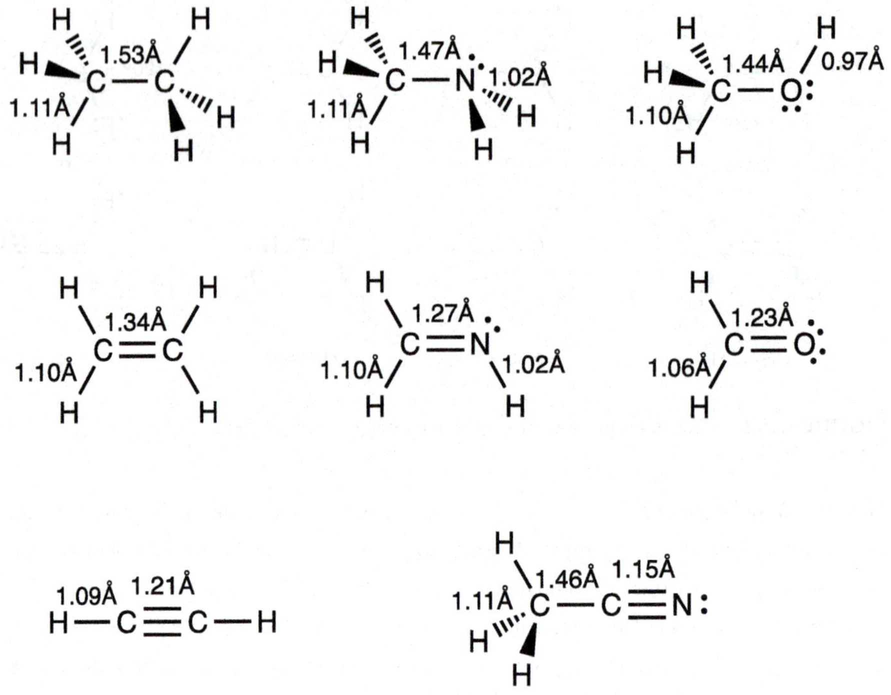

FIGURE 1.13. Bond lengths in organic molecules.

Problems

1.12. Ignoring any bond between carbon and hydrogen, indicate the shortest bond in the molecules shown below.

   a) $CH_3CCCH_2CH_3$ b) $CH_3COCH_2CH_2CHCH_2$

1.13. What is the approximate chlorine-carbon-carbon bond angle in vinyl chloride (CH_2=CHCl)?

1.14. Cyclopropane is a cyclic compound with the molecular formula C_3H_6. Explain, why the carbon-carbon-carbon bond angle in cyclopropane is significantly smaller than 109°.

1.5. POLARITY OF MOLECULES AND PHYSICAL PROPERTIES OF ORGANIC COMPOUNDS

Knowledge of molecular geometry is essential for identifying polar and nonpolar molecules. The polarity of a molecule is measured by the **molecular dipole moment** [μ, in Debye units (D)], which is defined as the vector sum of the individual bond dipoles (see Section 1.1) within the molecule. Any **polar molecule** is characterized by a dipole moment with a value that differs from zero, while a **nonpolar molecule** has $\mu = 0$ D, even if it has polar covalent bonds between some atoms within the molecule. Several polar and nonpolar molecules with their related dipole moment directions are shown in Figure 1.14. For example, a molecule of CO_2 is a nonpolar molecule overall, with $\mu = 0$ D, despite the presence of two highly polar C=O bonds. This is because of the linear geometry of CO_2, with equal dipole moments of C=O pointing in exactly opposite directions, and therefore cancelling each other out. A molecule of H_2O is bent (the angle H–O–H is 104.5°, see Figure 1.10), such that both H–O

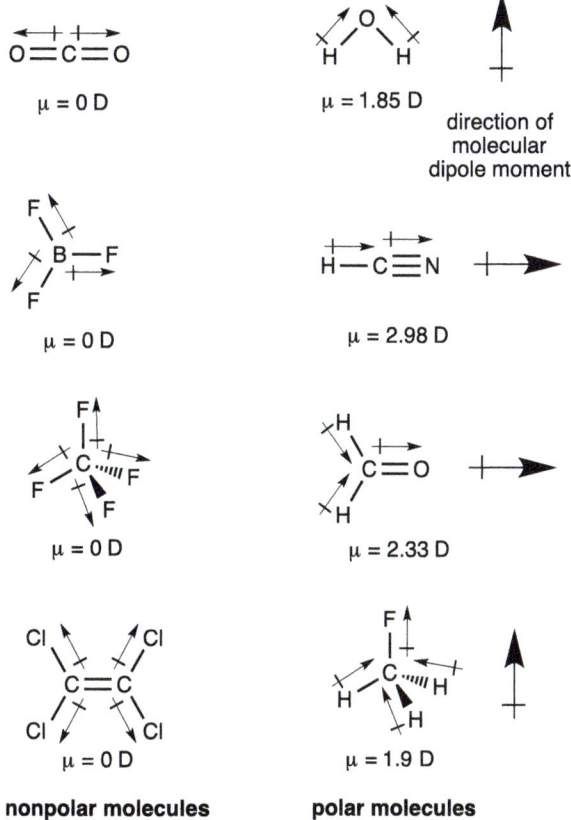

FIGURE 1.14. Examples of nonpolar and polar molecules with related directions of dipole moments.

dipoles are partially pointing in the same direction, creating a molecular dipole moment of 1.85 D. The molecular dipole moments of chemical compounds can be determined experimentally by physical measurement, or can be theoretically calculated.

Electrostatic attraction between polar molecules is important for understanding the physical properties of organic compounds. Chemical compounds exist as gases, liquids, or solids depending on the attractive forces between molecules or ions. The strongest electrostatic attractions exist between opposite ions in ionic compounds, the **ionic attractive forces**. For example, it takes 188 kcal/mol to separate the cation Na^+ from the anion Cl^-. The presence of these forces explains the extremely high melting and boiling points of ionic compounds; for example, NaCl melts at 801 °C and boils at 1,413 °C.

In general, polar organic compounds have higher boiling points (bp) and melting points (mp) compared to nonpolar compounds of similar size. For example, the nonpolar compound ethane (C_2H_6, μ = 0 D) has a bp of –89 °C and an mp of 183 °C, while the more polar compound chloromethane (CH_3Cl, μ = 1.9 D) has a bp of –24 °C and an mp of –97 °C. Polar compounds have higher boiling and melting points than nonpolar compounds due to electrostatic attraction between opposite charges of the dipole, known as the **intermolecular dipole-dipole attraction**. Dipole-dipole attraction is a much weaker attractive force (generally within 2-10 kcal/mol) compared to the ionic attraction in NaCl and other salts. A special case of the strongest dipole-dipole attraction (about 5-10 kcal/mol) is represented by **hydrogen bonding**, which is typical for molecules that have a hydroxyl functional group (OH) in their structure. Compounds with a hydroxyl functional group usually have much higher boiling points. For example, methanol (CH_3OH) has a bp of +65 °C, which is 89 degrees higher than chloromethane. Hydrogen bonding is explained by the high polarity of the O–H bond, in combination with the very small size of protons, which can be shared by oxygen atoms of neighboring molecules. Compounds with hydroxyl functional groups are also highly soluble in water because their molecules form hydrogen bonds with water molecules. Weaker hydrogen bonding exists for molecules with N–H groups (i.e., amines). For example, methylamine (CH_3NH_2) has a bp of –6° C, which is higher than CH_3Cl, but much lower than CH_3OH.

Nonpolar compounds usually have low boiling and melting points because of the absence of electrostatic attraction between molecules. Weak **dispersion forces** (or **van der Waals forces**) of about 0.02–2 kcal/mol exist between the nonpolar molecules due to the temporary dipoles induced in the adjacent atoms of molecules. Dispersion attraction depends on the size and the shape of a molecule. Methane (CH_4, bp = –161 °C, mp = –182 °C) has a much lower boiling point than the higher alkanes, such as pentane ($CH_3CH_2CH_2CH_2CH_3$, bp = 36 °C), because there is a larger area of contact between the bigger nonpolar molecules of pentane. The shape of molecules is also important. For example, the more rounded molecules of the isomeric neopentane (($CH_3)_4C$), have a smaller area of contact than normal pentane, and therefore neopentane has a lower boiling point (bp = 9 °C).

Problems

1.15 For each series of compounds, indicate the one with a molecular dipole moment of µ = 0.

a) [four chlorinated benzene structures: chlorobenzene; 1,4-dichlorobenzene; 1,2-dichlorobenzene; 1,3-dichlorobenzene]

b) Cl—C≡C—Br ; (Cl,H)C=C(H,Cl) ; Cl—CH₂—Cl (CH₂Cl₂) ; 1,2-dichlorocyclopropane

1.16 Sort the following compounds by increasing boiling point:

a) HO–CH₂CH₂CH₂CH₂CH₃ ; CH₃CH₂CH₂CH₂CH₃ ; HO–CH₂CH₂CH₂–OH

b) (CH₃)₃C–OH ; HO–CH₂CH₂CH₂CH₃ ; (CH₃)₂CH–OH

1.6. RESONANCE

For some organic molecules and molecular ions, a single Lewis structure does not provide an accurate representation of bonding and charge distribution. For example, the Lewis structure of the ionic organic compound sodium formate (HCO_2Na) consists of a negatively charged formate anion and sodium cation (Figure 1.15). The Lewis structure of the formate anion has two different bonds between carbon and oxygen atoms: the double C=O bond and a single bond linking carbon with the oxygen atom bearing a formal negative charge. As we know from Section 1.4, the double bond C=O is usually shorter than the single C–O bond; for example, the length of the C=O bond in formaldehyde ($H_2C=O$) is 1.23 Å and the length of the C–O bond in methanol ($H_3C–OH$) is 1.44 Å (see Figure 1.13 in Section 1.4). The real structure of the formate anion can be experimentally determined by physical methods, such as X-ray diffraction analysis, and it looks quite different from the Lewis structure: it has two equal carbon-oxygen bonds with a bond length of about 1.3 Å, and the negative charge is equally delocalized over two oxygen atoms (Figure 1.15). We do not see the lone pairs of unshared electrons in the experimentally determined real structure. However, the position of atoms and the distribution of electronic density within the molecule can be experimentally observed.

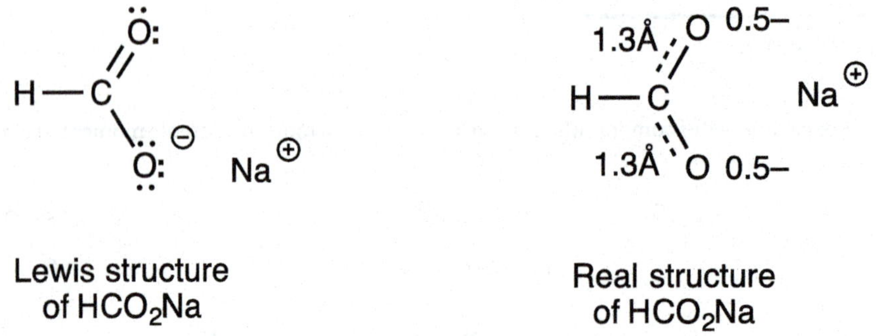

Lewis structure of HCO₂Na

Real structure of HCO₂Na

FIGURE 1.15. The Lewis formula and the real structure of sodium formate.

The failure of a single Lewis structure to show the position of electrons within the molecule is in principle explained by the wave-like behavior of electrons, which occupy with some degree of probability molecular orbitals within the molecule and are not firmly attached to each specific atom or a bond. Usually this problem occurs when an organic molecule has formal charges, multiple bonds, lone pairs, or atoms with incomplete octets of electrons. For such a molecule, several Lewis structures can be drawn, usually called **resonance contributing structures**, with nuclei in the same position, but different electron locations; the real structure, usually named a **resonance hybrid**, is a mixture of all contributing structures. The concept of resonance is of exceptional importance for the field of organic chemistry. It is widely used to describe the mechanisms of organic reactions, and to predict some properties of organic compounds (e.g., the higher acidity of carboxylic acids, compared alcohols (see Chapter 2). In general, the existence of several resonance contributors in a molecular ion leads to a significant stabilization of the ion due to the delocalization of electrical charge over several atoms. The concept of resonance brings elements of quantum mechanics to the simple Lewis structure representation of organic molecules.

Contributing structures can be drawn starting from one initial Lewis structure by changing the position of electrons from double bonds to lone pairs or vice versa. A **curved arrow** is commonly used to show the movement of a pair of electrons from one position to another, and straight **double-headed arrows,** also known as **resonance arrows**, indicate the relationship of Lewis structures as resonance contributing structures. It is important that each contributing structure satisfies the following requirements: 1) it is a correct Lewis structure with no more than 8 valence electrons on each non-hydrogen atom; 2) the position of nuclei can not be changed, only electrons can be moved; and 3) the overall number of electrons for each contributor, and therefore the overall charge, remains unchanged. Figure 1.16 presents the structure of a formate anion using resonance.

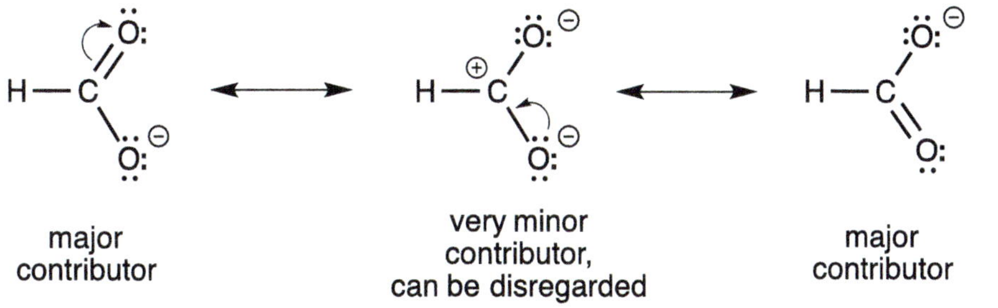

FIGURE 1.16. Resonance contributing structures of a formate anion.

The real structure of a molecule or a molecular ion is a hybrid of all contributing resonance structures but the input of each contributor to the hybrid can differ. For example, the middle contributor of the formate anion shown in Figure 1.16 has a very minor input in the real structure, because it has only six valence electrons on the carbon and two negative charges on both oxygens. The other two contributors have octets of electrons on every non-hydrogen atom and only one charged atom of oxygen. These two major contributors bring equal input to the real structure of the formate anion shown in Figure 1.15. Several simple rules that help identify the relative importance of resonance contributors are listed below.

The most important resonance contributors should have: 1) the largest number of covalent bonds and the smallest number of atoms with formal charges; 2) completely filled valence shells for all atoms; and 3) if there is a negative formal charge in the molecule, it should stay on the most electronegative atom.

In general, a reasonable minor resonance contributor can violate one of these three rules. Violation of two rules results in a minor contributor, which can often be disregarded. For example, the middle structure in Figure 1.16 violates rules one and two, which makes it a very minor, insignificant contributor. The structures that violate all three rules, and also structures with two charges on the same atom, can be disregarded as unreasonable. Figure 1.17 shows examples of resonance representation for several molecules.

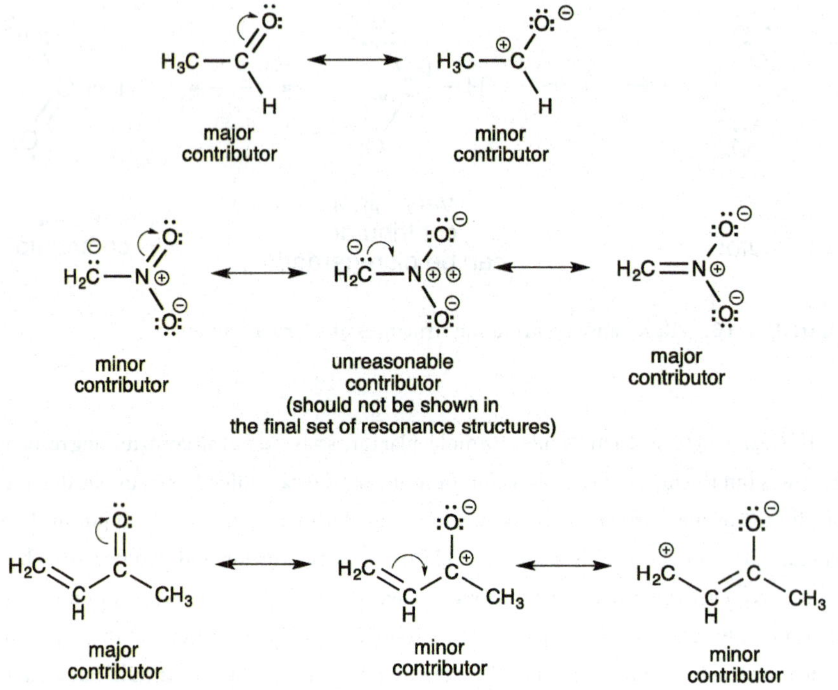

FIGURE 1.17. Examples of resonance representation for molecules and molecular ions.

Problems

1.17. Draw the most important resonance form for each of the species below.

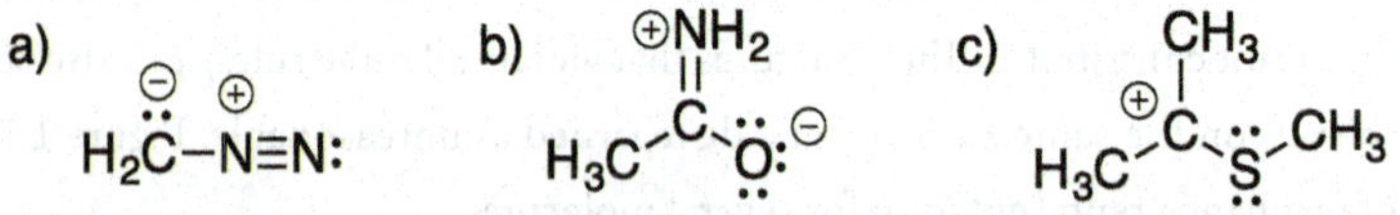

a) $H_2\overset{\ominus}{\underset{..}{C}}-N\overset{\oplus}{\equiv}N:$

b) $\overset{\oplus NH_2}{\underset{\|}{C}}$ $H_3C\overset{}{\underset{}{C}}\overset{..}{\underset{..}{\overset{\ominus}{O}:}}$

c) $H_3C-\overset{\overset{CH_3}{|}}{\underset{}{\overset{\oplus}{C}}}-\overset{..}{\underset{..}{S}}-CH_3$

1.18. Draw the contributing structures of the molecules below, as indicated by the curved arrows.

a)

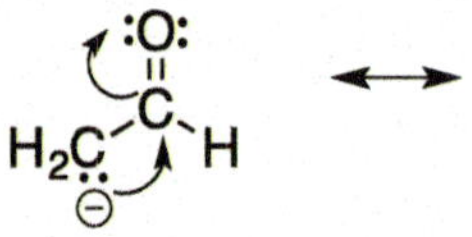

b)

1.19. Complete the resonance forms of the conjugate base of nitromethane ($^-CH_2NO_2$) by adding the missing charges.

1.20. Which of the following pairs of structures depict resonance structures?

1.7. INTRODUCTION TO MOLECULAR ORBITALS AND VALENCE BOND THEORY

Molecular orbital (MO) theory provides the basic methodology for theoretical chemistry. MO theory uses quantum mechanics and the wave theory to calculate the probability of finding electrons in different locations in atoms and molecules. In atoms, electrons occupy atomic orbitals (see Section 1.1); the most important for organic chemistry are orbitals 1s, 2s and the three orthogonal 2p orbitals (see Figure 1.1). MO theory is based on the assumption that electrons in molecules occupy orbitals that belong to the entire molecule, not just to a specific bond between atoms. Molecular orbitals are formed as a mathematical combination of all initial atomic orbitals (AO), and the number of MOs matches the number of initial AOs for each atom. Electrons fill the resulting MOs according to the same rules as for filling AOs (Pauli principle, Aufbau principle, and Hund's rule; see Section 1.1), starting from the MO lowest in energy. Pairs of electrons occupy the lowest energy MOs, which are called **bonding molecular orbitals**; the higher energy, unoccupied MOs are called **antibonding molecular orbitals**.

MO theory is a basic tool for computational studies, but it has only limited application for qualitative rationalization of bonding in organic molecules. The simplified version of MO theory commonly used in organic chemistry is called **valence bond (VB) theory**, and is an important tool for understanding and visualizing covalent bonding. According to VB theory, covalent bonds are formed when AOs of adjacent atoms overlap, and therefore, the bonds are localized between specific atoms, rather than spreading over the entire molecule, as assumed in MO theory. VB theory is in good agreement with Lewis structures, in which two shared electrons form a covalent bond between atoms.

Orbital hybridization (i.e., mixing several different orbitals to form the same number of identical hybrid orbitals) is a key concept of VB theory that explains the shape, reactivity, and electronic structure of organic compounds with single, double, and triple bonds. In particular, the ground state electronic configuration of carbon ($1s^2 2s^2 2p_x^1 2p_y^1$) does not explain the formation of molecules with tetrahedral carbon, such as a molecule of methane (CH_4; see Figure 1.9). Indeed, with this electronic configuration, we should expect the formation of a molecule :CH_2, by overlapping singly occupied $2p_x^1$ and $2p_y^1$ orbitals on carbon with the s^1 orbitals of two hydrogens. However, a molecule of :CH_2 with two single bonds and a lone pair formed by the $2s^2$ electrons will have an incomplete octet (six electrons) on carbon and therefore, will be lacking the stable, completely filled electronic configuration of a noble gas. Mixing the 2s orbital and three 2p orbitals on a carbon atom to produce four equivalent **sp³ hybrid orbitals** results in the formation of a tetrahedral molecule of CH_4 (Figure 1.18). All four sp³ orbitals have equal energy levels, closer to the 2p orbital level, but between that of the 2p and 2s orbitals, and are occupied by four single valence electrons. Each sp³ orbital consists of a larger lobe and a smaller back lobe with a node in the center (Figure 1.18). The core 1s orbital is not involved in orbital hybridization.

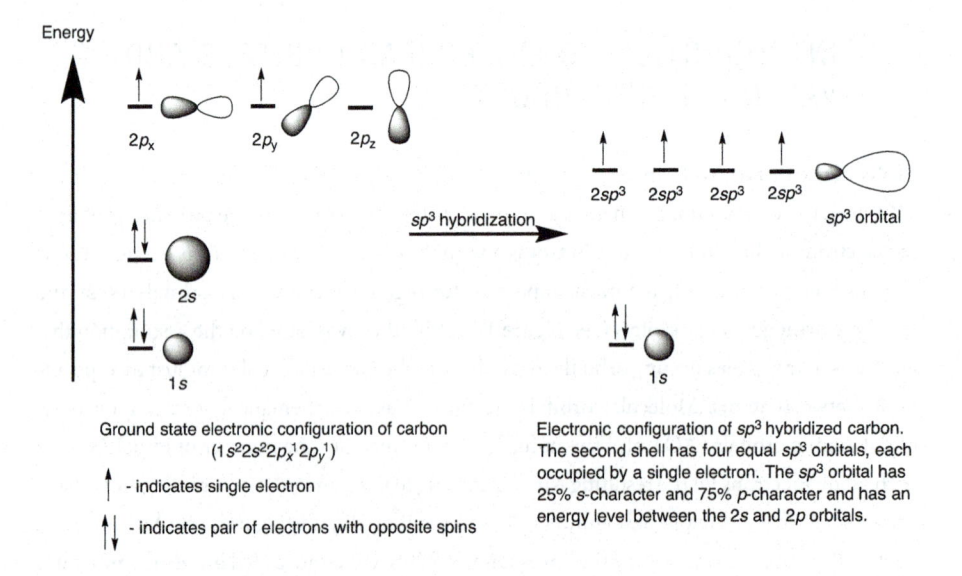

FIGURE 1.18. Formation of sp³ hybrid orbitals.

The four sp³ orbitals are arranged in space in the same direction as four bonds in a molecule of methane (Figure 1.19). The bonds to hydrogen atoms in a molecule of CH_4 are formed by the overlap of the singly occupied sp³ orbitals with singly occupied s orbitals of four hydrogen atoms. A pair of bonding electrons with opposite spins occupies each of the resulting four covalent bonds. Bonds formed by head-on overlapping of atomic orbitals with a cylindrical shape along the axis between the two connected nuclei are called **sigma (σ) bonds**. A molecule of ethane (CH_3CH_3) is formed in a similar fashion by overlapping sp³ orbitals of two sp³ hybridized carbon atoms, and s orbitals of six hydrogen atoms, as shown in Figure 1.19. A molecule of ethane has six C–H σ bonds and one C–C σ bond that is formed by the overlap of two sp³ orbitals on the carbon atoms. In general, a molecule of any alkane is built from sp³ hybridized carbon atoms and has C–C and C–H σ bonds between atoms.

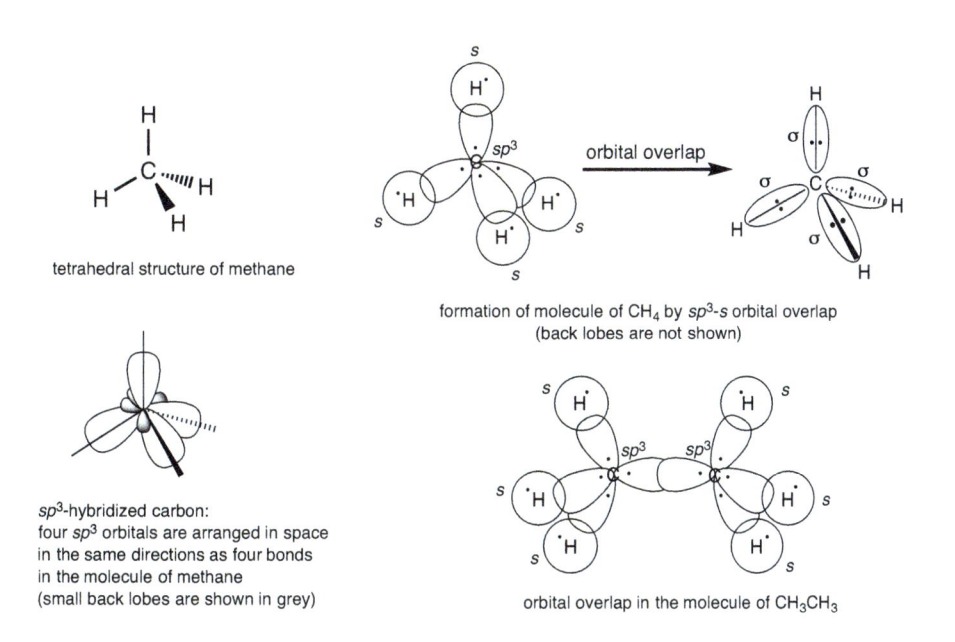

FIGURE 1.19. Formation of methane and ethane molecules from sp³ hybridized carbon atoms and hydrogen atoms.

Similar to alkanes, compounds of nitrogen and oxygen with tetrahedral electronic geometry (see Figure 1.10) also have sp³ hybridized central atoms with appropriate placement of valence electrons. For example, the molecule :NH_3 is formed by the overlap of three singly occupied sp³ orbitals on nitrogen with singly occupied s orbitals of three hydrogen atoms, and a lone pair occupying the fourth sp³ orbital. All single bonds in compounds of sp³ hybridized nitrogen or oxygen are σ bonds.

A different type of bonding is found in compounds with double and triple bonds. The formation of a trigonal planar molecule of ethylene ($CH_2=CH_2$) is explained by **sp² hybridization** of a carbon atom. In sp² hybridized carbon, the 2s orbital and two 2p orbitals are mixed, producing three equivalent **sp² hybrid orbitals**, while the third 2p orbital remains unchanged (Figure 1.20). All three sp² orbitals have equal energy, with energy levels between

2p and 2s orbitals, but lower than that of sp³ orbitals, and are occupied by single valence electrons. Each sp² orbital consists of a larger lobe, which is smaller in size compared to the sp³ orbital, and a smaller back lobe with a node in the center (Figure 1.20). A single electron also occupies the unchanged 2p orbital.

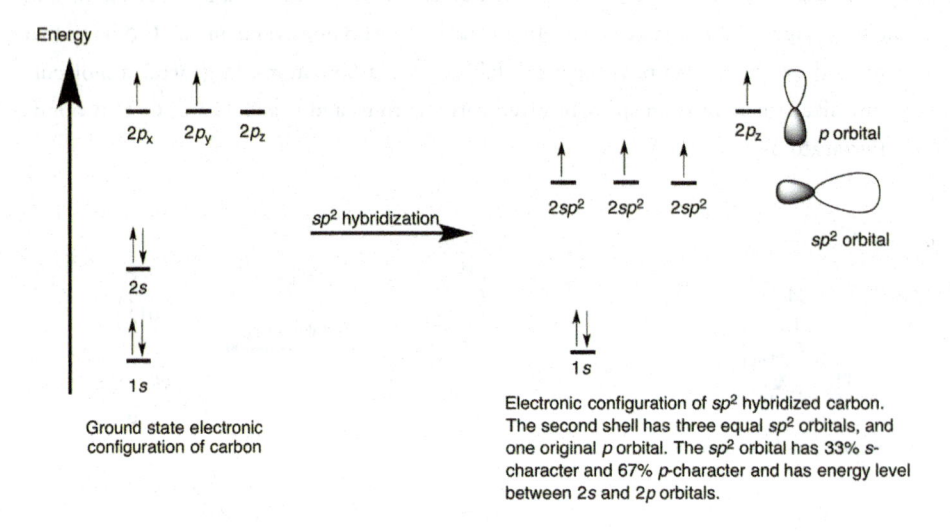

FIGURE 1.20. Formation of sp² hybrid orbitals.

The three sp² orbitals are arranged in the same plane as the carbon nucleus, in agreement with the trigonal planar geometry of the molecule $CH_2=CH_2$ (Figure 1.21). The σ bonds to hydrogen atoms in the molecule of $CH_2=CH_2$ are formed by the overlap of singly occupied sp² orbitals with singly occupied s orbitals of four hydrogen atoms, and one C–C σ bond is formed by the overlap of two sp² orbitals on the carbon atoms. A pair of bonding electrons with opposite spins occupies each of the resulting five σ bonds. In addition to the σ bond, the molecule has one π **bond**, where two lobes of one non-hybridized p orbital overlap side-by-side with two lobes of the p orbital on the second carbon. The π bond has two lobes, one above and one below the planar σ bond framework. In general, any double bond between carbon atoms, or C=N and C=O double bonds (see Figure 1.11 for examples of such molecules) is a combination of a σ bond and a π bond, and the atoms connected by a double bond have sp² hybridization. Other trigonal planar molecules, such as BF_3 (Figure 1.11) or carbocationic intermediates, such as H_3C^+, also have sp² hybridization of the central atom with an unoccupied p orbital.

trigonal planar structure of ethylene

orbital overlap

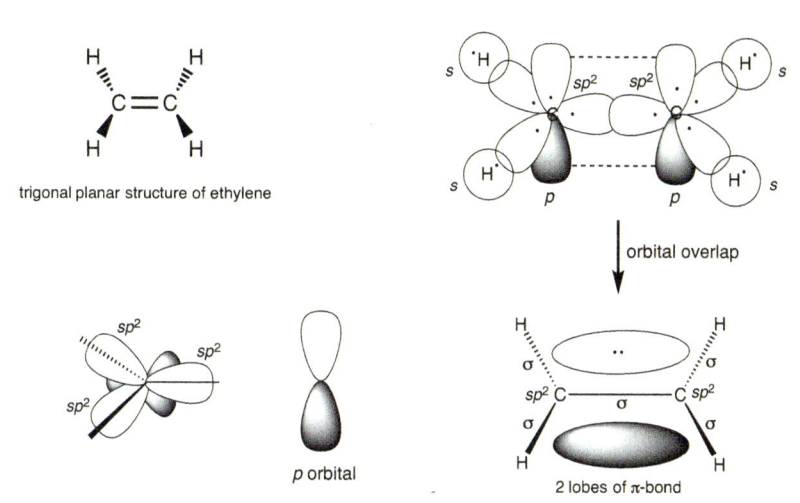

p orbital

sp²-hybridized carbon:
three *sp²* orbitals are arranged in space in the same directions as three bonds to carbon in the molecule of ethylene. (small back lobes are shown in grey)

2 lobes of π-bond

Formation of a molecule of ethylene from *sp²*-hybridized carbons. The bonds are formed by *sp²-sp²*, *s-sp²*, and *p-p* (side-by-side) orbital overlap. The σ-bonds are shown as lines, and the π-bond is shown as two lobes, one above and the second below the plane of the σ-bond framework.
(back lobes of *sp²* orbitals are not shown)

FIGURE 1.21. Formation of the molecule of ethylene from sp² hybridized carbon atoms and hydrogen atoms.

Finally, the formation of a linear molecule of acetylene (HCCH) is explained by **sp hybridization** of a carbon atom. In sp hybridized carbon, the 2s orbital and one 2p orbital are mixed, producing two equivalent **sp hybrid orbitals**, while the other two 2p orbitals remain unchanged (Figure 1.22). Both sp orbitals have equal energy (an energy level between those of the 2p and 2s orbitals, and lower compared to the sp² orbitals) and are occupied by single

Ground state electronic configuration of carbon

sp hybridization

sp orbital

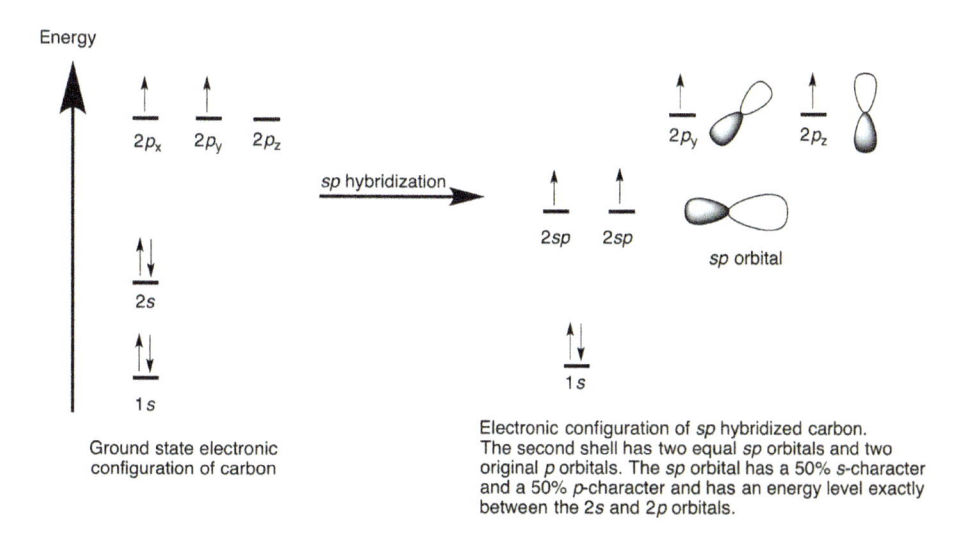

Electronic configuration of *sp* hybridized carbon. The second shell has two equal *sp* orbitals and two original *p* orbitals. The *sp* orbital has a 50% *s*-character and a 50% *p*-character and has an energy level exactly between the 2s and 2p orbitals.

FIGURE 1.22. Formation of sp hybrid orbitals.

COVALENT BONDING AND MOLECULAR STRUCTURE

valence electrons. Each sp orbital consists of a larger lobe, which is smaller in size than that of the sp² orbital, and a smaller back lobe with a node in the center (Figure 1.22). Single electrons also occupy the unchanged 2p orbitals.

The two sp orbitals are arranged in a line with the carbon nucleus in the center, in agreement with the linear geometry of a molecule of acetylene (Figure 1.23). The σ bonds to hydrogen atoms in the molecule of acetylene are formed by the overlap of singly occupied sp orbitals with the singly occupied s orbitals of two hydrogen atoms, and one C–C σ bond is formed by the overlap of two sp orbitals on carbon atoms. A pair of bonding electrons with opposite spins occupies each of the resulting three σ bonds. In addition to the C–C σ bond, the molecule has two orthogonal π bonds, in which two lobes of each non-hybridized p orbital overlap side-by-side with two lobes of the respective p orbital on the second carbon. One π bond is in the plane of the drawing, and the second π bond is in the perpendicular plane (Figure 1.23). Each of the two π bonds is occupied by a pair of bonding electrons with opposite spins. In general, any triple bond between carbon atoms, or the CN triple bond (see Figure 1.12 for examples of such molecules), is a combination of a σ bond and two orthogonal π bonds, and the atoms connected by a triple bond have sp hybridization. Other linear molecules, such as CO_2 (Figure 1.11) or allene ($H_2C=C=CH_2$) also have sp hybridization of the central atom.

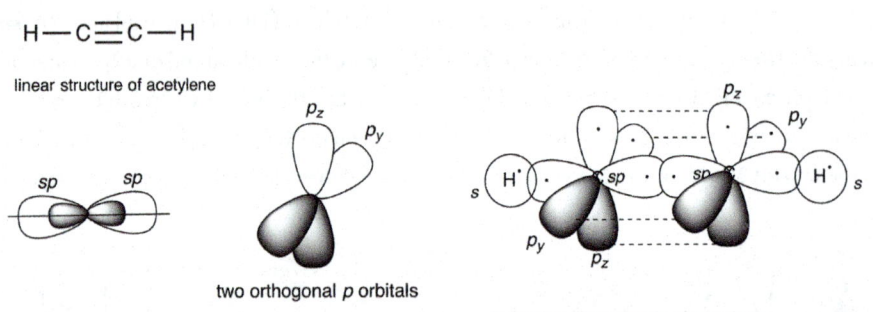

FIGURE 1.23. Formation of the molecule of acetylene from sp hybridized carbon atoms and hydrogen atoms.

The concept of orbital hybridization is important for understanding the structure and reactivity of organic molecules. In particular, the existence of different types of σ bonds account for the relative bond lengths of the C–H single bonds in organic compounds of different classes, as shown in Figure 1.13 (see Section 1.4). In alkynes, the C–H σ bond (1.09 Å) is formed by the overlap of the smallest sized sp orbital with the s orbital of hydrogen.

The main lobe of the sp² orbital is larger, and the sp³ orbital is the largest, which explains the longer C–H bond in alkenes (1.10 Å) and the longest C–H bond in alkanes (1.11 Å). **Bond strength**, defined as the amount of energy required to break a covalent bond, also known as the **bond dissociation energy (or BDE)**, is related to the bond length; in general, the longer covalent bond between two elements is weaker than the shorter bond between atoms of the same element. For example, the single C–C bond (BDE = 85 to 90 kcal/mol) is weaker than the shorter double bond (BDE = 150 to 170 kcal/mol), and the shortest triple bond is the strongest bond (BDE = 200-230 kcal/mol). This order of bond strengths is explained by combination of σ bond and π bonds in alkenes and alkynes, with a π bond being generally weaker than a σ bond. The presence of π bonds in alkenes and alkynes also explains the special reactivity of these compounds in electrophilic addition reactions discussed in Chapters 7 and 8.

Problems

1.21. Which atomic orbitals overlap to form the carbon-carbon bond of acetonitrile (CH_3CN)?

1.22. Why is the center carbon atom of an allene molecule ($H_2C=C=CH_2$) sp-hybridized?

1.23. Consider the molecule in the box and answer the following questions.

a) How many σ bonds are formed by the overlap of the two sp orbitals?
b) How many σ bonds are formed by the overlap of the sp and sp² orbitals?
c) How many σ bonds are formed by the overlap of the sp and sp³ orbitals?
d) How many σ bonds are formed by the overlap of the sp² and sp² orbitals?
e) How many σ bonds are formed by the overlap of the sp³ and sp³ orbitals?
f) How many σ bonds are formed by the overlap of the s and sp orbitals?
g) How many σ bonds are formed by the overlap of the s and sp² orbitals?
h) How many σ bonds are formed by the overlap of the s and sp³ orbitals?
i) How many σ bonds are formed by the overlap of the s and s orbitals?
j) What is the total number of π bonds in the whole molecule?

CHAPTER 2

Proton Transfer Reactions in Organic Chemistry

Reactions involving the transfer of protons from one molecule (an **acid**) to another (a **base**) are extremely important in nature. The proton (H⁺) has a vacant electronic shell and therefore it is a very small particle, with a diameter of about 0.88×10^{-15} m, as compared to the hydrogen atom's (H•) size of about 0.53×10^{-10} m. Because of its small size, a proton can jump from one molecule to another very quickly, almost like an electron. This sets proton transfer reactions apart from other chemical reactions involving movement of atoms or molecules. It is important to understand that protons can exist as single particles (H⁺ or p⁺) only in vacuum, in the absence of molecules. In solution or in solid state, a proton always combines with molecules; for example, in water, it exists mainly as the hydronium ion, H_3O^+. However, chemists commonly use a simplified symbol 'H⁺' to indicate a proton combined with water or other molecules in a solution.

2.1. DEFINITION OF ACIDS AND BASES

Several definitions of acids and bases exist. The Swedish scientist **Svante Arrhenius** proposed one of the first definitions in 1884. According to Arrhenius, an acid is a substance that dissolves in water, releasing protons, while a base is a source of hydroxide anions (HO⁻) in an aqueous solution. However, many organic reactions do not involve water, and so this definition has limited applicability.

The **Brønsted-Lowry theory**, which was proposed independently by Johannes Nicolaus Brønsted and Thomas Martin Lowry in 1923, provides a more general description of acid-base reactions. According to this theory, an acid (A–H) is a donor of protons and a base (:B⁻) is an acceptor of protons, as shown in the following **proton-transfer** equilibrium (Figure 2.1).

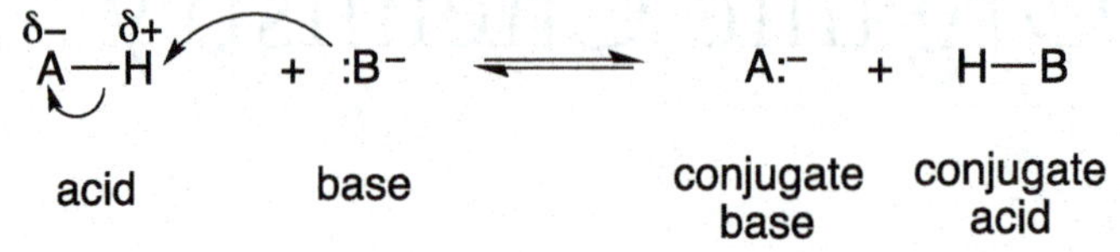

$$\overset{\delta-\;\;\delta+}{A-H} \;+\; :B^- \;\rightleftharpoons\; A:^- \;+\; H-B$$

acid   base   conjugate base   conjugate acid

FIGURE 2.1. Brønsted-Lowry definition of acids and bases.

The proton is transferred from an acid (A–H) to a base (:B⁻) producing a **conjugate base** (A:⁻) and a **conjugate acid** (H–B). It is essential that the acid has a polar A–H bond with a partial positive charge on the hydrogen atom; hydrogen compounds of elements that are less electronegative than hydrogen cannot donate protons. In fact, metal hydrides such as lithium hydride, Li⁺:H⁻, act as strong bases, but not acids. The base (:B⁻) must have a lone electron pair in order to accept a proton by forming a covalent bond H–B. The basic atom B may have a full negative charge, or a partial negative charge in a covalent molecule.

The curved arrows shown in Figure 2.1 indicate movement of electrons from the non-bonded position on the basic atom B to form the covalent bond H–B and a simultaneous cleavage of the bond A–H to give the conjugate base :A⁻. It is important to keep in mind that in any acid-base reaction, the base acts as a **donor of electrons** and the acid is an **acceptor of electrons**, as shown by the curved arrows. This idea is the basis for the most general definition of acids and bases, proposed by Gilbert N. Lewis. A **Lewis acid** is defined as any compound acting as an acceptor of a pair of electrons from a base. The Lewis definition of acids includes the Brønsted-Lowry acids, also called **protic acids**, and compounds of elements with an incomplete octet of electrons, such as boron fluoride BF_3 and aluminum chloride $AlCl_3$ (see also Section 1.2). In related terminology (see Chapters 5 and 7), the acceptors of electrons (i.e., Lewis acids) are commonly called **electrophiles**, meaning "electron loving" and the donors of electrons are called **nucleophiles**, meaning "nucleus loving".

In modern organic chemistry, the Brønsted-Lowry definition of acids and bases is most commonly used for the description of proton transfer reactions, while the term Lewis acid is reserved for compounds of elements with incomplete octet of electrons.

2.2. ACID-BASE EQUILIBRIUM AND RELATIVE STRENGTH OF ACIDS

Relative strength of an acid (A–H) is defined by the position of acid-base equilibrium shown in Figure 2.1. For a strong acid, this equilibrium will be almost completely shifted to the right side, producing a conjugate base (:A⁻) and a conjugate acid (H–B) as major components of the mixture. **Strong inorganic acids,** such as sulfuric acid (H_2SO_4), hydrogen chloride (HCl), hydrogen bromide (HBr), and hydrogen iodide (HI) are the most powerful donors of protons that always act as an acid in the acid-base equilibrium. On the other side, some ionic inorganic compounds, such as sodium hydroxide (NaOH), sodium amide ($NaNH_2$), and sodium hydride (NaH) always act as **strong bases** and never serve as proton donors in an acid-base equilibrium. In contrast, most organic compounds can act either as proton acceptors or proton donors in the presence of strong inorganic acids or bases. For example, methanol (CH_3OH) functions as a proton acceptor in the presence of HCl, but functions as a proton donor in the presence of NaOH (Figure 2.2). It should be noted that in this equilibrium hydroxide anion (HO⁻) acts as the actual base, while sodium cation (Na⁺) stays unchanged in the solution, and thus, is considered a **spectator cation**.

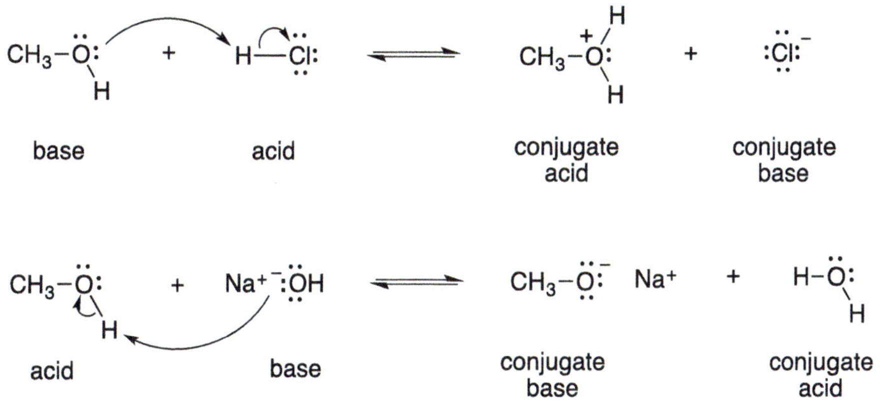

FIGURE 2.2. Methanol reactivity as a base or as an acid.

The strength of an acid (A–H) can be quantitatively characterized by measuring the position of acid-base equilibrium in water solution (Figure 2.3). The equilibrium in this reversible reaction is characterized by an equilibrium constant, K_{eq}, which can be calculated using the experimentally measured molar concentrations of each component (shown as [AH], [H_2O], [A⁻], and [H_3O^+] in Figure 2.3). For dilute aqueous solutions, the molar concentration of water is almost constant, [H_2O] = 55.6 mol/L, which allows simplification of the equation using the combined constant $K_{eq}[H_2O]$, called an **acid dissociation constant** (K_a). The value of K_a provides a quantitative measure of an acid's strength in aqueous solution. Stronger acids have larger K_a values, because A^- and H_3O^+ are the predominant species in the equilibrium. For weak organic acids, the values of K_a usually are very small numbers with a negative exponent; for example, the K_a of acetic acid CH_3CO_2H is 1.74×10^{-5}. It is more convenient to use a logarithm of this constant with a negative sign: **$pK_a = -\log_{10} K_a$**. For acetic acid, the pK_a value is 4.76. Stronger acids have lower pK_a than weaker acids. Strong acids A–H, for which acid-base equilibrium (Figure 2.3) is significantly shifted to the right side, have negative pK_a values.

$$A-H + H_2O \xrightleftharpoons{K_{eq}} A{:}^- + H_3O^+$$

$$K_{eq} = \frac{[A{:}^-][H_3O^+]}{[AH][H_2O]}, \text{ where [AH], [H}_2\text{O], [A:}^-\text{], [H}_3\text{O}^+\text{] are concentrations in mol/L, for dilute solutions [H}_2\text{O] = 55.6 mol/L}$$

$$K_a = K_{eq}[H_2O] = \frac{[A{:}^-][H_3O^+]}{[AH]}$$

$$pK_a = -\log_{10} K_a$$

FIGURE 2.3. Definition of acid dissociation constant (K_a) and pK_a.

Representative pK_a values for some organic and inorganic compounds are listed in Table 2.1. It should be noted that these numbers are not very accurate for the weakest acids like alkanes and for the strongest acids, which are almost completely ionized in the water producing the hydronium ion H_3O^+.

TABLE 2.1. pK_a values for some organic and inorganic compounds.

COMPOUND (A–H)	pK_a	CONJUGATE BASE
CH_4	50–60	$^-:CH_3$
$CH_2=CH_2$	44	$^-:CH=CH_2$
NH_3	38	$^-:NH_2$
H_2	35	$^-:H$
$HC\equiv CH$	25	$^-:C\equiv CH$
CH_3CH_2OH	15.9	$CH_3CH_2O^-$
H_2O	15.7	HO^-
NH_4^+	9.24	$:NH_3$
CH_3CO_2H	4.76	$CH_3CO_2^-$
HF	3.2	F^-
H_3O^+	-1.74	H_2O
H_2SO_4	-5	HSO_4^-
HCl	-7	Cl^-
HBr	-8	Br^-
HI	-9	I^-

Relative **basicity** of a Brønsted-Lowry base is defined as the tendency of a compound to act as a proton acceptor in the acid-base equilibrium. Basicity of compound A: or anion A:⁻ is quantitatively measured by the pK_a of the corresponding conjugate acid AH⁺ or AH. Larger pK_a values of conjugate acids correspond to higher basicity of the base. The anions of the strongest acids, such as I⁻, Br⁻, Cl⁻, and HSO_4^- are very weak bases, while the anions of weak acids, such as ⁻:CH_3, ⁻:NH_2, $CH_3CH_2O^-$, and HO⁻ are strong bases. In general, the negatively charged bases have much higher basicity than the related uncharged molecules. For example, ⁻:NH_2 is a much stronger base than :NH_3 as measured by the pK_a values of respective conjugate acids (pK_a of NH_3 = 38 and pK_a of NH_4^+ = 9.24). Likewise, HO⁻ is a much stronger base than H_2O (pK_a of H_2O = 15.7 and pK_a of H_3O^+ = −1.74). It should be noted that the acidity of positively charged acids (such as H_3O^+ and NH_4^+) is always much higher than the acidity of related neutral molecules (H_2O and NH_3), while the negatively charged species (HO⁻ and NH_2^-) can never act as a proton donor.

Values of pK_a for the acid and the conjugated acid can be used to predict the direction of a proton transfer reaction and evaluate the position of acid-base equilibrium. The general rule is that protons are transferred from a stronger acid (i.e., lower pK_a value) to make a weaker conjugate acid (i.e., higher pK_a value). For example, ethanol (pK_a = 15.9) reacts with sodium hydride Na⁺:H⁻, transferring a proton to the hydride anion (:H⁻) to produce a molecule of hydrogen H_2 as the conjugate acid (Figure 2.4, equation 1). Hydrogen (pK_a = 35) is a much weaker acid than ethanol (pK_a = 15.9), with a difference in the acidity constant (K_a) of about

10^{19}. In principle, any acid transfer reaction is an equilibrium; however, with such a huge difference in acidity between an acid and a conjugate acid, this reaction (equation 1 below) can be considered a one-way process. Note that Na⁺ is a spectator ion and not involved in the proton transfer process.

$$CH_3CH_2OH + NaH \longrightarrow CH_3CH_2ONa + H_2 \quad \text{(eq. 1)}$$
$$pK_a = 15.9 \hspace{6cm} pK_a = 35$$

$$CH_3CH_2OH + NH_3 \rightleftharpoons CH_3CH_2O^- + {}^+NH_4 \quad \text{(eq. 2)}$$
$$pK_a = 15.9 \hspace{6cm} pK_a = 9.24$$

$$CH_3CH_2OH + NaOH \rightleftharpoons CH_3CH_2ONa + H_2O \quad \text{(eq. 3)}$$
$$pK_a = 15.9 \hspace{6cm} pK_a = 15.7$$

$$CH_3C{\equiv}CH + NaNH_2 \longrightarrow CH_3C{\equiv}CNa + NH_3 \quad \text{(eq. 4)}$$
$$pK_a = 25 \hspace{6cm} pK_a = 38$$

FIGURE 2.4. Examples of proton transfer reactions.

Reaction of ethanol with ammonia NH_3 is an equilibrium in which molecules of CH_3CH_2OH and NH_3 are the dominating species because ${}^+NH_4$ is a stronger acid than ethanol (Figure 2.4, eq. 2). In this case, the difference in pK_a is just about 6 pK_a units, so the reaction can be viewed as an uneven equilibrium (eq. 2). In the third case (eq. 3), the reaction is almost even equilibrium because pK_a values for the acid and the conjugate acid are very close (15.9 and 15.7). In the last case (eq. 4), the reaction is almost completely shifted to the right. In this case, the molecule of acid (propyne) has two different types of hydrogens: those in the CH_3-group (pK_a about 50, the same as in methane CH_4) and the acetylenic hydrogen ($pK_a = 25$, the same as in acetylene C_2H_2). Only the sufficiently acidic acetylenic hydrogen can participate in an acid transfer reaction with the amide $^-{:}NH_2$ to give NH_3 ($pK_a = 38$) as the conjugate acid (eq. 4).

Problems

2.1. For each of the following compounds, draw the structures of its conjugate base and its conjugate acid:

A) $H_2C{=}CH_2$ B) CH_3CH_2OH C) NH_3 D) HF

2.2. Use table 2.1 to identify the stronger base within each pair.

a) $^-$:CH$_3$ and $^-$:C≡CH

b) CH$_3$CH$_2$O$^-$ and CH$_3$CO$_2^-$

c) F$^-$ and I$^-$

d) $^-$:CH$_3$ and $^-$:NH$_2$

2.3. For each reaction, identify the strongest base, the strongest acid, the weakest acid, and the weakest base; indicate the side of equilibrium.

a) H$_3$C-C(=O)-CH$_3$ + HO-C(=O)-O$^-$ ⇌ H$_3$C-C(=O)-CH$_2^-$ + HO-C(=O)-OH

pk$_a$ = 19.3   pk$_a$ = 6.1

b) H$_2$C=CH$_2$ + H-I ⇌ H$_3$C-CH$_2^+$ + I$^-$

pk$_a$ = -9   pk$_a$ = -3

c) H$_3$C-C(=O)-O$^-$ + H-C≡N ⇌ H$_3$C-C(=O)-OH + $^-$C≡N

pk$_a$ = 9   pk$_a$ = 4.76

2.3. EFFECTS OF MOLECULAR STRUCTURE ON ACIDITY

As a general rule, the acidity of an acid H–A is predetermined by the relative stability of the conjugate base :A$^-$. Any factor stabilizing the anion :A$^-$ will shift the acid-base equilibrium to the right (Figure 2.3).

The position of element A in the periodic table is one of the most important factors determining acidity of the H–A bond. The first three elements of the second period, Li, Be, and B, are not able to form acidic A–H compounds because these elements have lower electronegativity than hydrogen. The acidity of simple hydrogen derivatives of the next elements in the second period, CH$_4$ (pK$_a$ about 50), NH$_3$ (pK$_a$ = 38), H$_2$O (pK$_a$ = 15.7), and HF (pK$_a$ = 3.2), is significantly increasing from carbon to fluorine owing to the increased electronegativity of these elements. When the size of atoms is about the same, the stronger tendency of an element to attract electrons has a significant stabilizing effect on the anions. The carbanion (H$_3$C:$^-$) is less stable than the fluoride anion (F$^-$), and therefore CH$_4$ is a weaker acid than HF. The size of the atom bearing a negative charge has an even stronger effect on the stability of an anion. Within a group in the periodic table, the hydrogen derivatives of the lower elements have significantly higher acidity: HF (pK_a = 3.2) is a much weaker acid than HCl (pK_a = –7), HBr (pK_a = –8), and HI (pK_a = –9). Larger atom sizes allow better delocalization of the negative charge over a larger volume of space, resulting in anions with higher stability. Figure 2.5 illustrates the general pattern of acidity for simple hydrogen derivatives of some elements.

Acidity within the 2nd period:

stronger acid ⟶

Acid:	H₃C–H	H₂N–H	HO–H	F–H
pKₐ:	50-60	38	15.7	3.2
Conjugate base:	H₃C:⁻	H₂N⁻	HO⁻	F⁻

⟵ stronger base

Acidity within group 17:

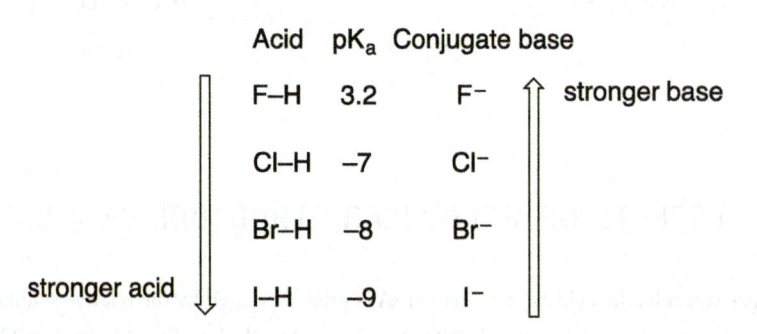

	Acid	pKₐ	Conjugate base	
	F–H	3.2	F⁻	↑ stronger base
	Cl–H	–7	Cl⁻	
	Br–H	–8	Br⁻	
stronger acid ↓	I–H	–9	I⁻	

FIGURE 2.5. Acidity of simple hydrogen derivatives of elements in the periodic table.

The relative acidity of the H–A bond in complex molecules can be significantly different from element A's simple hydrogen derivatives. Hybridization of carbon atoms has a very strong effect on the C–H acidity. Methane and other alkanes, which have hydrogen directly connected to the sp³ hybridized carbon atoms, are much weaker acids (pK_a about 50–60) than alkenes with sp² hybridized C–H bonds (pK_a about 44); alkynes with sp hybridized C–H bonds have the highest acidity (pK_a about 25). This difference in acidity reflects the greater stability (lower energy level) of the alkynyl anion in which the lone electron pair occupies the sp hybridized orbital (50% s character), as compared to the less stable alkenyl anion (sp² hybridized orbital, 33% s character), and the least stable alkyl anion (sp³ hybridized orbital, 25% s character). Hybridized orbitals with higher s characters are lower in energy and therefore, more stable (see Section 1.7).

Acidity of the O–H bond can be strongly affected by the **resonance stabilization** of the anion. For example, methanol CH_3OH (pK_a = 15.5) is a much weaker acid compared to formic acid HCOOH (pK_a = 3.75) because its anion (CH_3O^-) has no resonance contributors, whereas the negative charge in a formate anion (HCOO⁻) is delocalized over two oxygen atoms, (represented by two equal resonance contributors in Figure 1.16, in Section 1.6). In general, charge delocalization in the anion :A⁻ leads to increased acidity of the acid A–H.

Finally, the relative acidity of compounds bearing the same acidic functional group can be slightly influenced by a remote substituent, which can cause polarization of the chain of bonds leading to the charged atom in the anion. The example alcohols in Figure 2.6 illustrate this **inductive effect** of a substituent.

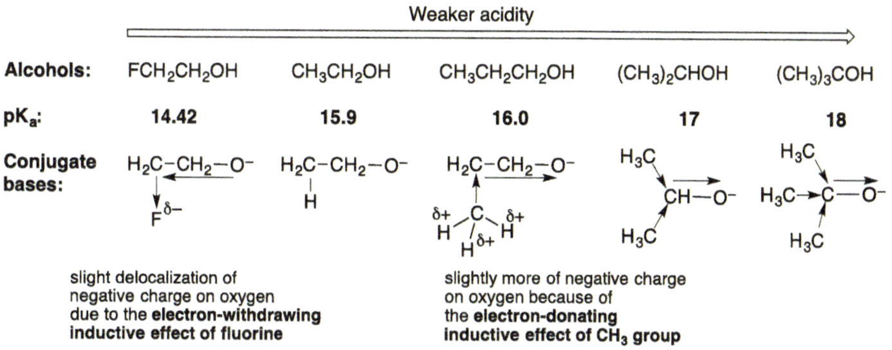

FIGURE 2.6. Inductive effect of a remote substituent on acidity of alcohols.

The first alcohol, 2-fluoroethanol (pK_a = 14.42) is slightly more acidic than ethanol (pK_a = 15.9) because the negative charge on the anion is slightly delocalized, owing to the shift of electronic density to the more electronegative fluorine atom (indicated by the arrows in Figure 2.6). Such a shift of electronic density toward the electronegative substituent is termed an **electron-withdrawing inductive effect**. The third alcohol, 1-propanol (pK_a = 16) is slightly less acidic than ethanol, because the additional CH_3-group has a weak **electron-donating inductive effect**, due to a very weak shift of electronic density from the less electronegative hydrogen atoms. This electron-donating effect results in a slight increase of negative charge on the oxygen atom in 1-propanol, compared to ethanol, which has destabilizing effect on the anion. The presence of additional CH_3-groups in the molecules of isopropanol $(CH_3)_2CHOH$ (pK_a = 17) and *tert*-butanol $(CH_3)_3COH$ (pK_a = 18) explains the further reduction of acidity in this group of compounds (Figure 2.6). In general, the inductive effect has a very small influence on the acidity (0.1-3 units of pK_a), and it sharply decreases with increased distance between the substituent and the atom bearing a negative charge in the anion. Inductive effect also explains differences in the acidity of substituted carboxylic acids (see Chapter 13) and some other classes of organic compounds.

Problems

2.4. Sort each set of compounds according to increasing relative acidity (least acidic to most acidic).

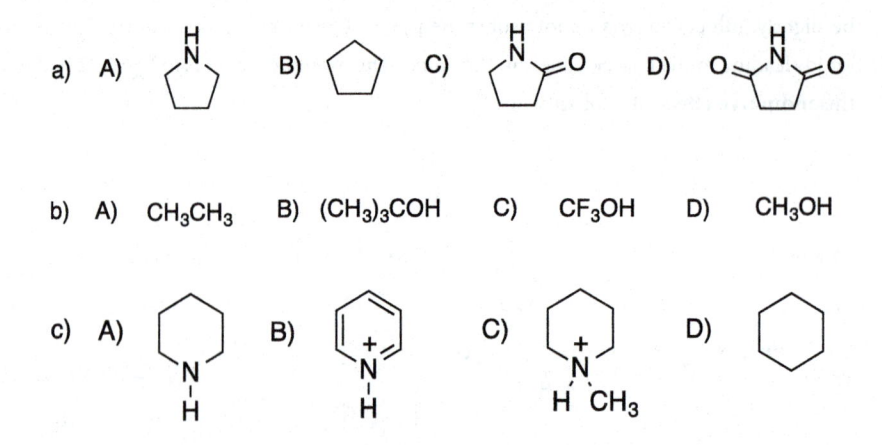

a) A) pyrrolidine (N-H) B) cyclopentane C) 2-pyrrolidinone (N-H, C=O) D) succinimide (O=C-N(H)-C=O)

b) A) CH_3CH_3 B) $(CH_3)_3COH$ C) CF_3OH D) CH_3OH

c) A) piperidine (N-H) B) pyridinium (N-H, aromatic) C) N-methyl piperidinium (H, CH₃) D) cyclohexane

2.5. Sort the following set of compounds according to increasing relative basicity (least basic to most basic):

A) CH_3S^- B) $CH_3CH_2^-$ C) $(CH_3)_2N^-$ D) CH_3O^-

2.6. All acid base reactions proceed as written. For each reaction indicate the strongest base, the strongest acid, the weakest base and the weakest acid. Use curved arrows to indicate the flow of electrons.

a) H₃C-Li + H-Ö-H ⟶ CH₄ + Li⁺ :Ö-H

b) $CH_3-C(=O)-CH_3$ + H-Ö⁺H-H ⟶ $CH_3-C(=O^+H)-CH_3$ + H-Ö-H

c) $H-\overset{\overset{\displaystyle :\ddot{O}:}{\|}}{C}-\ddot{\underset{\ddot{}}{O}}:^{-}$ + $CH_3-\ddot{\underset{\ddot{}}{O}}-H$ ⟶ $H-\overset{\overset{\displaystyle :\ddot{O}:}{\|}}{C}-\ddot{\underset{\ddot{}}{O}}-H$ + $CH_3-\ddot{\underset{\ddot{}}{O}}:^{-}$

d) $H_3C-C\equiv C-H$ + $^-:\ddot{N}H_2$ ⟶ $H_3C-C\equiv C:^-$ + $\ddot{N}H_3$

2.7. For each compound indicate the most acidic proton or group of protons.

a) $CH_3\overset{\overset{\displaystyle O}{\|}}{C}CH_2\overset{\overset{\displaystyle O}{\|}}{C}CH_3$ b) $H_3C-C\equiv CH$ c) $CH_3\overset{\overset{\displaystyle O}{\|}}{C}OCH_2CH_3$

d) $CH_3CH=CH\overset{\overset{\displaystyle O}{\|}}{C}C(CH_3)_3$ e) $H_2NCH_2\overset{\overset{\displaystyle O}{\|}}{C}OH$ f) ascorbic acid structure (HO–CH(OH)–[lactone ring with HO, OH, =O])

2.8. Indicate the most basic site in each of the following molecules:

a) $H_2\ddot{N}-CH_2CH_2-\ddot{\underset{\ddot{}}{O}}-H$ b) $H_3C-\overset{\overset{\displaystyle :\ddot{O}:}{\|}}{C}-\ddot{\underset{\ddot{}}{O}}-CH_2CH_3$

c) $\underset{\underset{\ddot{N}H_2}{|}}{H_2C}-\overset{\overset{\displaystyle :\ddot{O}:}{\|}}{C}-\ddot{\underset{\ddot{}}{O}}-H$ d) $H_3C-\overset{\overset{\displaystyle :\ddot{O}:}{\|}}{C}-\underset{\underset{\displaystyle CH_2CH_3}{|}}{\ddot{N}}-CH_2CH_3$

CHAPTER 3

Alkanes and Cycloalkanes

Simple alkanes, and especially methane (CH_4), are the most abundant organic compounds on earth and in the solar system. The most important sources of alkanes on earth are natural gas and oil. Natural gas contains primarily methane and ethane, with small amounts of propane and butane; oil is a liquid mixture of higher alkanes and other hydrocarbons. Naturally occurring alkanes are of irreplaceable importance for our civilization as a primary source of energy and feedstock for the chemical industry. This chapter provides an overview of nomenclature, structural features of alkanes and cycloalkanes, and provides a background for understanding their reactions, such as oxidation and halogenation of alkanes.

3.1. STRUCTURE OF ALKANES AND CONSTITUTIONAL ISOMERISM

Alkanes are classified as saturated hydrocarbons with a general molecular formula of C_nH_{2n+2}. Alkanes are built from sp^3 hybridized carbon atoms and have only C–C and C–H σ bonds between atoms (see Figure 1.19 in Chapter 1). All carbon atoms in alkanes have tetrahedral geometry (see Section 1.4 and Figure 1.9 in Chapter 1). The names and condensed formulas of the first few representatives of the straight-chain, unbranched alkanes are listed in Table 3.1.

TABLE 3.1. Names and condensed formulas of unbranched alkanes.

NAME	MOLECULAR FORMULA	CONDENSED FORMULA
methane	CH_4	CH_4
ethane	C_2H_6	CH_3CH_3
propane	C_3H_8	$CH_3CH_2CH_3$
butane	C_4H_{10}	$CH_3(CH_2)_2CH_3$
pentane	C_5H_{12}	$CH_3(CH_2)_3CH_3$
hexane	C_6H_{14}	$CH_3(CH_2)_4CH_3$
heptane	C_7H_{16}	$CH_3(CH_2)_5CH_3$
octane	C_8H_{18}	$CH_3(CH_2)_6CH_3$
nonane	C_9H_{20}	$CH_3(CH_2)_7CH_3$
decane	$C_{10}H_{22}$	$CH_3(CH_2)_8CH_3$
undecane	$C_{11}H_{24}$	$CH_3(CH_2)_9CH_3$
dodecane	$C_{12}H_{26}$	$CH_3(CH_2)_{10}CH_3$
tridecane	$C_{13}H_{28}$	$CH_3(CH_2)_{11}CH_3$
tetradecane	$C_{14}H_{30}$	$CH_3(CH_2)_{12}CH_3$

For the first three alkanes (methane, ethane, and propane) only single structural formulas are possible. Alkanes with more than three carbon atoms can be assembled from atoms in different ways, forming structural isomers (see Section 1.2). Structural formulas of isomeric butanes and pentanes with the corresponding common name of each isomer are shown in Figure 3.1. The simplest isomer of an alkane in which the carbon atoms are connected in a straight-chain with no branches is called *n*-alkane (n is for "normal"). In other isomers (branched alkanes), the chain of carbon atoms is branched at one or more points. The number of possible isomers sharply increases with the number of carbon atoms (e.g. C_6H_{14} has five isomers, C_7H_{16} has nine isomers, C_8H_{18} has 18 isomers, C_9H_{20} has 25 isomers, $C_{10}H_{22}$ has 75 isomers, and $C_{30}H_{62}$ has over 4 billion isomers). Each of the isomeric alkanes can be isolated as an individual chemical substance characterized by a specific boiling point, melting point, and other physical properties.

Boiling points of isomeric butanes and pentanes are listed in Figure 3.1. It should be noted that the more highly branched isomers have lower boiling points compared to the less branched isomers. For example: *n*-pentane (no branches) boils at 36 °C, isopentane (one branch) boils at 28 °C, and neopentane (two branches) boils at 9 °C. These differences in boiling points reflect the shape of molecules. The more branched alkanes are more compact and have a smaller area of attractive contact between molecules (see Section 1.5).

CH₃CH₂CH₂CH₃   CH₃CH(CH₃)₂

n-butane   isobutane

boiling point 0 °C   boiling point −11 °C

CH₃CH₂CH₂CH₂CH₃   CH₃CH(CH₃)CH₂CH₃   CH₃C(CH₃)₃

n-pentane   isopentane   neopentane

boiling point 36 °C   boiling point 28 °C   boiling point 9 °C

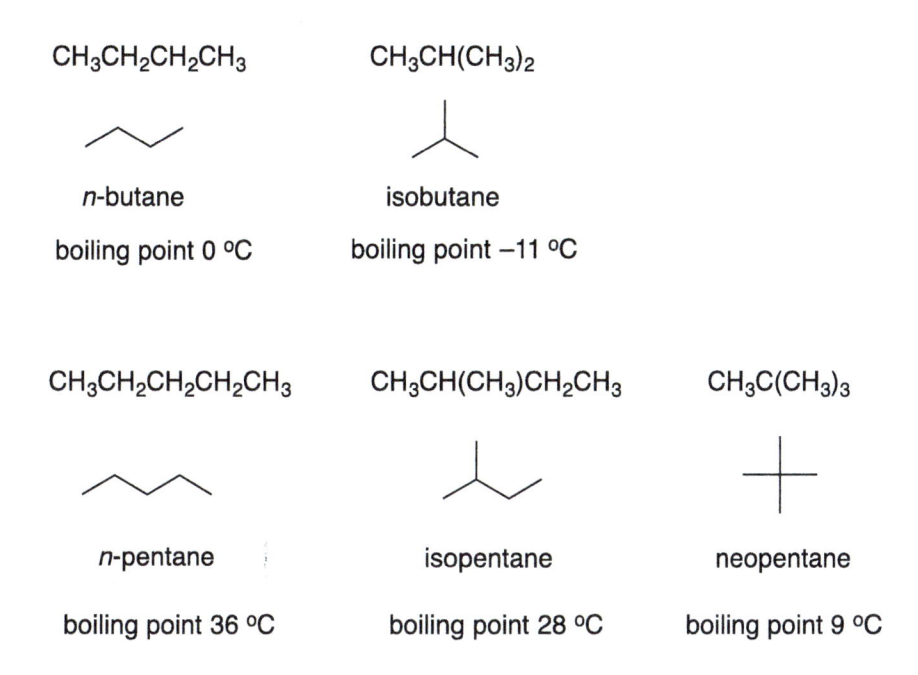

FIGURE 3.1. Isomeric butanes (C₄H₁₀) and pentanes (C₅H₁₂).

Each carbon atom in an alkane molecule can be classified as **primary (1°), secondary (2°), tertiary (3°), or quaternary (4°)** depending on the number of C–C bonds with this atom. A 1° carbon atom is bonded directly to one carbon and three hydrogen atoms, a 2° carbon is bonded to two carbons and two, a 3° carbon is bonded to three carbons and one hydrogen, while a 4° carbon has only carbon atoms bonded directly to it. Hydrogen atoms connected to the 1° carbon atom are called primary hydrogens; secondary and tertiary hydrogens are bonded to 2° and 3° carbons respectively. For example, an isopentane molecule (Figure 3.1) has three 1° carbons, one 2° carbon and one 3° carbon, while neopentane has four 1° carbons and one 4° carbon atom. There are 12 primary hydrogen atoms in neopentane. A molecule of *n*-butane has six 1° hydrogens and four 2° hydrogen atoms.

It is important to understand the difference between isomers and resonance contributors (Section 1.6). Isomers are different compounds with the same molecular formula. It is impossible to convert one isomer into another without breaking the strong σ bonds between atoms. In contrast, resonance contributors show the same chemical compound with different distributions of electronic density in the molecule.

Problems

3.1. Write condensed and line structures for isomeric hexanes and heptanes. Identify the types of carbon and hydrogen atoms in each molecule.

3.2. A molecule of isooctane has five primary, one secondary, one tertiary, and one quaternary carbon atom. Draw the line structure of isooctane.

3.2. NOMENCLATURE OF ALKANES

Common names of the simplest branched alkanes such as isobutane, isopentane, and neopentane are retained in the IUPAC nomenclature and used as the preferred names in industry and in academic research. For larger alkanes, the systematic IUPAC names are used (see Section 1.3 for general rules of IUPAC naming). The IUPAC name of any branched alkane includes the following two parts: 1) root name that indicates the longest carbon chain (the parent chain); and 2) prefixes in front of the root name identifying substituents attached to specific carbons of the main chain. The substituents (branches) in the parent chain are called **alkyl groups**. An alkyl group has the general molecular formula C_nH_{2n+1}, and the name of a specific alkyl group is derived from the corresponding parent alkane by changing the suffix –ane to suffix –yl. Similar to the parent chain, the substituents can be unbranched or branched. For the unbranched substituents, the names methyl ($-CH_3$), ethyl ($-CH_2CH_3$), propyl ($-CH_2CH_2CH_3$), butyl ($-CH_2CH_2CH_2CH_3$), and so on, are used. The most important names for the branched alkyl groups include the following: isopropyl for $-CH(CH_3)_2$, tert-butyl for $-C(CH_3)_3$, isobutyl for $-CH_2CH(CH_3)_2$, and sec-butyl for $-C(CH_3)CH_2CH_3$. The following abbreviations are often used for alkyl groups: Me (methyl), Et (ethyl), Pr (propyl), Bu (butyl), iPr (isopropyl), t-Bu (tert-butyl), iBu (isobutyl), and s-Bu (sec-butyl). The following rules are used to assign IUPAC names for branched alkanes:

1. Name the longest chain of carbon atoms (the parent chain) using the standard names of unbranched alkanes (see Table 3.1). This is the root name of the branched alkane. If there are two or more chains of the same length, choose the chain with the greater number of substituents.
2. Put numbers on the carbons of the parent chain starting from the end that has the nearest substituent.
3. Name each substituent as the appropriate alkyl group. Assign a number to each substituent according to the number of carbon atoms of the parent chain to which the substituent is attached.

4. List the numbered alkyl groups in alphabetical order before the root name.
5. Use prefixes di, tri, tetra, penta for the same substituents – these prefixes and the prefixes *sec-* and *tert-* are not alphabetized. For punctuation, see examples in Figure 3.2.

According to IUPAC rules, the systematic name of isobutane is 2-methylpropane, that of isopentane is 2-methylbutane, and neopentane is 2,2-dimethylpropane. Figure 3.2 shows additional examples of alkanes illustrating the process of correct selection and numbering of the parent chain (both the line-angle structure and condensed formula are shown).

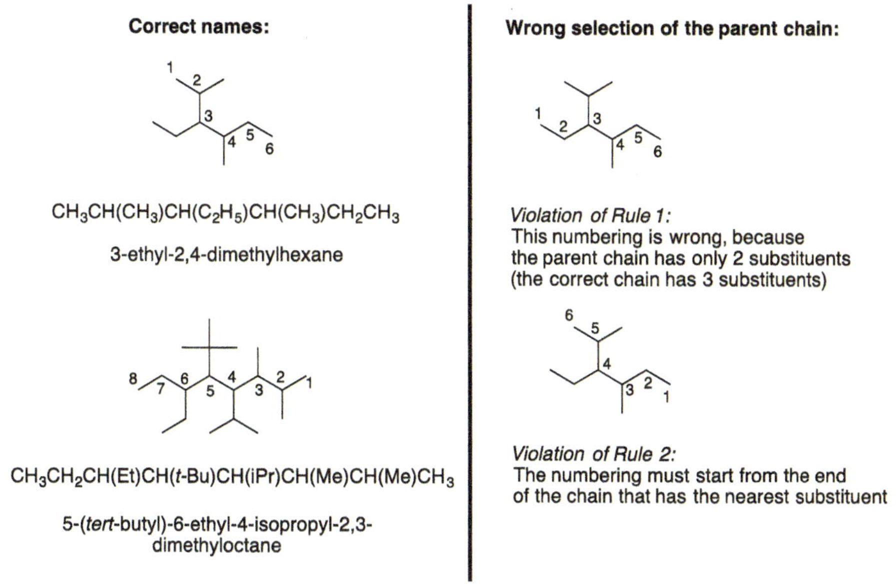

Correct names:

CH₃CH(CH₃)CH(C₂H₅)CH(CH₃)CH₂CH₃

3-ethyl-2,4-dimethylhexane

CH₃CH₂CH(Et)CH(*t*-Bu)CH(iPr)CH(Me)CH(Me)CH₃

5-(*tert*-butyl)-6-ethyl-4-isopropyl-2,3-dimethyloctane

Wrong selection of the parent chain:

Violation of Rule 1:
This numbering is wrong, because the parent chain has only 2 substituents (the correct chain has 3 substituents)

Violation of Rule 2:
The numbering must start from the end of the chain that has the nearest substituent

FIGURE 3.2. IUPAC naming of branched alkanes.

Problems

3.3 Provide the IUPAC name for each of the following compounds:

a)

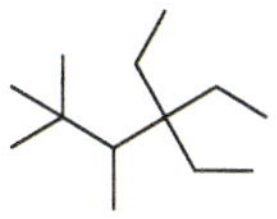

b) (CH₃)₂CH(CH₂)₃CH₃

c)

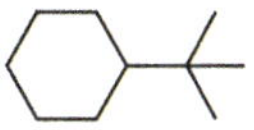

3.3. CONFORMATIONS OF ALKANES

Alkane molecules with two or more carbon atoms are flexible at room temperature and can change their shape by rotation about carbon-carbon single bonds. Such three-dimensional representations of flexible molecules are called **conformations** (or **conformers**, or **rotamers**, or **conformational isomers**). Several conformations of a molecule of *n*-pentane are shown in Figure 3.3. Knowledge of the preferred conformations of flexible molecules is essential for understanding the physical, chemical, and biological properties of organic compounds.

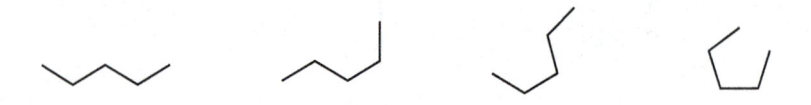

FIGURE 3.3. Conformations of a molecule of *n*-pentane.

It is important to understand the difference between conformers and structural isomers. All conformers of a chemical compound have the same connectivity of atoms in the molecule and have the same IUPAC name, and mutual conversion of conformers does not involve breaking of a chemical bond. In principle, it is possible to freeze a single conformer at a very low temperature, but under normal conditions all conformations are in fast equilibrium. Some conformers are more stable (less **strained**) than the others and dominate in the equilibrium. For example, the least strained conformation of *n*-pentane (the lowest in energy, most stable conformation) is the zigzag conformation (shown as the first drawing in Figure 3.3). Molecular **strain** is defined as the extra energy accumulated in a molecule due to structural distortions in comparison to the unstrained molecule.

Methane does not have any conformers because there are no C–C bonds in this molecule. The molecule of ethane has two principal conformations: a **staggered** and an **eclipsed** conformation (Figure 3.4). The conformations of ethane in Figure 3.4 are shown as a three-dimensional drawing (see Section 1.4) and as a **Newman projection**, which visualizes the molecule along the C–C bond from front to back, with the front carbon represented by a dot and the back carbon as a circle. In the least strained staggered conformation, all hydrogen atoms at the adjacent carbons are separated as far apart as possible. The eclipsed conformation is about 3 kcal/mol higher in energy, mainly due to the repulsion between the C–H bonds on the adjacent carbons. This type of molecular strain caused by the repulsion of electrons in neighboring bonds is called **torsional strain**.

3-D drawings:

Newman projections:

60° angle

60° rotation

staggered conformation
(least strained)

eclipsed conformation
(most strained,
3 kcal/mol higher in energy)

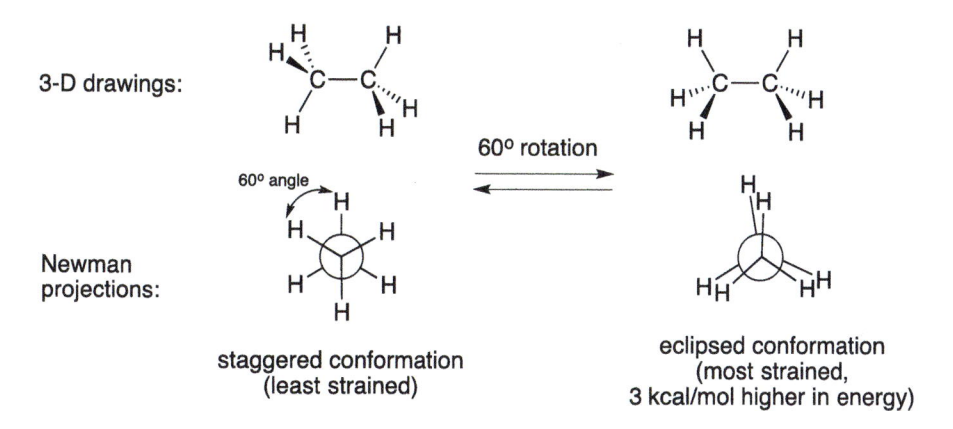

FIGURE 3.4. Staggered and eclipsed conformations of ethane.

The change in molecular strain during the course of rotation about the C–C bond in ethane is illustrated by the energy diagram shown in Figure 3.5. The 60° rotation changes a molecule of ethane from a staggered to an eclipsed conformation, and the next 60° rotation brings it back to staggered conformation. Rotation through 360° completes the cycle. Under normal conditions, molecules of ethane remain in the most stable staggered conformation.

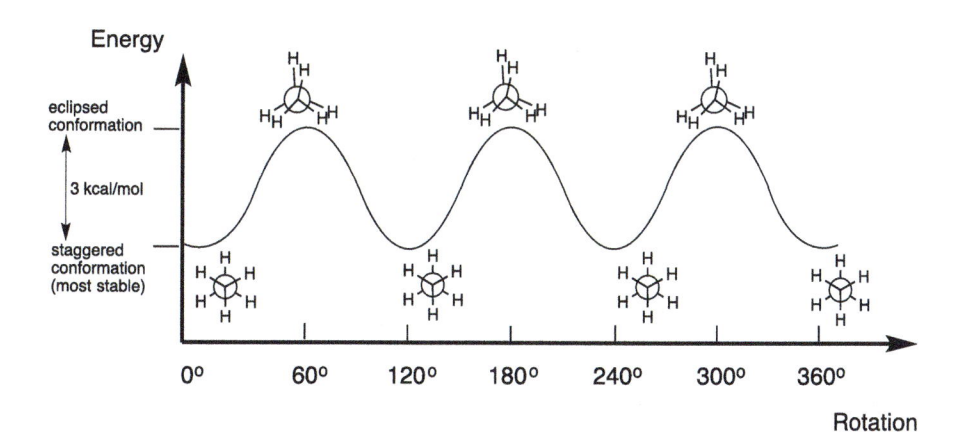

FIGURE 3.5. Energy diagram showing change in molecular strain during the course of rotation about the C–C bond in ethane.

For larger molecules, some additional important conformations involving different types of molecular strain are possible. In Figure 3.6, the conformations of butane ($CH_3CH_2CH_2CH_3$) are shown as Newman projections for the rotation about the central C–C bond between carbons 2 and 3. The most stable anti conformation of butane is completely unstrained because there is no torsional strain and the bulky methyl groups are on opposite sides of the bond, far away from each other. The next staggered conformation of butane, the gauche conformation, has two methyl groups in close proximity and the repulsion between these bulky groups increases molecular strain by about 0.9 kcal/mol. This type of strain, due to the spatial proximity of atoms or groups, is called **steric strain**. The least stable conformation of butane, the Me-Me eclipsed conformation, is highly destabilized by both torsional and steric strain. Under normal conditions, butane molecules remain mostly in the anti conformation and less in the gauche conformation, only passing quickly through the eclipsed conformations during the process of molecular rotation.

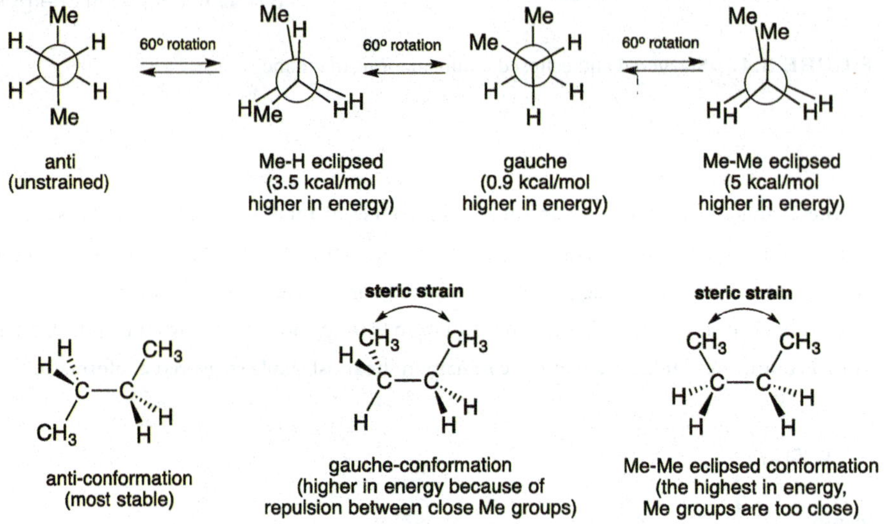

FIGURE 3.6. Conformations of butane for rotation about the C_2–C_3 bond.

Problems

3.4. Convert the Newman projections into a line angle structure and provide the IUPAC name of each compound.

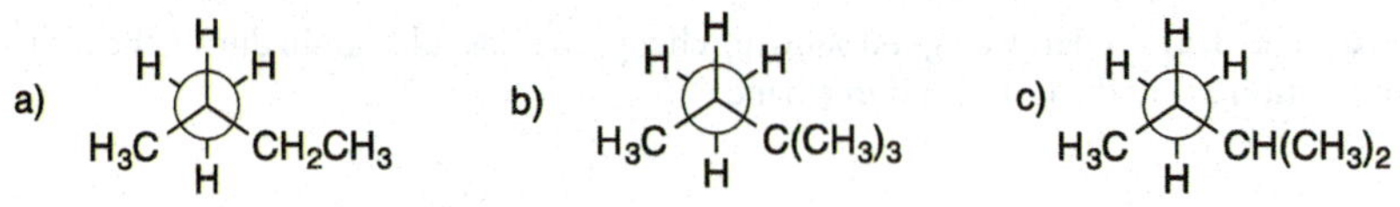

3.5. Which of the following Newman projections depicts the least stable conformer of 1-bromo-2-methyl-propane?

A) [Newman projection with Br, CH₃ on front; H, H on front sides; CH₃ on back bottom, H, H on back]
B) [Newman projection with H on top front; H₃C, H on front sides; Br on back bottom, CH₃, H on back]
C) [Newman projection with Br on top front; H₃C, H on front sides; CH₃ on back bottom, H, H on back]
D) [Newman projection with H₃C on top front; Br, H on front sides; H on back bottom, H₃C, H on back]

3.4. CYCLOALKANES AND RING CONFORMATIONS

Cycloalkanes are hydrocarbons with a ring of carbon atoms in their structure. Their general molecular formula is C_nH_{2n}. Cycloalkanes have only C–C and C–H σ bonds between atoms. A molecule of cycloalkane (C_nH_{2n}) can add one molecule of hydrogen to give a saturated alkane (C_nH_{2n+2}) and therefore it has one unit of unsaturation (see Section 1.3). Compounds with two rings of carbon atoms in a molecule formed by single bonds are called **bicycloalkanes** and have two units of unsaturation.

Figure 3.7 shows the first four representatives of cycloalkanes (cyclopropane, cyclobutane, cyclopentane, and cyclohexane). An unsubstituted cycloalkane is named by adding the prefix cyclo to the name of the parent alkane to indicate the number of carbon atoms in the ring. For substituted cycloalkanes, the name of the substituent is included as a prefix; for example, methylcyclobutane and *tert*-butylcyclohexane. When two or more substituents are present, the carbons of the ring are numbered so that the substituents get the lowest possible numbers, e.g., 1-ethyl-2-methylcyclopentane and 1-*isopropyl*-3-*tert*-butyl-4-methylcyclooctane (see additional examples in Figure 3.10).

Cycle:	△	□	⬠	⬡
C–C–C angle in planar cycle:	60°	90°	108°	120°
IUPAC name:	cyclopropane	cyclobutane	cyclopentane	cyclohexane
3D view of actual molecule:	[eclipsed structure]	[butterfly structure]	[envelope structure]	[chair structure]
	rigid triangle (no conformations)	butterfly conformation	envelope conformation	chair conformation

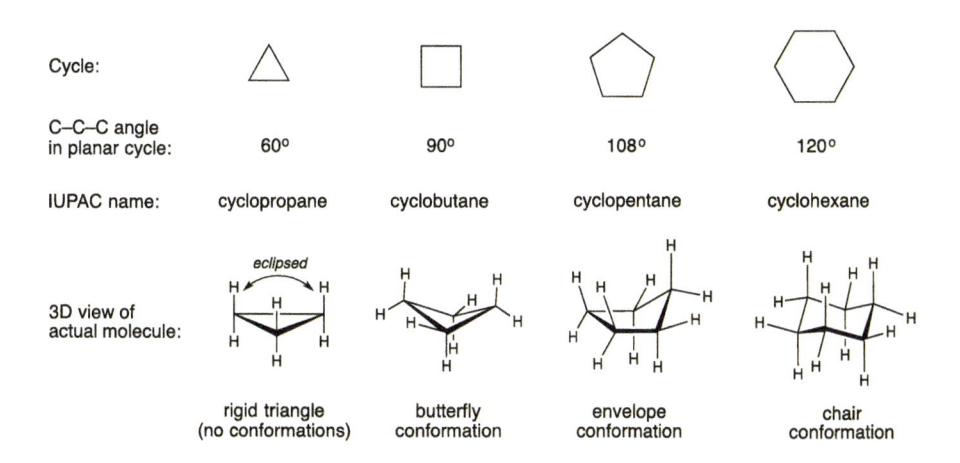

FIGURE 3.7. Unsubstituted cycloalkanes.

The actual three-dimensional structures of cycloalkanes and their stabilities strongly depend on the size of the carbon ring. A molecule of cyclopropane is rigid and has no conformations (Figure 3.7). It is a highly strained compound, mainly because of the considerable deviation of bond angles at carbon atoms. The normal C–C–C bond angle for a tetrahedral carbon is equal to 109.5° (see Section 1.4), while the formal bond angle in cyclopropane is 60° because of the standard geometry of an equilateral triangle. Such a deviation from the normal tetrahedral angle introduces significant **angle strain** to the molecule. Furthermore, because of the rigid, flat structure of cyclopropane, hydrogen atoms at all three carbons are forced to stay in an eclipsed conformation (Figure 3.7), adding torsional strain to the molecule. The overall strain energy of cyclopropane is about 28 kcal/mol, which is very significant excessive energy compared to the unstrained alkanes. Because of the high strain energy, cyclopropane has a higher chemical reactivity and produces more heat upon combustion than alkanes.

Cyclobutane also has significant angle strain because of the deviation of the square angle (90°) from the normal tetrahedral angle (109.5°). However, the torsional strain in cyclobutane is reduced owing to the greater flexibility of the four-membered ring. The four carbon atoms in cyclobutane do not stay in one plane; instead the ring has a folded or "puckered" conformation (Figure 3.7). One of the carbon atoms moves out of the plane formed by the other three carbons, which results in reduced eclipsing interactions. At room temperature, there is a quick interconversion between the puckered conformations of cyclobutane, known as the "butterfly" conformations.

Cyclopentane does not have any significant angle strain. In order to relieve the torsional strain caused by eclipsing hydrogens, the ring adopts a so-called "envelope" conformation (Figure 3.7). A molecule of cyclopentane exists in a dynamic equilibrium between five equal envelope conformations in which each of the carbon atoms consistently moves out of plane. Because of their low steric strain and high stability, five-membered rings are very common in nature.

Carbon atoms in cyclohexane are forced to stay out of plane in order to avoid a significant angle strain (120° in a flat hexagonal ring versus the unstrained tetrahedral angle of 109.5°) and the torsional strain caused by 6 pairs of eclipsing hydrogens. At room temperature, the overwhelming majority of the molecules of cyclohexane stay in the most stable "chair" conformation (Figure 3.7). In the chair conformation, carbon atoms 2, 3, 5, and 6 lie in the same plane, atom 1 is below the plane, and atom 4 is above the plane (Figure 3.8). In this conformation, each carbon adopts the almost perfect tetrahedral geometry with C–C–C bond angles of 110.9°. All the carbon-hydrogen bonds are fully staggered, eliminating the torsional strain. Because of the overall low strain, molecules with a six-membered ring in a chair conformation are very common in natural compounds such as carbohydrates (see Chapter 17).

The C–H bonds at each carbon atom in the chair conformation have two different arrangements: **axial bonds** (C–H_a) pointing straight up or straight down, and **equatorial bonds** (C–H_e) pointing away from the ring. There are three axial bonds pointing straight up (at carbons 2, 4, and 6) and three axial bonds pointing straight down (at carbons 1, 3, and 5). The equatorial bonds at carbons 2, 4, and 6 are tilted slightly down, while equatorial bonds at carbons 1, 3, and 5 are pointing slightly up. When drawing the chair conformation, it is important to keep in mind that there are three sets of parallel bonds in this structure: 1) bonds H_e-C_1, C_2-C_3, C_5-C_6, C_4-H_e; 2) bonds H_e-C_2, C_3-C_4, C_1-C_6, C_5-H_e; and 3) bonds H_e-C_3, C_1-C_2, C_4-C_5, C_6-H_e.

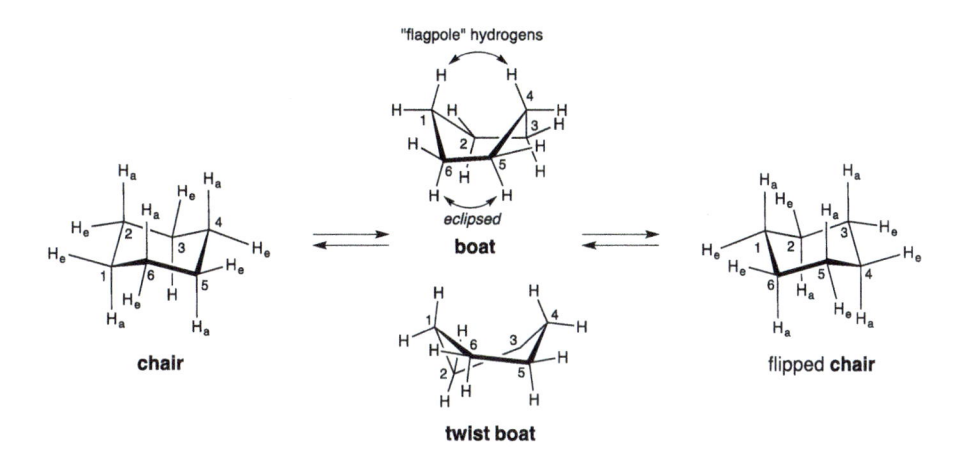

FIGURE 3.8. Conformations of cyclohexane.

Chair conformation of cyclohexane is flexible and the molecule exists in a dynamic equilibrium between several conformations. In the process of "flipping" the chair, one of the carbon atoms (atom C_1 in Figure 3.8) moves up, resulting in the so called **boat conformation**, which is about 6.5 kcal/mol higher in energy than the chair conformation. In the boat conformation, four carbons (C_2, C_3, C_5, and C_6) are forced to stay in an eclipsed conformation, which adds torsional strain to the molecule. In addition, the "flagpole" hydrogens at C_1 and C_4 are located too close to each other, adding steric strain to the molecule. Strain in the boat conformation can be partially relieved by a slight twisting of the ring, resulting in the **twist boat conformation** (Figure 3.8). The twist boat is about 1.5 kcal/mol more stable than the boat conformation because of some relief in the torsional strain and in the flagpole strain. The process of flipping is finalized by moving carbon atom C_4 downward, resulting in the flipped chair (Figure 3.8). It is important to note that all axial hydrogens exchange their positions with equatorial hydrogens during flipping. In the case of substituted cyclohexanes, flipping of the chair results in a similar change of orientation of substituents from axial to equatorial and vice versa.

Figure 3.9 illustrates conformational equilibrium for the chair conformation of methylcyclohexane. In the process of flipping, the methyl substituent changes orientation from axial to equatorial. The chair conformation with methyl group in equatorial position is more stable (lower in energy by about 1.7 kcal/mol) and dominates in the equilibrium (Figure 3.9). The higher energy of the axial conformation is explained by the closeness of the methyl group (at C_1) and two axial hydrogens at C_3 and C_5; this type of steric strain is called **axial-axial (or diaxial) interaction**. The extra strain caused by diaxial interaction strongly depends on the size of the axial substituent. For bulky substituents such as the *tert*-butyl group, the axial conformation is so high in energy that the more stable equatorial conformation almost completely predominates at equilibrium.

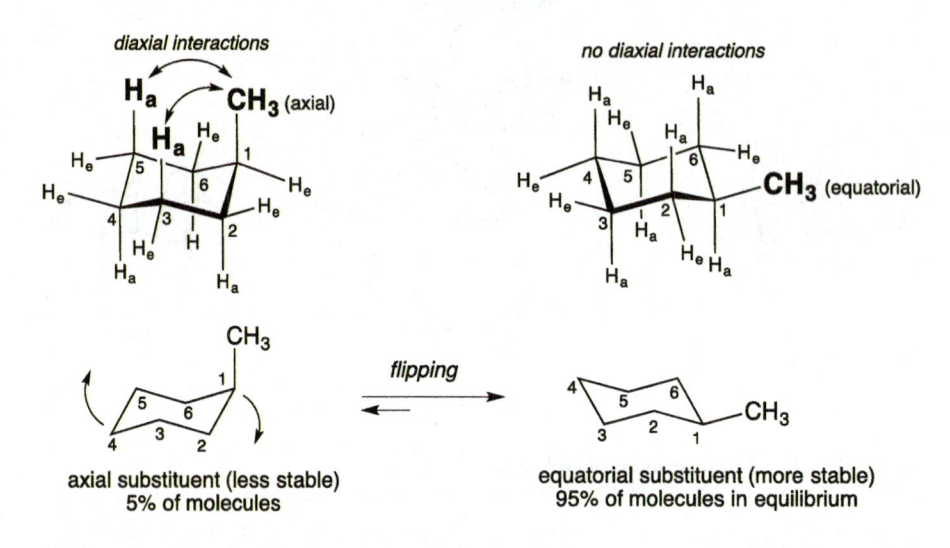

FIGURE 3.9. Conformations of methylcyclohexane.

Problems

3.6. Draw chair representations of each of the compounds below. Which compound contains a quaternary (4°) carbon atom?

 a) cyclohexane b) methylcyclohexane
 b) 1,1-dimethylcyclohexane d) *trans*-1,3-dimethyl cyclohexane

3.7. Sort the following structures by increasing stability (least stable to most stable):

60 ORGANIC CHEMISTRY

3.5. *CIS*, *TRANS* ISOMERISM IN SUBSTITUTED CYCLOALKANES

Cycloalkanes with two substituents on the same side of the ring are referred to as **cis isomers** and cycloalkanes with substituents on the opposite sides of the ring are **trans isomers**. Several examples of such isomers with the corresponding IUPAC names are shown in Figure 3.10. Atoms in *cis* and *trans* isomers have the same connectivity (as can be seen from their IUPAC names), but substituents are oriented of differently in space. For example, the ethyl group is attached to C_1 and the methyl group is attached to C_2 of the cyclobutane ring in both cis-1-ethyl-2-methylcyclobutane and trans-1-ethyl-2-methylcyclobutane; but the spatial arrangement of substituents at the carbon atoms (the **configuration** of carbon atoms) is different, indicated by the italicized prefixes *cis* and *trans* at the beginning of the IUPAC name. *Cis* and *trans* isomers in cycloalkanes are classified as **stereoisomers** (also named **configurational isomers**). According to the general definition, two molecules are stereoisomers if they are made of the same atoms and connected in the same sequence, but the atoms are positioned differently in space. In contrast to conformers (see Sections 3.3 and 3.4), stereoisomers cannot be interconverted without breaking covalent bonds. *Cis* and *trans* isomers exist as different chemical compounds with distinctly different physical and chemical properties. Additional types of stereoisomers include enantiomers (Chapter 4) and *E/Z*-isomers of alkenes (see Chapter 6).

cis-1-ethyl-2-methylcyclobutane   *trans*-1-ethyl-2-methylcyclobutane

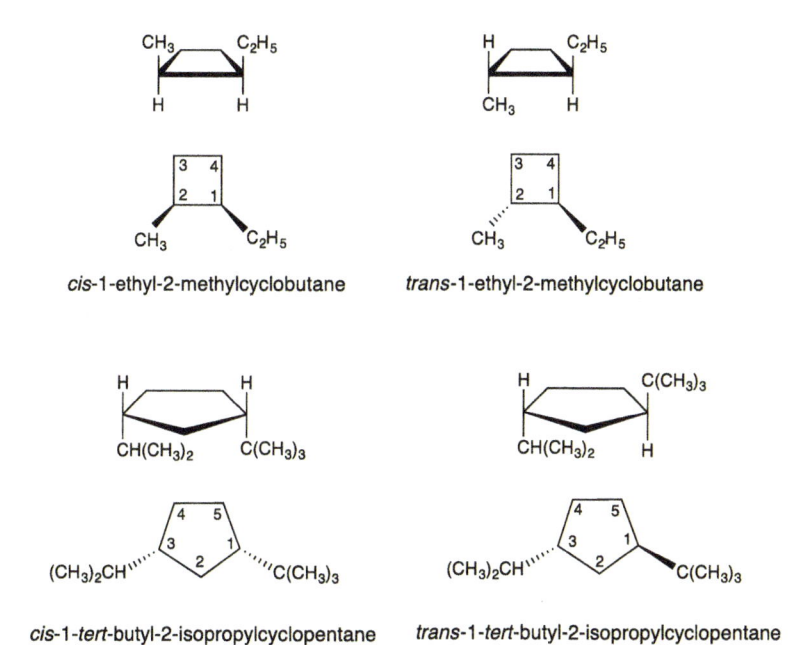

cis-1-*tert*-butyl-2-isopropylcyclopentane   *trans*-1-*tert*-butyl-2-isopropylcyclopentane

FIGURE 3.10. Examples of *cis* and *trans* isomers of disubstituted cycloalkanes.

The *cis/trans* isomerism in substituted cyclohexanes is particularly important because of the presence of six-membered rings in naturally occurring compounds such as carbohydrates. The major, most stable conformations of the stereoisomeric disubstituted cyclohexanes are shown in Figure 3.11. The difference between *cis* and *trans* stereoisomers can be easily noticed when cyclohexane is shown as a planar six-membered ring. In reality, as we know from the previous section, cyclohexane ring mainly exists in the chair conformation with substituents either at the axial or equatorial position. For example, the stable, low energy conformer of *trans*-1,2-dimethylcyclohexane has both substituents in equatorial positions, and flipping the chair forces these substituents to the less stable axial positions (Figure 3.11). The *trans*-configuration can be easily identified in the axial conformation: one axial substituent is pointing straight up, and the other is on the opposite side of the ring pointing straight down. It is more difficult to distinguish between *cis/trans* isomers when substituents occupy equatorial positions. However, as indicated by the arrows in Figure 3.11, one equatorial bond is tilted slightly up (above the ring), and the other is pointing slightly down (below the ring).

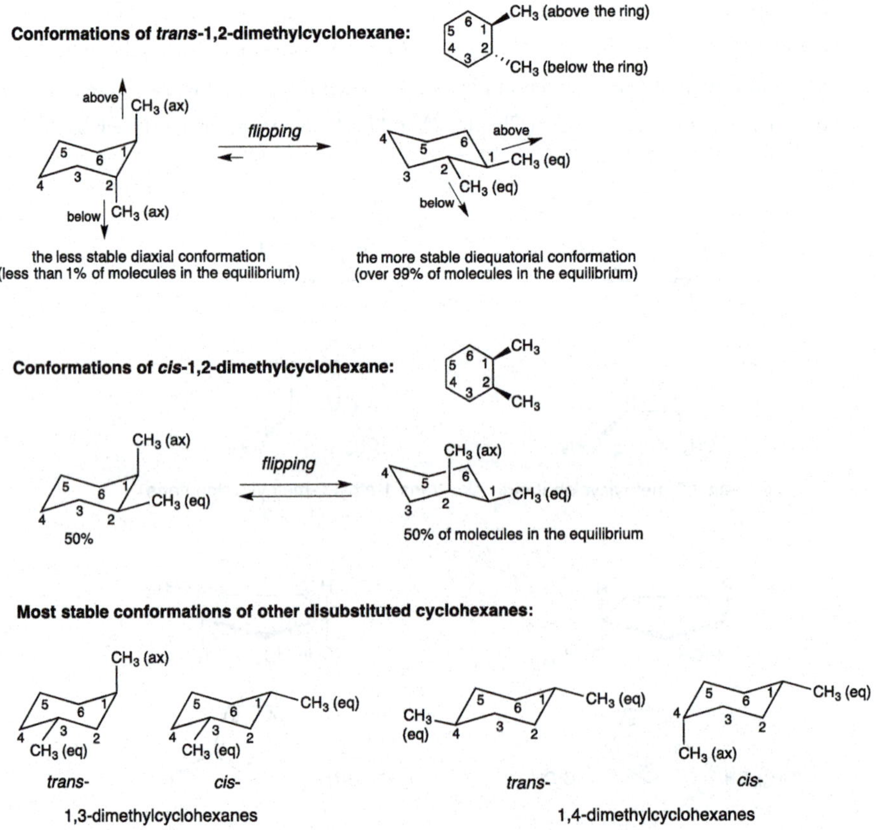

FIGURE 3.11.   *Cis* and *trans* isomers of disubstituted cyclohexanes.

In *cis*-1,2-dimethylcyclohexane, one substituent is in the axial position and the other is in the equatorial position. Figure 3.11 also illustrates the most stable conformations of *cis* and *trans* isomers of 1,3-dimethylcyclohexane and 1,4-dimethylcyclohexane. When two different substituents are present in the cyclohexane ring, the larger one will always stay in the equatorial position in order to reduce the diaxial interactions (see previous section).

Problems

3.8. Consider following pairs of structures and determine if they are in all respects identical, *cis*/*trans* isomers, conformational isomers, or constitutional isomers.

a), b), c), d), e), f)

3.6. RADICAL REACTIONS: OXIDATION AND HALOGENATION OF ALKANES

Alkanes and cycloalkanes have very low chemical reactivity at room temperature. At higher temperatures, they react with oxygen from the air, producing a flame and forming carbon dioxide and water as the final combustion products. Combustion of natural alkanes (natural gas and higher alkanes distilled from crude oil) provides our civilization's main source of energy. Alkanes also serve as a primary feedstock for the chemical industry. Partial oxidation, halogenation, and thermal conversion of alkanes are basic industrial processes that supply essential chemical products. All chemical reactions of alkanes proceed with cleavage of the very strong C–C or C–H single bonds and involve an intermediate formation of **radicals** (also known as **free radicals**). Radicals are defined as chemical species that have unpaired valence electrons. Radical intermediates (R•) play a key role in many chemical processes including combustion, halogenation, and polymerization, and they are also important in atmospheric chemistry, plasma chemistry, and biochemistry.

Halogenation of alkanes is an important chemical reaction used for industrial preparation of haloalkanes (also commonly called **alkyl halides**), which have many practical applications, including as important solvents. Chloroalkanes and bromoalkanes are prepared by direct reaction of alkanes with Cl_2 or Br_2 in the presence of heat or light. Fluoroalkanes and iodoalkanes are usually prepared by indirect methods because fluorine is an extremely reactive element that reacts explosively with alkanes, while iodine has a very low reactivity and in general does not react with alkanes.

When a mixture of methane and chlorine is heated or exposed to light, a fast reaction occurs, forming chloromethane (methyl chloride) and hydrogen chloride as the products (Figure 3.12). Further reaction with excessive Cl_2 may lead to the replacement of additional hydrogens, with chlorine forming dichloromethane (methylene chloride), trichloromethane (chloroform), and tetrachloromethane (carbon tetrachloride) as the products.

$$CH_4 + Cl_2 \xrightarrow{\text{heat or light}} CH_3Cl + HCl$$
$$\text{chloromethane}$$

$$CH_3Cl + Cl_2 \xrightarrow{\text{heat or light}} CH_2Cl_2 + HCl$$
$$\text{dichloromethane}$$

$$CH_2Cl_2 + Cl_2 \xrightarrow{\text{heat or light}} CHCl_3 + HCl$$
$$\text{trichloromethane}$$

$$CHCl_3 + Cl_2 \xrightarrow{\text{heat or light}} CCl_4 + HCl$$
$$\text{tetrachloromethane}$$

FIGURE 3.12. Chlorination of methane.

These reactions are generally classified as **substitution reactions** (reactions resulting in the substitution of one atom or functional group with another atom or group) and they proceed by the **radical chain mechanism**. A **reaction mechanism** in general is a description of a chemical reaction showing the movement of electrons with drawing structures of intermediate products at each step. A detailed reaction mechanism may also include the analysis of transition states and energy changes during the course of a reaction. Figure 3.13 shows the radical chain mechanism for the formation of chloromethane.

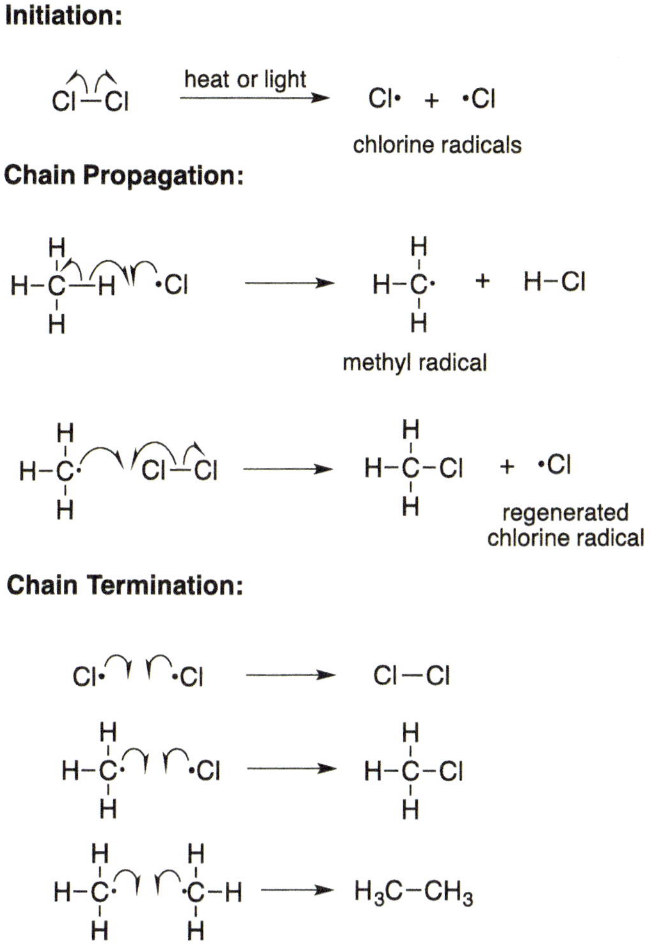

Initiation:

$$Cl-Cl \xrightarrow{\text{heat or light}} Cl\cdot + \cdot Cl$$
chlorine radicals

Chain Propagation:

[CH₄ + ·Cl → ·CH₃ + H-Cl]
methyl radical

[·CH₃ + Cl-Cl → CH₃-Cl + ·Cl]
regenerated chlorine radical

Chain Termination:

$$Cl\cdot + \cdot Cl \longrightarrow Cl-Cl$$

[·CH₃ + ·Cl → CH₃-Cl]

[·CH₃ + ·CH₃ → H₃C-CH₃]

FIGURE 3.13. Radical chain mechanism of chlorination reaction.

In general, the mechanism of a **radical chain reaction** consists of three consecutive processes: 1) chain initiation; 2) chain propagation; and 3) chain termination. At the **initiation step**, a molecule of chlorine breaks into two chlorine atoms (or chlorine radicals) in reaction to heat or light. This is the so-called **homolytic bond cleavage** in which the bonding electron pair of the covalent bond splits evenly to form two uncharged radicals. The homolytic bond cleavage is shown in Figure 3.13 by two "fishhook" arrows, each indicating the movement of a single electron. The cleavage of the Cl–Cl bond requires external energy (endothermic reaction), which is supplied by a light or heat source.

Chain propagation is a sequence of steps in which the initial Cl• radical reacts with a molecule of CH_4, producing final products and regenerating the initial radical. The regenerated Cl• radical reacts with another molecule of CH_4, repeating the cycle, and this continues the chain. The process of chain propagation consists of two different steps: 1) hydrogen abstraction from CH_4 by Cl• resulting in intermediate formation of a methyl radical; and 2) reaction of the methyl radical with Cl_2 regenerating Cl•. Both of these steps involve homolytic cleavage of single bonds, as illustrated by the fishhook arrows in Figure 3.13. This chain is initiated by a single Cl• radical and can repeat thousands of times until one of the participating radicals is removed from the chain by some different reaction (chain termination).

Chain termination is a chemical reaction that ceases radical regeneration in the chain propagation process. Examples of chain termination reactions may include the combination of any two radicals present in the reaction mixture as shown in Figure 3.13. For example, a chlorine radical may combine with a methyl radical, forming a molecule of CH_3Cl and terminating the chain. It should be noted that concentration of radicals in the reaction mixture is very low, which makes the combination of two radicals a very rare event.

Other alkanes or cycloalkanes can be chlorinated or brominated similar to methane; however, the formation of isomeric products of halogenation (the so-called **regioisomers** or **positional isomers**, which differ by the position of substituent in the carbon chain), is possible in these reactions. For example, chlorination of isobutane produces two regioisomers: 2-chloro-2-methylpropane as the major product and 1-chloro-2-methylpropane as the minor product (Figure 3.14). The major product in this reaction is formed by radical chlorination of the tertiary (3°) hydrogen, which is indicative of a relatively faster radical substitution reaction of the tertiary H, compared to the primary (1°) H atoms. Likewise, bromination of propane predominantly yields the substitution product of the secondary (2°) hydrogen atom, despite the fact that there are more 1° hydrogens (six) than 2° hydrogens (two) in a propane molecule, making the involvement of 1° hydrogens more likely. Reactions that give predominantly one regioisomer out of several possible isomers are called **regioselective reactions**. In general, the order of reactivity of C–H bonds in radical substitution reactions is: 3° C–H (most reactive) > 2° C–H > 1° C–H > CH_4 (least reactive).

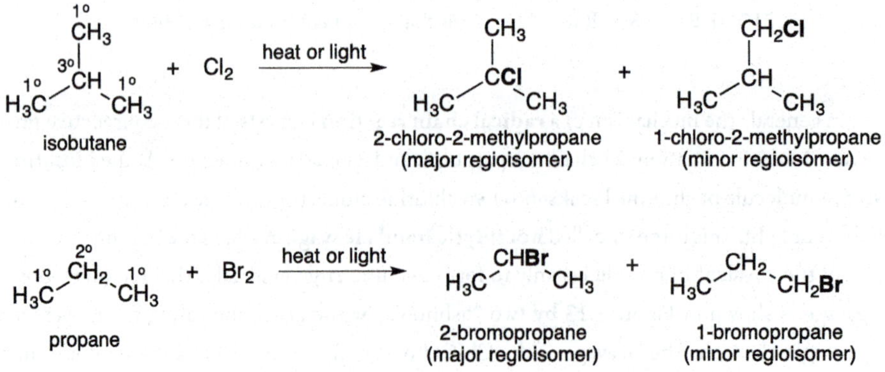

FIGURE 3.14. Regioselective chlorination and bromination of alkanes.

The regioselectivity of halogenation reactions is explained by the relative stability of radical intermediates formed in these reactions (Figure 3.15). Tertiary radicals are the most stable and are generated faster than secondary and primary radicals. The electronic structure of radicals involves the sp² hybridized carbon atom with a trigonal planar arrangement of three σ bonds and the single electron occupying an unhybridized 2p orbital. In principle, the electronic structure of radicals is similar to the electronic structure of carbocations (Chapter 5). Like carbocations, the greater stability of the more highly substituted radicals is due to increased delocalization of electronic density resulting from the electron releasing inductive effect of alkyl groups. Radicals that have a possibility of resonance delocalization have the greatest stability (Chapter 7).

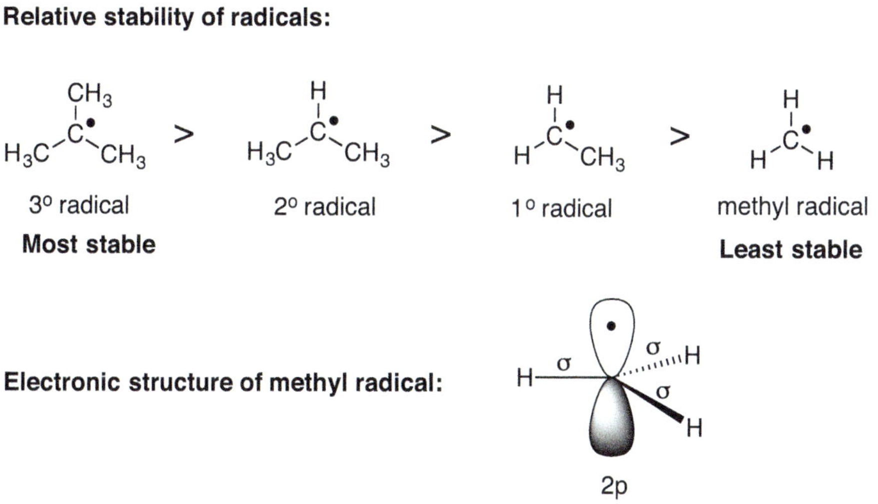

Relative stability of radicals:

3° radical — Most stable > 2° radical > 1° radical > methyl radical — Least stable

Electronic structure of methyl radical:

FIGURE 3.15. Electronic structure and relative stability of radicals.

Problems

3.9 Sort the following set of radicals according to increasing relative stability:

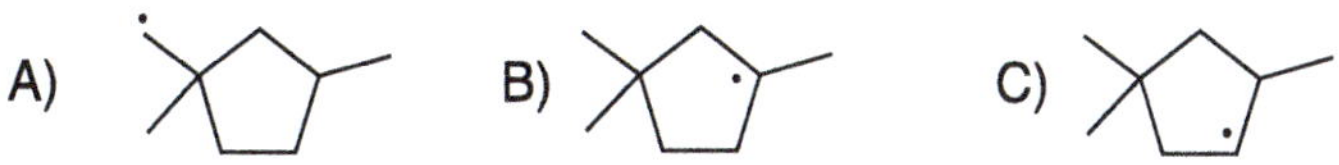

A)   B)   C)

3.10. Provide major products in each of the radical reactions below. Propose a mechanism for each reaction.

a) 2,3-dimethylbutane + Br$_2$, heat or light →

b) tert-butylcyclopentane + Cl$_2$, light →

CHAPTER 4

Stereochemistry

Stereochemistry is the study of molecules in three-dimensional space. Understanding the three-dimensional shapes of molecules is essential for understanding the physical, chemical, and biological properties of organic compounds. This chapter introduces the fundamental concept of chirality and provides an overview of different types of stereoisomers.

4.1. CONSTITUTIONAL ISOMERS AND STEREOISOMERS

From earlier chapters, we know that isomers can be divided into two general groups – **constitutional isomers** and **stereoisomers**. Constitutional isomers are a set of molecules with the same molecular formula but different connectivity of atoms. Constitutional isomers have different chemical and physical properties. Examples of constitutional isomers with a molecular formula C_3H_6O are shown in Figure 4.1.

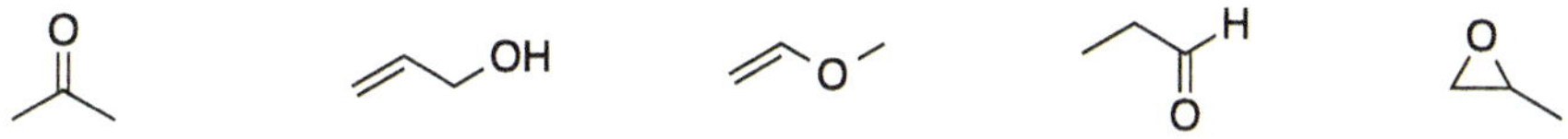

FIGURE 4.1. Examples of constitutional isomers with a molecular formula C_3H_6O. All molecules have different connectivity of atoms.

The definition of constitutional isomers ignores any reference to bond multiplicity (for example single versus double bond) as well as position of the non-bonding electrons and/or charges. Structures that have the same connectivity of atoms but differ by the position of the non-bonding electrons and/or charges are called resonance contributors (see Chapter 1). The structures of resonance contributors are neither constitutional isomers nor stereoisomers. Resonance structures are shown in an attempt to fully describe charge delocalization and bond multiplicity, which affects the physical and chemical properties of a compound (see Chapter 1). The three structures shown in Figure 4.2 represent the same compound. The connectivity for all of those structures is the same because the atoms are in the same position.

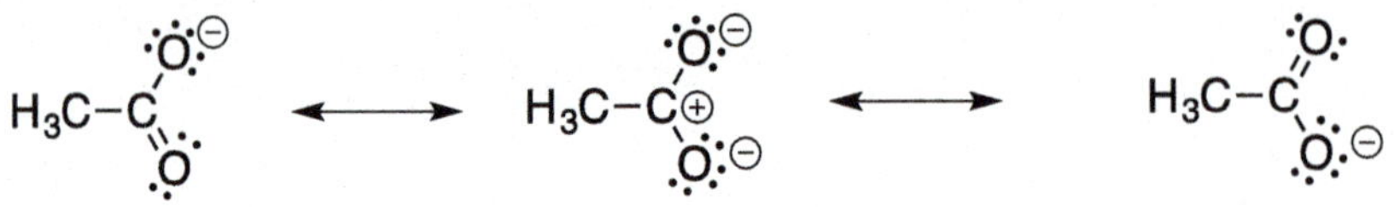

FIGURE 4.2. Resonance structures are not isomers.

Stereoisomers have the same molecular formula and the same connectivity of atoms, but differ in the orientation of atoms in space. **Conformational stereoisomers**, or **conformers** (see Chapter 3), differ by rotation around one or more C–C single bonds. Eclipsed and staggered conformation of ethane and the chair and boat conformation of cyclohexane (Figure 4.3) are examples of conformers.

FIGURE 4.3. Conformational isomers are stereoisomers that differ by rotation around one or more single C–C bonds.

Stereoisomers that cannot be interconverted by rotation around single bonds are classified as **configurational isomers**. So far we have discussed configurational isomers involving *cis/trans* isomers of cycloalkanes (Chapter 3). Other examples are *cis/trans* isomers of alkenes, which will be discussed in Chapter 6, and stereoisomers with tetrahedral carbon atoms with four different substituents, which are the focus of this chapter. Examples of configurational isomers are shown in Figure 4.4. Stereoisomers cannot be interconverted without breaking one or more covalent bonds.

FIGURE 4.4. Examples of configurational isomers.

4.2. CHIRAL MOLECULES AND R/S-NOMENCLATURE OF ENANTIOMERS

The word **chirality** is derived from the Greek word χειρ (kheir), for hand. Familiar chiral objects are hands, feet, and also shoes (Figure 4.5). Your left foot will not fit in your right shoe, which means that your left and right feet are not superposable. In other words, they are related as mirror images that can not occupy the same space. All chiral objects have the property that they cannot be superposed on their mirror images. In contrast, objects that can be superposed on their mirror image are called **achiral**. All achiral objects are characterized by the presence of internal symmetry, such as a plane of symmetry that cuts the object into two halves that are mirror images of each other. For example, the human body is an achiral object because it has an internal plane of symmetry in the center of the body. Other examples of achiral objects include pants and jackets.

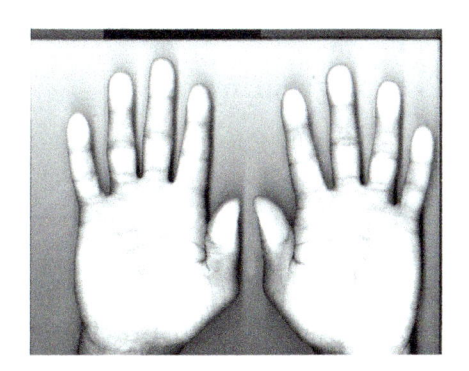

FIGURE 4.5. Examples of chiral objects: left and right hands and shoes.

In organic chemistry all molecules that contain a tetrahedral carbon atom with four different substituents are chiral. The carbon atom that provides this property is called a **chiral center** or an **asymmetrical carbon atom**. The two configurational stereoisomers of a chiral molecule that are related as mirror images are called **enantiomers**. Enantiomers are stereoisomers that are non-superposable mirror images. Figure 4.6 shows an example of a pair of enantiomers.

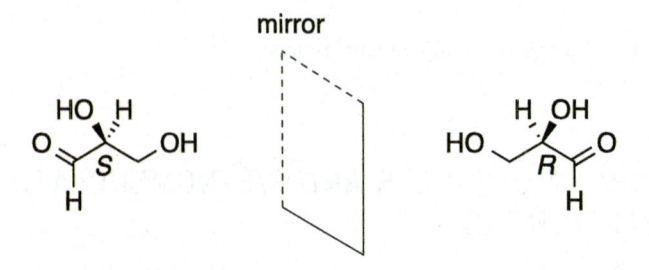

FIGURE 4.6. Enantiomers are non-superposable mirror images.

In order to distinguish between enantiomers, we need to name them by assigning the chiral center with a stereochemical descriptor. Two nomenclature systems for naming enantiomers exist. The general **Cahn-Ingold-Prelog R/S system** for naming enantiomers has been officially approved by IUPAC. Biologists on the other hand prefer the older **D/L system** which is commonly used to describe the chiral centers of amino acids and carbohydrates. We will discuss the D/L system in Chapter 17; the rules of the R/S system are explained below.

4.2.1 Rules for assignment of R/S configuration of chiral centers

The assignment of the configuration of a chiral center is accomplished in three steps:

1. Assign numbers for the priority of each of the four substituents on the chiral center using the priority rules (see below), with #1 as the highest priority and #4 as the lowest priority.
2. The structure needs to be rotated in space in such a way that the lowest priority substituent (#4) points to the back - away from the observer.
3. If the three groups projecting towards you are ordered from highest priority #1 to lowest priority #3 clockwise, then the configuration of the chiral center is R. If the three groups projecting toward you are ordered from highest priority #1 to lowest priority #3 counterclockwise, then the configuration of the chiral center is S. The labels R and S are derived from the Latin words for right (rectus) and left (sinister) respectively.

The priority of substituents is established according to **Cahn-Ingold-Prelog Priority Rules**:

1. Prioritize the atoms or groups of atoms based on the atomic number of the atoms directly attached to the chiral center. The higher the atomic number, the higher the priority.

-I	>	-Br	>	-Cl	>	-F
-SH	>	-OH	>	$-NH_2$	>	$-CH_3$

2. If two or more atoms directly attached to chiral center are the same, then priority is assigned based on the next set of atoms (i.e., atoms attached to the directly-bonded atoms). Continue if needed until priority can be assigned at the first point of difference.

CH_2Cl	>	$-CH_2SH$	>	$-CH_2F$
$-CH_2OH$	>	$-CH_2NH_2$	>	CH_2CH_3

3. If two atoms have substituents of the same priority, higher priority is assigned to the atom with more of these substituents.

   $-C(CH_3)_3$ > $-CH(CH_3)_2$ > $-CH_2CH_3$ > $-CH_3$

4. Atoms connected by double/triple bonds are considered to be bonded to an equivalent number of similar "phantom" atoms by single bonds, as shown in the following schematic. Note: "phantom" atoms are bonded to no other atoms.

STEREOCHEMISTRY 73

Examples of Cahn-Ingold-Prelog priority rules to assign configuration of chiral centers

Example 1: Molecule of bromofluoromethanol

1. Prioritize the atoms or groups of atoms based on the atomic number of the atoms directly attached to the chiral center. The higher the atomic number, higher the priority.
   Number the four atoms or groups of atoms so that #1 has the highest priority and #4 has the lowest priority

   atomic number
   Br   35
   F   9
   O   8
   H   1

2. Position the molecule in space so that the lowest priority group (#4) points away from the observer. There are three possibilities as shown in the following schematic; all of them represent the same structure. Either of them can be used in the next step.

3. If the three groups projecting toward you are ordered from highest (#1) to lowest priority (#3) clockwise, then the configuration is R. If the three groups projecting toward you are ordered from highest priority (#1) to lowest priority (#3) counter-clockwise, then the configuration is S.

   Reads 1,2,3 in clockwise direction => R

   (*R*)-bromofluoromethanol

Repeating these steps for the other enantiomer will result in the following:

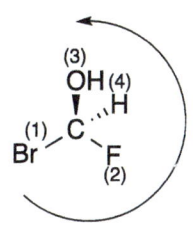

Reads 1,2,3 in counterclockwise direction => S

(*S*)-bromofluoromethanol

Example 2: Molecule of 3-amino-2-methylpropan-1-ol

If two or more atoms directly bonded to a chiral center are the same, then priority is assigned based on the next set of atoms (i.e., atoms adjacent to the directly bonded atoms). Continue until the first point of difference where a priority can be assigned.

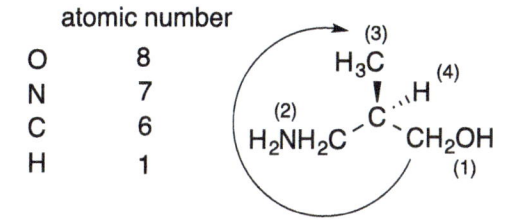

atomic number

O	8
N	7
C	6
H	1

Reads 1,2,3 in clockwise direction => R

(*R*)-3-amino-2-methylpropan-1-ol

Example 3: Molecule of 4-bromo-1-chloro-2-methylbutane

Since priority is assigned at the first point of difference, this may result in a group containing an atom with higher atomic number having lower priority. In this example, even though bromine has a higher atomic number than chlorine, it is beyond the first point of difference.

	atomic number
Br	35
Cl	17
C	6
H	1

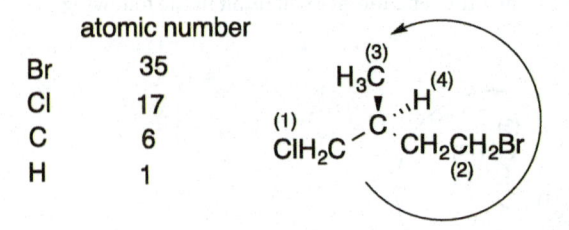

Reads 1,2,3 in counterclockwise direction => S

(S)-4-bromo-1-chloro-2-methylbutane

Example 4: Molecule of 4-(2,2,2-trifluoroethyl)-4-(trifluoromethyl)nonan-1-ol

A large group with more atoms does not necessarily have a higher priority than a smaller group with fewer atoms.

	atomic number
F	9
O	8
C	6
H	1

Reads 1,2,3 in counterclockwise direction => S

(S)-4-(2,2,2-trifluoroethyl)-4-(trifluoromethyl)nonan-1-ol

Example 5: Molecule of 1-ethyl-1-isopropyl-3,3-dimethylcyclohexane

If two atoms have substituents of the same priority, higher priority is assigned to the atom with more of these substituents.

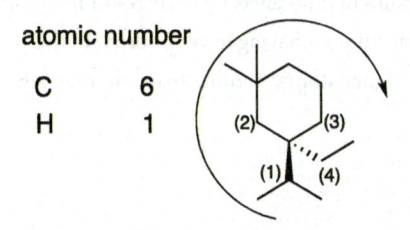

	atomic number
C	6
H	1

Reads 1,2,3 in clockwise direction => R

(R)-1-ethyl-1-isopropyl-3,3-dimethylcyclohexane

Example 6: Molecule of 4-(*tert*-butyl)-4-ethynyl-5-hexenoic acid

Atoms connected by double/triple bonds are considered bonded to an equivalent number of similar "phantom" atoms by single bonds. Note: "phantom" atoms are bonded to no other atoms.

atomic number

O	8
C	6
H	1

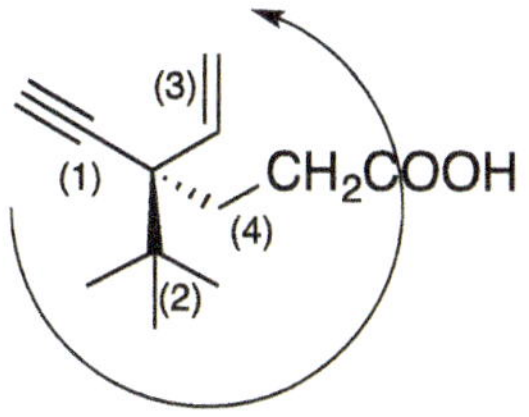

Reads 1,2,3 in counterclockwise direction => S

(*S*)-4-(*tert*-butyl)-4-ethynyl-5-hexenoic acid

Problems

4.1. Determine if the following compounds are chiral or achiral:

A) 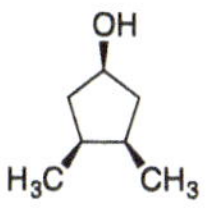 B) 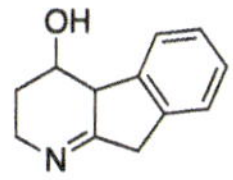 C) 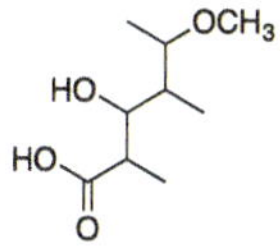 D)

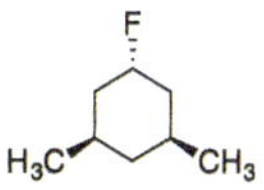

4.2. How many chiral centers do each of these compounds have?

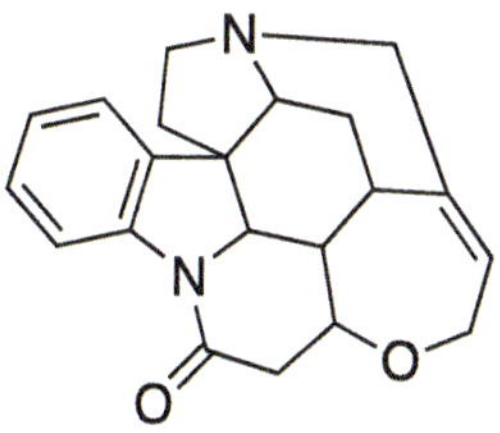

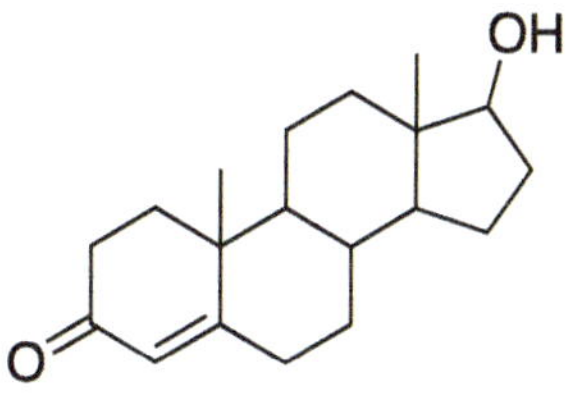

4.3. Assign priorities to the following sets of substituents:

a) –H, –CH(CH$_3$)$_2$, –B(OCH$_3$)$_2$, –CH$_2$CH$_2$OCH$_3$

b) –CO$_2$H, –CH$_2$SH, –CH=CH$_2$, –OH

c) –CN, –CH$_2$NH$_2$, –CONH$_2$, –CON(CH$_3$)$_2$

d) –Br, –CH$_2$CH$_2$Br, –C(CH$_3$)$_3$, –CH$_2$Cl

4.4. Consider each of the following orders of priority (highest to lowest within each order). Which order is incorrect?

I) –OH > –CH$_2$OH > –CH$_3$ > –H

II) –CH=CH$_2$ > –CH$_2$COOH > –CH$_2$CH=CH$_2$ > –CH$_3$

III) –NH$_3^+$ > –COO$^-$ > –CH$_3$ > –H

IV) –NH$_3^+$ > –CHO > –CH$_2$SH > –CH$_3$

4.3. MOLECULES WITH MORE THAN ONE CHIRAL CENTER

4.3.1. Enantiomers and diastereomers

Regardless of how many chiral centers are present in a molecule, enantiomers differ in configuration at all chiral centers. To distinguish between stereoisomers that are enantiomers and stereoisomers that are not enantiomers the term **diastereomers** (sometimes called diastereoisomers) is introduced. The definition of diastereomers is quite simple: they are stereoisomers that are not enantiomers. Note that this definition is very broad and therefore does not require chiral centers to be present for stereoisomers to be diastereomers. While a pair of enantiomers are mirror images, two diastereomers are not related as mirror images. Furthermore, if two configurational stereoisomers are related as diastereomers, they need to differ at least at one stereocenter, but not at all stereocenters. Figure 4.7 describes the relationship between diastereomers and enantiomers in a compound with two chiral centers.

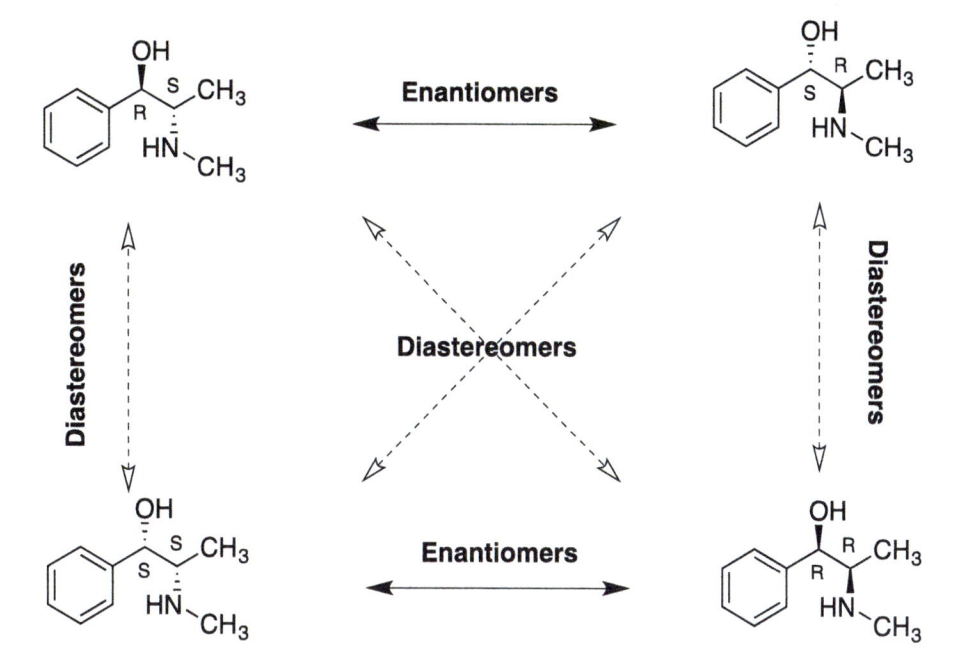

FIGURE 4.7. Relationship between diastereomers and enantiomers in a molecule with two chiral centers.

4.3.2. Meso compounds

Meso compounds are achiral molecules with two or more chiral centers. These molecules are characterized by the presence of internal symmetry. For example, an imaginary plane of symmetry may cut the molecule in two halves that are related to each other as mirror images. Because of the internal symmetry, mirror images of meso compounds are superposable and thus are identical molecules. Several examples of meso compounds are shown in Figure 4.8. Note: In order to see the plane of symmetry in a chair conformation of cis-1,2-dibromocyclohexane you need to convert it to a planar six-membered ring.

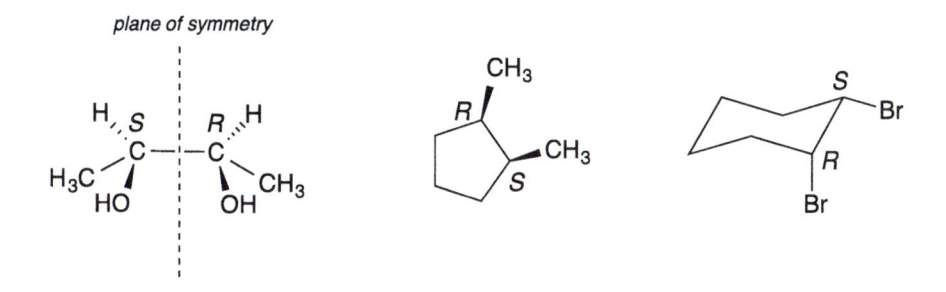

FIGURE 4.8. Examples of meso compounds.

STEREOCHEMISTRY 79

4.3.3. Fischer projections

The Fischer projection, introduced by Hermann Emil Fischer in 1891, is a two-dimensional representation of a three-dimensional organic molecule by projection. Figure 4.9 shows an example of a Fischer projection.

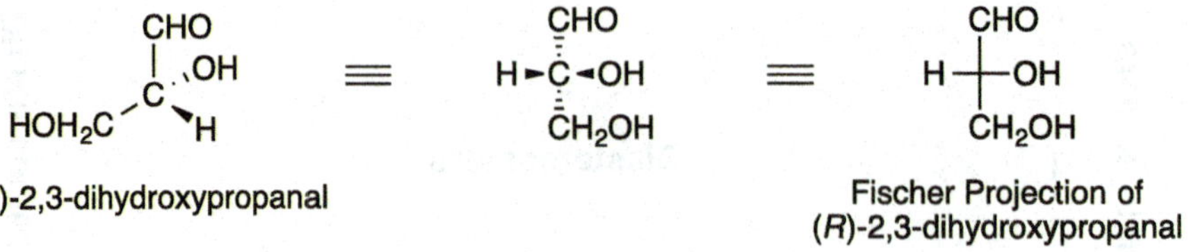

FIGURE 4.9. Fischer projection of (R)-2,3-dihydroxypropanal.

General concept of a Fischer projection: All nonterminal bonds are depicted as horizontal or vertical lines. The longest carbon chain is depicted vertically, with its carbon atoms represented by lines crossing in the center. The orientation of the carbon chain is such that the vertical bonds point to the back away from the observer and the horizontal bonds point towards the observer.

> ***Memory Aid*** – *the relative arrangement of the bonds can be remembered by the fact that the horizontal bonds are coming out of the plane of the paper to hug you.*

Note the difference between Fischer projections and Lewis structures (Figure 4.10):

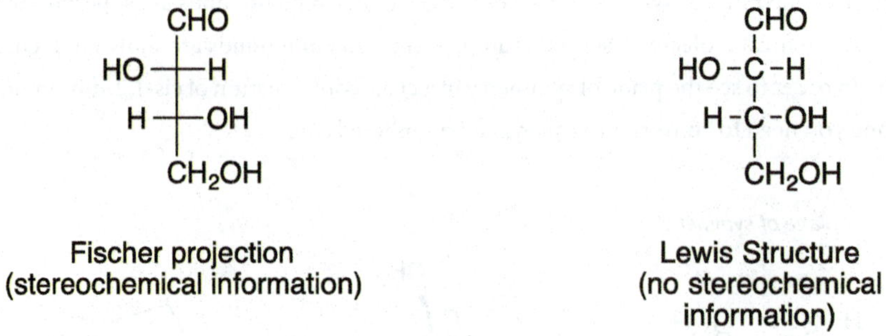

FIGURE 4.10. Fischer projection vs. Lewis structure.

ORGANIC CHEMISTRY

Fischer projections can be easily confused with Lewis structures. Lewis structures are not intended to provide any stereochemical information! Fischer projections are widely used to illustrate the special orientation of groups in biological molecules with more than one chiral center, such as carbohydrates (Figure 4.11).

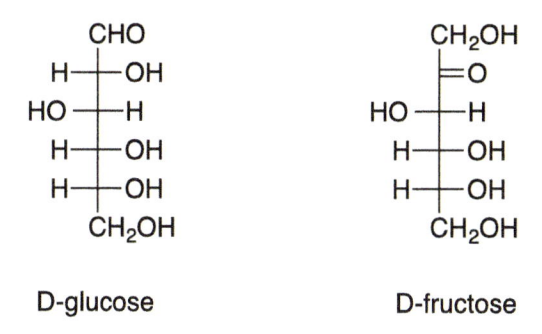

```
        CHO                    CH₂OH
   H ──┬── OH                    ║O
   HO ──┼── H             HO ──┬── H
   H ──┼── OH             H ──┼── OH
   H ──┴── OH             H ──┴── OH
        CH₂OH                  CH₂OH

      D-glucose              D-fructose
```

FIGURE 4.11. Fischer projections of naturally occurring carbohydrates.

When relating one Fischer projection to another, it is important to realize that Fisher projections can only be manipulated within the two-dimensional plane in which it is drawn. This means it cannot be arbitrarily rotated in three-dimensional space. Even in the plane it is drawn, Fischer projections can only be rotated by 180°. Rotations of the structure by 90° or 270° will result in the inversion of all chiral centers in the molecule.

Problems

4.5. In each of the following compounds, assign the R/S configurations to each chiral center. Which are enantiomers, which are diastereomers?

4.6. In which of the following structures are both chiral centers in R-configuration?

STEREOCHEMISTRY   81

4.7. Determine the configuration (R or S) at the carbon atoms 2 and 3 in each compound.

a)
```
       CHO
        |2
   H ---|--- OH
        |3
   H ---|--- OH
        |
      CH₂OH
```

b)
```
       CHO
        |2
   F ---|--- H
        |3
   H ---|--- F
        |
      CH₂OH
```

c)
```
       CHO
        |2
   H ---|--- OH
        |3
   HO---|--- H
        |
      CH₂SH
```

4.8. Which of the following are meso compounds?

I) cis-1,3-dimethylcyclohexane

II) (1R,2R)-1,2-dimethylcyclohexane

III) (3R,4S)-3,4-dimethylhexane

IV) cis-1,4-dimethylcyclohexane

4.4. PROPERTIES OF ENANTIOMERS AND DIASTEREOMERS

Enantiomers for the most part have identical physical and chemical properties. For example, a pair of enantiomers will have identical melting point, boiling point, energies, and reactivities. However, enantiomers behave differently with respect to plane-polarized light and in their interaction with other chiral molecules and materials. The ability of enantiomers to rotate the plane of plane-polarized light is called optical activity. Chemically, the different behavior of enantiomers in a chiral environment provides a very useful tool to separate enantiomers from one another. Since biological systems are chiral, enantiomers may have different biological effects.

4.4.1. Optical activity

A sample of material that is able to rotate the plane of polarization of a beam of plane-polarized light is said to possess an ability called **optical activity** (or to be **optically active**). This optical rotation is a classical distinguishing characteristic of systems containing unequal amounts of corresponding enantiomers. An enantiomer causing rotation in a clockwise direction (when viewed in the direction facing the oncoming light beam) is called **dextrorotatory** and its chemical name or formula is designated by the prefix (+)-; one causing rotation in the opposite direction is called **levorotatory** and designated by the prefix (−)-. A pair of enantiomers of the same compound will have optical rotation identical in magnitude, but opposite in sign. It is important to note that the sign and magnitude of the optical rotation is independent of the configuration of a chiral center. This means molecules with R configuration could be dextrorotatory or levorotatory.

Thus, in a mixture of enantiomers, an excess of one of the enantiomers is required for optical activity of the whole system. The larger the excess, the greater the magnitude of optical activity for the same amount of compound (with the obvious limit of 100% of one enantiomer). Mixtures with an equal quantity of each enantiomer will not exhibit optical activity as their magnitudes would sum to zero. Such mixtures are called **racemic mixtures** or **racemates**. The **enantiomeric excess** (**ee**) is used to specify the purity of an enantiomer.

4.4.2. Separation of enantiomers (resolution of racemates)

1. Manual separation of enantiomers: During the crystallization of a racemic mixture, the enantiomers crystallize separately and form two macroscopically different kinds of crystals with a mirror-image relationship. These crystals can be separated manually with a pair of tweezers. Less than 10% of all racemates are known to crystallize as mixtures of **enantiopure** crystals.
2. Resolution of racemates by conversion into diastereomers: Enantiomers can be converted into diastereomeric salts by treatment with chiral acid or base, which can then be separated by fractionated crystallization. Alternatively, covalent derivatization of enantiomers with chiral reagents also yields diastereomers. These diastereomers can be separated by common separation techniques as they possess different physical and chemical properties. In addition, the separation of enantiomers can be carried out by chromatographic methods using chiral stationary phases. In this approach, the movement of one enantiomer is retarded relatively more than the other due to the formation of a more stable chiral complex with the stationary phase.
3. Kinetic resolution of enantiomers: Different rates of reaction of enantiomers with chiral reagents allows for the separation of enantiomers. The reaction of enantiomers with enzymes is a special case of this approach.

In general, resolution of a racemic mixture is wasteful when only one enantiomer is needed. Therefore, it is more economical to synthesize that single enantiomer stereoselectively. Asymmetric synthesis involving reagents derived from nature's pool of enantiomerically pure compounds will avoid any resolution processes completely.

4.4.3. Enantiomers and biology

Numerous proteins and enzymes, most of the amino acids, carbohydrates, nucleosides, and a number of alkaloids and hormones are chiral. Furthermore, all of these molecules are present just as a single enantiomer. In other words, they are **homochiral**. Thus, our bodies are remarkably effective chiral selectors, being composed of these compounds. Chiral compounds we ingest may act different from each other within our bodies, resulting in different biological effects.

Distinctive tastes of certain foods may be due to the presence of different enantiomers. For example, dill contains the R-enantiomer of carvone and therefore tastes different than spearmint which contains the S-enantiomer of the same compound (Figure 4.12).

Our bodies may also metabolize each enantiomer via separate pathways to generate different pharmacological activity. Thus, one enantiomer may produce a desired therapeutic effect, while another may be inactive or, in the worst case, produce undesired or toxic effects. In the 1960's, the drug thalidomide was sold as racemic mixture to relieve the symptoms of morning sickness. Shortly after marketing of this drug, it was observed that women who had taken it prior to the third trimester of their pregnancy were giving birth to children with malformed extremities. It was later found that while R-enantiomer had the desired antiemetic pharmacological effect, the S-enantiomer was teratogenic, damaging the fetus.

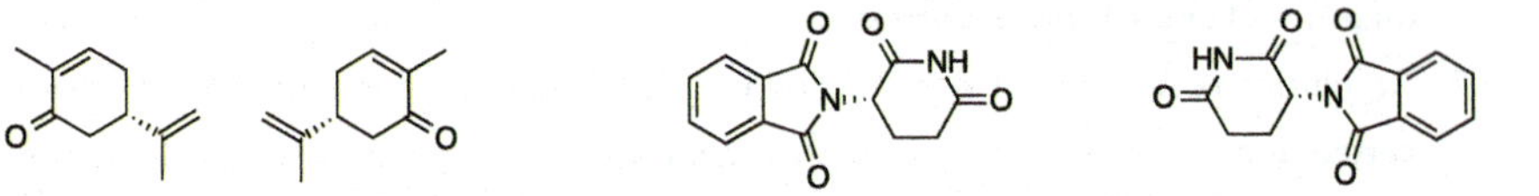

S-carvone (dill taste)

R-carvone (spearmint taste)

S-Thalidomide (causes birth defects)

R-Thalidomide (relieves symptoms of morning sickness during pregnancy)

FIGURE 4.12. Enantiomers may have different biological effects.

CHAPTER 5

Nucleophilic Substitution and β-Elimination Reactions

Nucleophilic substitution is very common in nature and widely employed in the chemical industry and research. Thus, it is one of the most fundamental types of chemical reactions. In a nucleophilic substitution reaction, an electron-donating reagent (a **nucleophile**) replaces a suitable substituent at a carbon atom of an organic molecule. The substituent is referred to as a **leaving group** and the organic molecule as an organic **substrate**. Nucleophilic substitution can be further classified according to the nature of organic substrate as: 1) nucleophilic substitution occurring at the sp^3 hybridized carbon atom in substituted alkanes or cycloalkanes (also known as **aliphatic nucleophilic substitution**); 2) **acyl** nucleophilic substitution reactions occurring at the sp^2 hybridized carbon atom of a carbonyl group (see Chapter 13); and 3) **aromatic** nucleophilic substitution reactions occurring at the sp^2 hybridized carbon atom of aromatic compounds – derivatives of benzene (see Chapter 15). When the nucleophilic reagent is a strong Brønsted base, aliphatic nucleophilic substitution reactions are accompanied by the formation of alkenes, which are products of β-elimination reactions. This chapter describes the general features and the mechanisms of aliphatic nucleophilic substitution and β-elimination reactions. Understanding these mechanisms is essential for predicting the structure of reaction products and for optimizing reaction conditions.

5.1. OVERVIEW OF NUCLEOPHILIC SUBSTITUTION AND β-ELIMINATION REACTIONS

A typical nucleophilic substitution reaction involves the replacement of a halogen atom (X, the leaving group) in haloalkane (RX, the substrate) with an anionic or an uncharged nucleophile (Nu:) according to the general schemes shown in Figure 5.1. The actual reagents that are sources of anionic nucleophiles include the spectator cations, usually Na⁺ or K⁺, which are not involved in the process of nucleophilic substitution and are not part of the organic product. In the case of neutral nucleophiles, the product of the substitution reaction is an organic salt, with a positively charged organic ion and a negatively charged halide ion (the leaving group X:⁻). The anionic nucleophiles in general are better donors of electrons compared to the uncharged nucleophiles and therefore are classified as **strong nucleophiles**. Neutral molecules of water and alcohols are **weak nucleophiles**, while amines, sulfides, and phosphines are **moderate nucleophiles**. The leaving group X can also be classified as a **good leaving group,** for the most stable anions of strong acids (Cl⁻, Br⁻, I⁻; see Section 2.3 in Chapter 2), and a **poor leaving group,** for the less stable anions of weaker acids (F⁻, HO⁻, etc.). Only substrates with good leaving groups in their structures can effectively participate in nucleophilic substitution.

Reactions with anionic nucleophiles:

RX + Nu:⁻ M⁺ —solvent→ RNu + X:⁻ M⁺
substrate anionic   product
           nucleophile

R = alkyl or cycloalkyl; X = Cl, Br, or I; M = Na, K, or Li

Examples of anionic nucleophiles:

Nu:⁻	Reagent (Name)
HO⁻	NaOH (sodium hydroxide)
CH₃O⁻	CH₃ONa (sodium methoxide)
HS⁻	NaSH (sodium hydrosulfide)
CH₃S⁻	CH₃SNa (sodium methanethiolate)
I⁻	NaI (sodium iodide)
⁻N=N=N⁻	NaN₃ (sodium azide)
N≡C⁻	KCN (potassium cyanide)
HC≡C⁻	HCCLi (lithium acetylide)

Reactions with uncharged nucleophiles:

RX + Nu: —solvent→ RNu⁺ X:⁻
substrate uncharged   product
           nucleophile   (organic salt)

Examples of uncharged nucleophiles:

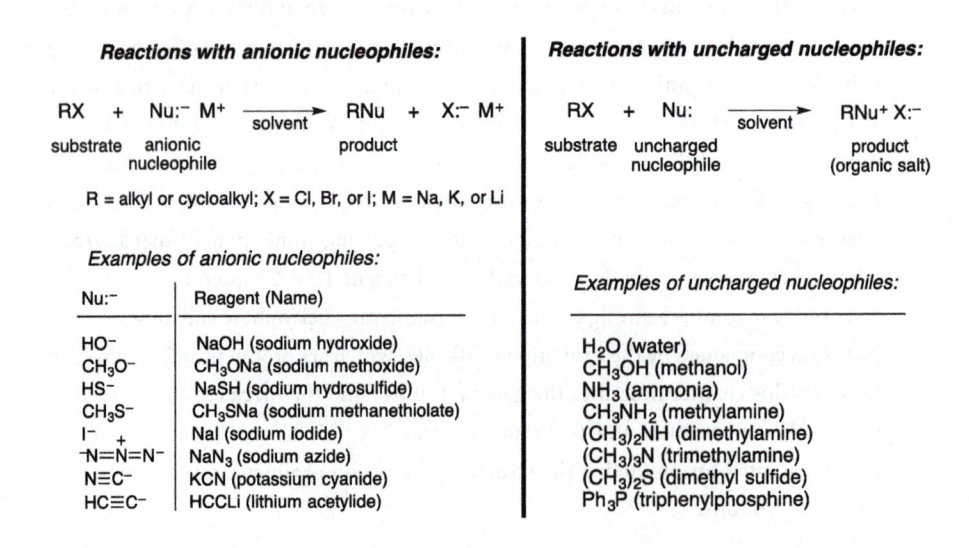

H₂O (water)
CH₃OH (methanol)
NH₃ (ammonia)
CH₃NH₂ (methylamine)
(CH₃)₂NH (dimethylamine)
(CH₃)₃N (trimethylamine)
(CH₃)₂S (dimethyl sulfide)
Ph₃P (triphenylphosphine)

FIGURE 5.1. General description of nucleophilic substitution reactions in haloalkanes.

During nucleophilic substitution, the pair of non-bonding electrons from the nucleophile (Nu:) moves to form a bond with a carbon atom of the substrate, while the bond between the carbon atom and the leaving group X breaks in such a way that a pair of electrons stays on

X:⁻. This process can be generally illustrated by using curved arrows as shown in Figure 5.2. A detailed description of the mechanisms of nucleophilic substitution is provided in the next section of this chapter.

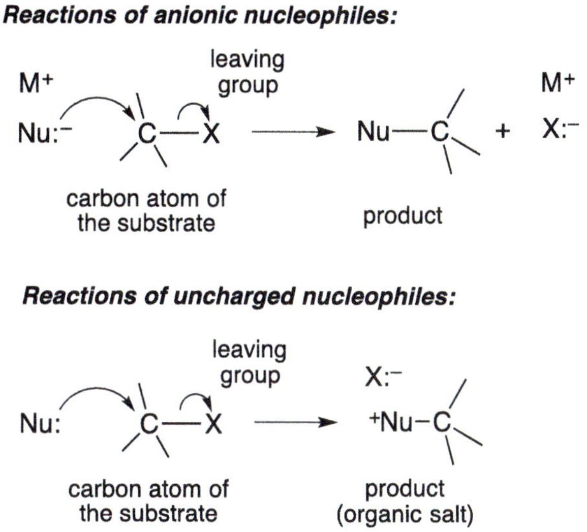

Reactions of anionic nucleophiles:

$$M^+ \; Nu:^- \;\; \overset{\text{leaving group}}{C-X} \longrightarrow Nu-C \; + \; X:^- \;\; M^+$$

carbon atom of the substrate → product

Reactions of uncharged nucleophiles:

$$Nu: \;\; \overset{\text{leaving group}}{C-X} \longrightarrow \;^+Nu-C \; \; X:^-$$

carbon atom of the substrate → product (organic salt)

FIGURE 5.2. Movement of electrons in the process of nucleophilic substitution.

Several examples of nucleophilic substitution reactions are shown in Figure 5.3. Reaction of sodium azide with chlorocyclopentane in methanol as the solvent produces azidocyclopentane as the organic product and sodium chloride (Na⁺Cl⁻) as the inorganic by-product. Please note that the inorganic by-product is usually omitted from schematics of organic reactions. The reaction of iodomethane with organic amine (CH$_3$CH$_2$NH$_2$) as an uncharged nucleophile produces ethylmethylammonium iodide, an organic salt. The reactions of a bromoalkane with cyanide or acetylide anions as nucleophiles yield a nitrile or a substituted alkyne as the respective organic products. It should be noted that there are several different ways of writing a schematic of an organic reaction. The nucleophilic reagent can be placed above the arrow or in line with the substrate; the solvent can be shown above or below the arrow, along with other reaction conditions such as temperature, presence of light, and catalyst; and the insignificant by-product (such as inorganic salt) can be omitted. The substrates and the products can be presented as line structures or condensed Lewis structures.

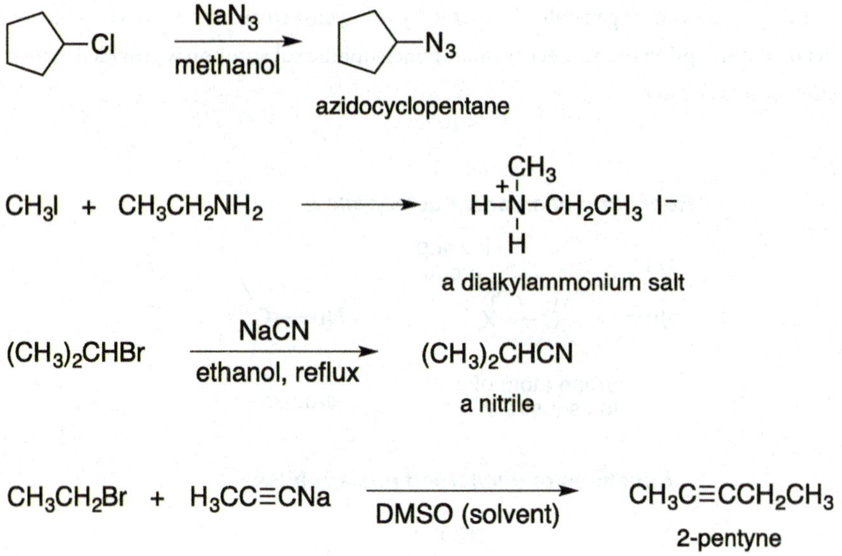

$$\text{cyclopentyl-Cl} \xrightarrow[\text{methanol}]{\text{NaN}_3} \text{cyclopentyl-N}_3$$

azidocyclopentane

$$CH_3I + CH_3CH_2NH_2 \longrightarrow \overset{\overset{CH_3}{|}}{\underset{\underset{H}{|}}{H-\overset{+}{N}-CH_2CH_3}} \; I^-$$

a dialkylammonium salt

$$(CH_3)_2CHBr \xrightarrow[\text{ethanol, reflux}]{\text{NaCN}} (CH_3)_2CHCN$$

a nitrile

$$CH_3CH_2Br + H_3CC\equiv CNa \xrightarrow{\text{DMSO (solvent)}} CH_3C\equiv CCH_2CH_3$$

2-pentyne

FIGURE 5.3. Examples of nucleophilic substitution reactions.

In some reactions of haloalkanes with strongly basic nucleophilic reagents (such as sodium hydroxide NaOH, sodium ethoxide EtONa, or potassium *tert*-butoxide *t*-BuOK), alkenes are formed as the by-products or as the primary reaction products. For example, the reaction of chlorocyclohexane with aqueous sodium hydroxide gives a mixture of cyclohexanol and cyclohexene (Figure 5.4). Two competing processes explain the formation of these products: the nucleophilic substitution of chlorocyclohexane with HO⁻ as a nucleophile, and the β-elimination reaction in which HO⁻ acts as a base toward chlorocyclohexane. In the course of the elimination reaction, the base removes a proton from the β-carbon atom while the chloride anion leaves from the α-carbon atom resulting in formal elimination of HCl and formation of a double bond (Figure 5.4).

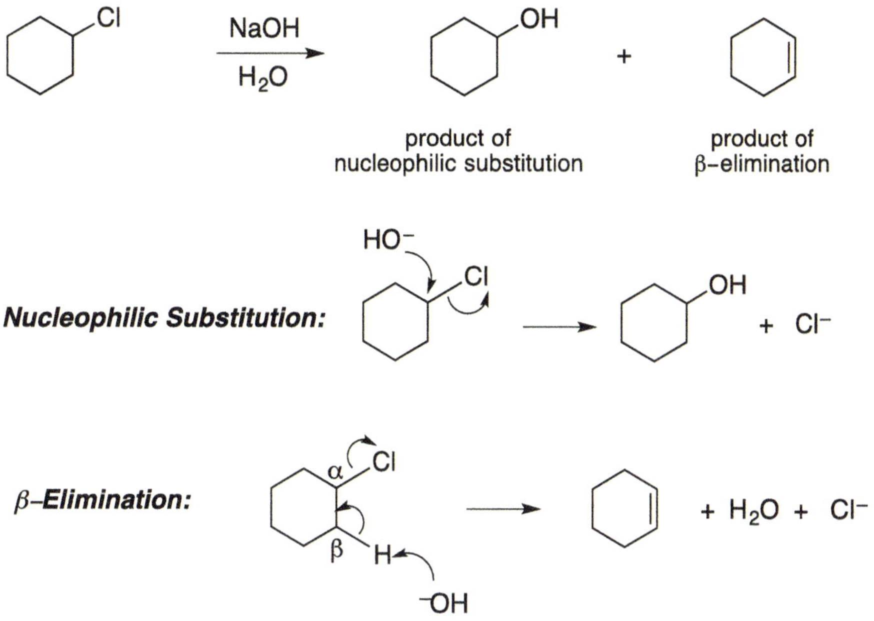

FIGURE 5.4. Competing nucleophilic substitution and β-elimination.

The reactions of 2°- or 3°-haloalkanes with stronger bases (EtONa or t-BuOK) give exclusively products of β-elimination, while weaker bases such as NaOH give a mixture of substitution and β-elimination products. A detailed description of the mechanisms of a β-elimination reaction is provided in Section 5.4 of this chapter.

Problems

5.1. Draw the products of nucleophilic substitution in the reactions below. Identify the substrate, leaving group, and nucleophile, and use curved arrows to show the movement of electrons in each reaction.

a)  (CH₃)₂CHBr + NaSH, ethanol

b) CH₃CH₂I + Ph₃P $\xrightarrow{\text{pentane}}$

c) 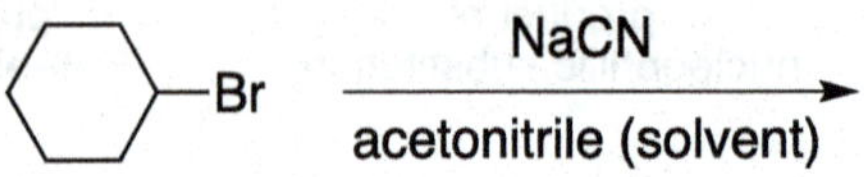 [cyclohexyl]–Br $\xrightarrow[\text{acetonitrile (solvent)}]{\text{NaCN}}$

5.2. Draw the products of elimination in the reactions below. Identify the substrate, leaving group, and base, and use curved arrows to show the movement of electrons in each reaction.

a) 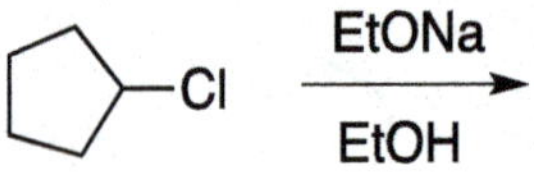 [cyclopentyl]–Cl $\xrightarrow[\text{EtOH}]{\text{EtONa}}$

b) 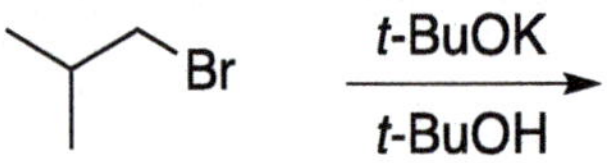 (isopropyl)CH–Br (2-bromo-3-methylbutane) $\xrightarrow[\text{t-BuOH}]{\text{t-BuOK}}$

5.3. Provide the products of the following reactions:

a) 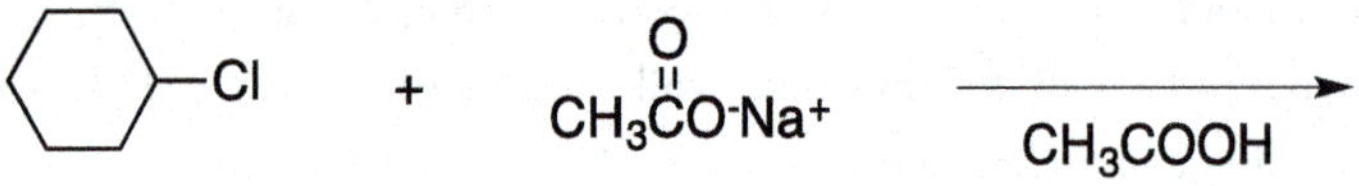 [cyclohexyl]–Cl + CH₃CO⁻O⁻Na⁺ $\xrightarrow{\text{CH}_3\text{COOH}}$

b) 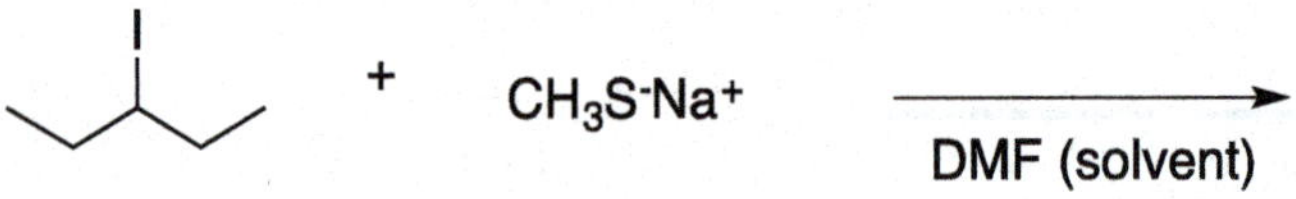 (2-iodo-3-methylbutane) + CH₃S⁻Na⁺ $\xrightarrow{\text{DMF (solvent)}}$

c) 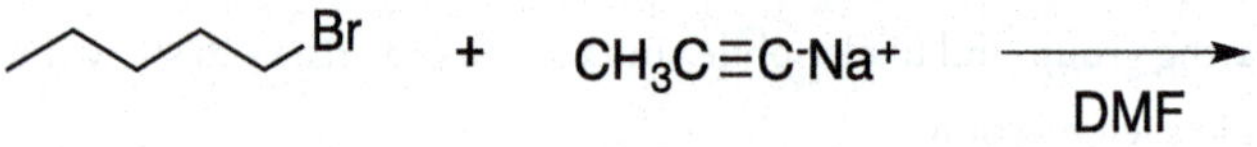 CH₃CH₂CH₂CH₂–Br + CH₃C≡C⁻Na⁺ $\xrightarrow{\text{DMF}}$

d) 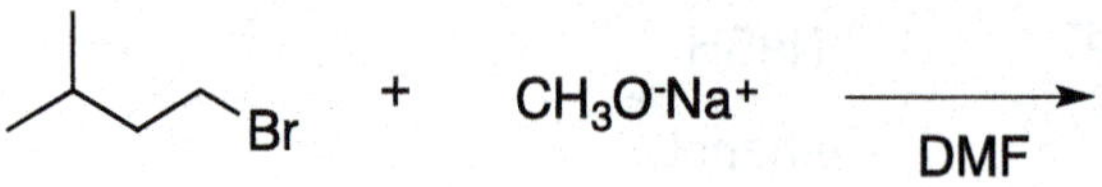 (2-bromo-3-methylbutane) + CH₃O⁻Na⁺ $\xrightarrow{\text{DMF}}$

e) 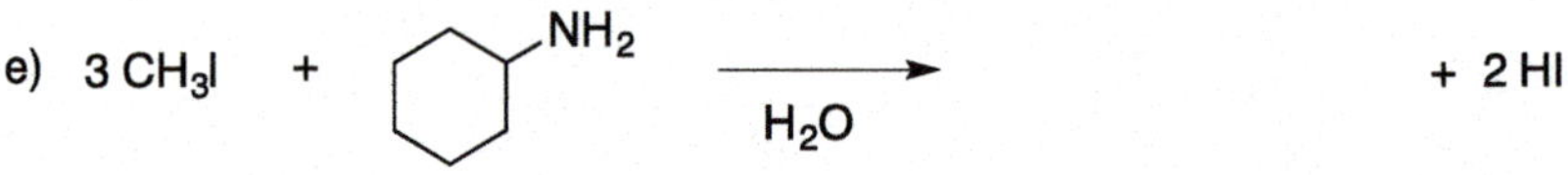 3 CH₃I + [cyclohexyl]–NH₂ $\xrightarrow{\text{H}_2\text{O}}$ + 2 HI

90 ORGANIC CHEMISTRY

5.2. S_N2 AND S_N1 MECHANISMS OF NUCLEOPHILIC SUBSTITUTION IN HALOALKANES

All nucleophilic substitution reactions result in the movement of electrons from a nucleophile to a substrate, and in the departure of a leaving group as, shown in Figure 5.2. However, the exact timing of these two events can be different depending mainly on the structure of the substrate. In the reactions of methyl halides (CH_3X) and 1° haloalkanes, the new C–Nu bond formation and breaking of the C–X bond occurs simultaneously according to the **substitution nucleophilic bimolecular mechanism (S_N2 mechanism)**. In the case of 3° haloalkanes, **substitution nucleophilic unimolecular mechanism (S_N1)** occurs in which the C–X bond breaks first, forming the carbocationic intermediate R^+, which then combines with a nucleophile in the second step of the reaction. 2° Haloalkanes undergo nucleophilic substitution by the S_N2 mechanism in the presence of strong and moderate nucleophiles and by S_N1 mechanism in the presence of weak nucleophiles.

5.2.1. S_N2 mechanism

A detailed description of an S_N2 mechanism for the reaction of bromomethane with a hydroxide anion is shown in Figure 5.5. The initial approach of the negatively charged nucleophile to the substrate is driven by the electrostatic attraction between HO^- and a partially positively charged carbon atom at the C–Br bond. Note that the nucleophile approaches from the side opposite to the leaving group (Br), which bears a partial negative charge. At some point in the process, the electrons from HO^- start forming a new bond to carbon with the simultaneous breaking of the C–Br bond. Such an intermediate arrangement of reactants is called the **transition state** (indicated by a double dagger (‡) in Figure 5.5) and it corresponds to the highest energy of the system in the course of a reaction (referred to as **activation energy**). After reaching the transition state, the transfer of electrons finalizes quickly to form the products. The energy diagram in Figure 5.5 shows the energy change during the course of an S_N2 reaction. An S_N2 reaction is called a **bimolecular reaction** because two reactants are involved in the transition state (the nucleophile and the substrate). The overall rate of an S_N2 reaction depends on the concentrations of both of the participating reactants and is mathematically expressed by the following second order kinetics equation: Rate = k $[CH_3Br]$ $[HO^-]$.

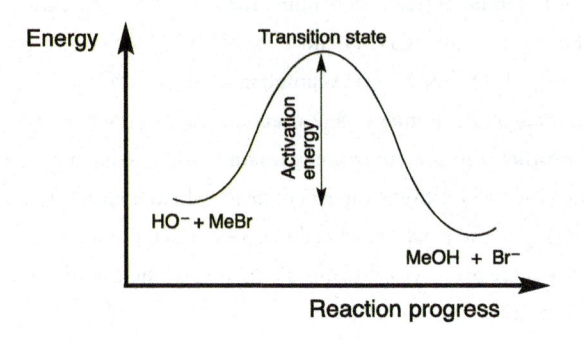

FIGURE 5.5. S_N2 mechanism and energy diagram.

S_N2 reactions can proceed with various types of solvents. Water and alcohols such as CH_3OH and CH_3CH_2OH are classified as **protic solvents** because they can serve as sources of protons due to the presence of a relatively acidic hydroxyl function. They are readily available and inexpensive solvents, and thus, they are commonly used in these reactions. However, the reactivity of nucleophiles in protic solvents is rather low because protic solvents can form hydrogen bonds to anionic nucleophiles, blocking them from interaction with the substrate molecule. The best solvents for S_N2 reactions are the so-called **polar aprotic solvents,** which are highly polar molecules incapable of forming hydrogen bonds to the nucleophiles. Commonly used polar aprotic solvents include acetone (CH_3COCH_3), DMSO (dimethyl sulfoxide CH_3SOCH_3), acetonitrile (CH_3CN), and DMF (N,N-dimethylformamide $HCONMe_2$).

The structure of alkyl group R in the substrate molecule RX has a particularly strong effect on the rate of S_N2 reactions. The nucleophile can easily approach the electrophilic carbon atom of the methyl substrate (CH_3X), which has only small hydrogen atoms connected to it. In the case of 1° and 2° haloalkanes, the additional alkyl groups hinder access to the carbon atom, as shown in Figure 5.6. The approach of nucleophiles to the electrophilic carbon atoms of 3° haloalkanes is completely blocked by the alkyl groups, making a substitution reaction by the S_N2 mechanism impossible. In general, steric hindrance within the substrate is the major factor that determines the rate of an S_N2 reaction.

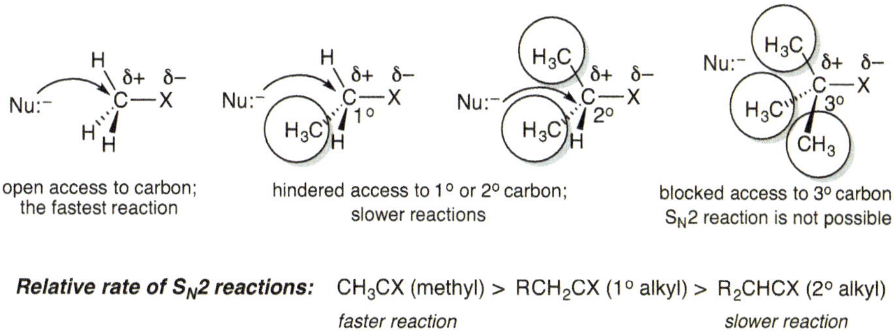

open access to carbon; the fastest reaction

hindered access to 1° or 2° carbon; slower reactions

blocked access to 3° carbon; S$_N$2 reaction is not possible

Relative rate of S$_N$2 reactions: CH$_3$CX (methyl) > RCH$_2$CX (1° alkyl) > R$_2$CHCX (2° alkyl)
faster reaction slower reaction

FIGURE 5.6. Effect of steric hindrance on S$_N$2 reactions.

5.2.2. S$_N$1 mechanism

Nucleophilic substitution in 3° haloalkanes can occur only by the unimolecular S$_N$1 mechanism. During the course of an S$_N$1 reaction, the R–X bond breaks first, forming a carbocationic intermediate R$^+$, which subsequently combines with the nucleophile during a fast step of the reaction. The overall rate of reaction is determined by the first, slow step of the R–X bond breaking, which requires higher activation energy. When the nucleophile (Nu:) reacts with the carbocationic intermediate R$^+$ in the next step, it does not affect the overall reaction rate. This is why the most typical S$_N$1 reactions occur with weak nucleophiles, such as water, alcohols, or carboxylic acids, which also serve as solvents in these reactions. Nucleophilic substitution reactions with protic solvents acting as nucleophiles are commonly called **solvolysis reactions**. A detailed description of the S$_N$1 mechanism for the solvolysis of 3°-haloalkane in water (a hydrolysis reaction) is shown in Figure 5.7.

In the first step of the process, the bromine anion departs from the substrate molecule with a pair of electrons (referred to as the **heterolytic bond cleavage**). During this process, electrons from the C–Br bond start shifting to the bromine atom and the bond stretches almost, but not quite to the point of breaking. This point in the process corresponds to the first transition state, which is characterized by the highest energy of the system (the activation energy) over the course of Step 1. After reaching the first transition state, the C–Br bond finally breaks, forming a carbocation R$^+$. In the second step of the reaction, the carbocation quickly combines with a water molecule (the uncharged nucleophile) to form an oxonium ion. This step requires smaller activation energy and is also characterized by a transition state. The overall rate of an S$_N$1 reaction is determined by the rate of the first step, which involves only molecules of t-BuBr, and therefore, it is mathematically expressed by the following first order kinetics equation: Rate = k [t-BuBr].

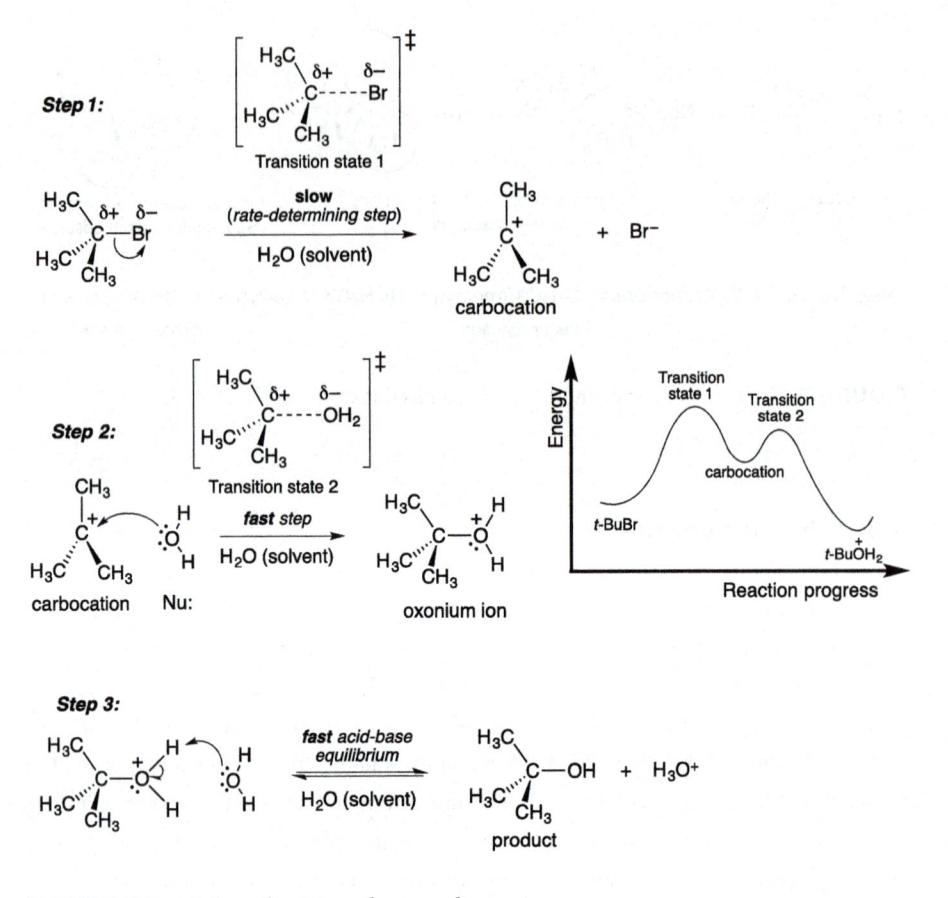

FIGURE 5.7. S$_N$1 mechanism and energy diagram.

Carbocationic intermediates in S$_N$1 reactions are highly unstable species that cannot be isolated but can be observed by spectroscopic methods at low temperatures. The carbon atom in a carbocation has only 6 valence electrons and, therefore, can react as a Lewis acid (or an electrophile, see Section 2.1). Carbocations have trigonal planar geometry with an sp^2 hybridized central atom and vacant p orbital (Figure 5.8). The relative stability of carbocations is predetermined by the nature of substituents attached to the central atom; substituents that help to delocalize the overall positive charge will increase the stability of a carbocation. In general, the most stable carbocations have three electron-donating alkyl groups (see Section 2.3 in Chapter 2) attached to the central carbon (3° carbocations), or they have a resonance delocalization of the positive charge due to the neighboring double bond (allylic or benzylic carbocations), or an element with a lone pairs of electrons, such as in the resonance contributor protonated carbonyl (Figure 5.8). Formation of a stabilized carbocation is essential for S$_N$1 reactions and therefore, S$_N$1 reactions of methyl halides (CH$_3$X) and 1° haloalkanes are highly unlikely. Reactions of the most stable 3° carbocations or the carbocations stabilized by resonance generally follow an S_N1 mechanism. The S_N1 solvolysis reactions of 2° haloalkanes are, in principle, possible in the absence of strong nucleophiles.

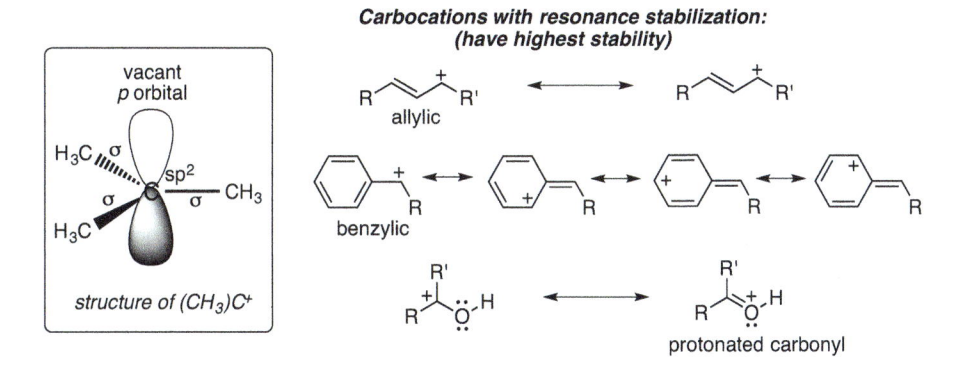

FIGURE 5.8. Electronic structure and relative stability of carbocations.

In some reactions involving carbocationic intermediates, rearrangement of a less stable carbocation to the more stable carbocation can be observed. A **rearrangement** is generally defined as a process resulting in the conversion of a molecule to an isomeric molecule via bond breaking and new bond formation. Rearrangement of a 2° carbocation to the more stable 3° carbocation can occur either by migration of a hydrogen atom with a pair of electrons (hydride anion, H:⁻) to the neighboring carbocationic center, or by a similar migration of an alkyl group, as shown in Figure 5.9. The first process is named a 1,2-hydride shift and the second process is generally called a 1,2-alkyl shift. More specifically, the process involves a 1,2-methyl shift when a methyl group with a pair of electrons (H_3C:⁻) migrates to the neighboring carbocationic center. Specific examples of solvolysis reactions involving 1,2-hydride shift and 1,2-methyl shift are shown in Figure 5.9.

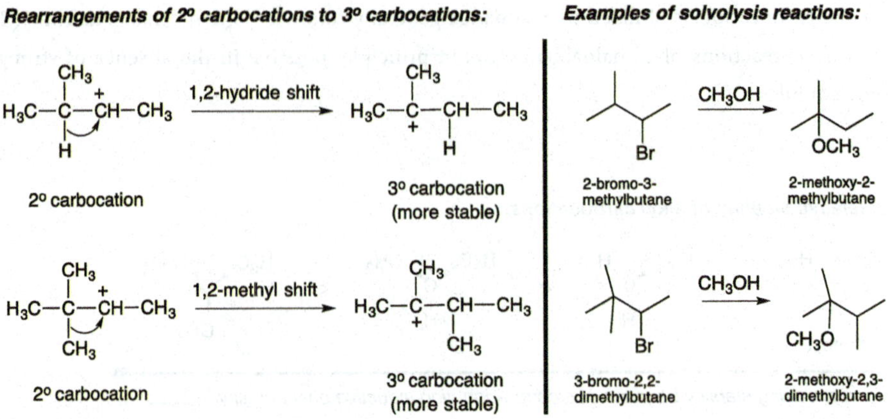

Rearrangements of 2° carbocations to 3° carbocations:

$$H_3C-\underset{H}{\overset{CH_3}{\underset{|}{\overset{|}{C}}}}-\overset{+}{C}H-CH_3 \xrightarrow{\text{1,2-hydride shift}} H_3C-\underset{H}{\overset{CH_3}{\underset{|}{\overset{|}{\overset{+}{C}}}}}-CH-CH_3$$

2° carbocation → 3° carbocation (more stable)

$$H_3C-\underset{CH_3}{\overset{CH_3}{\underset{|}{\overset{|}{\overset{+}{C}}}}}-CH-CH_3 \xrightarrow{\text{1,2-methyl shift}} H_3C-\underset{CH_3}{\overset{CH_3}{\underset{|}{\overset{|}{C}}}}-\overset{+}{C}H-CH_3$$

2° carbocation → 3° carbocation (more stable)

Examples of solvolysis reactions:

2-bromo-3-methylbutane $\xrightarrow{CH_3OH}$ 2-methoxy-2-methylbutane

3-bromo-2,2-dimethylbutane $\xrightarrow{CH_3OH}$ 2-methoxy-2,3-dimethylbutane

FIGURE 5.9. Carbocationic rearrangements.

Problems

5.4. Identify the S_N2 or S_N1 mechanisms in the reactions below. Draw each mechanism with curved arrows and show carbocationic intermediates when appropriate.

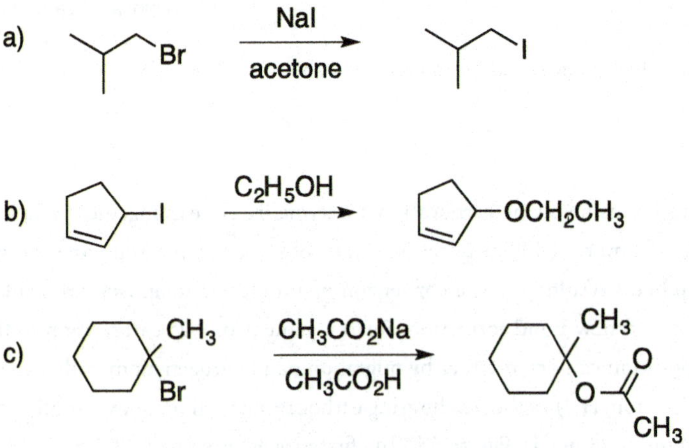

a) (CH₃)₂CHBr $\xrightarrow[\text{acetone}]{NaI}$ (CH₃)₂CHI

b) 3-iodocyclopentene $\xrightarrow{C_2H_5OH}$ 3-ethoxycyclopentene

c) 1-bromo-1-methylcyclohexane $\xrightarrow[CH_3CO_2H]{CH_3CO_2Na}$ 1-methylcyclohexyl acetate

5.5. The reaction of 1-bromopropane ($CH_3CH_2CH_2Br$) with the azide ion is a typical S_N2 reaction. Answer the following questions about the rate of this reaction:

    a) What happens to the rate of the reaction if the concentration of 1-bromopropane is doubled, while the concentration of the azide ion stays the same?

    b) What happens to the rate of the reaction if the concentration of the azide ion is doubled, while the concentration of 1-bromopropane stays the same?

c) What happens to the rate of the reaction if the concentration of 1-bromopropane is halved, while the concentration of the azide ion is doubled?

d) What happens to the rate of the reaction if the concentration of both 1-bromopropane and the azide ion is doubled?

5.6. The reaction of *tert*-butylbromide $((CH_3)_3CBr)$ with the azide ion (N_3^-) in methanol is a typical S_N1 reaction. Answer the following questions about the rate of this reaction:

a) What happens to the rate of the reaction if the concentration of *tert*-butylbromide is doubled, while the concentration of the azide ion stays the same?

b) What happens to the rate of the reaction if the concentration of the azide ion is doubled, while the concentration of *tert*-butylbromide stays the same?

c) What happens to the rate of the reaction if the concentration of *tert*-butylbromide is halved, while the concentration of the azide ion is doubled?

d) What happens to the rate of the reaction if the concentration of both *tert*-butylbromide and the azide ion is doubled?

5.7. Sort the following compounds least reactive to most reactive according to their relative reactivity in a nucleophilic substitution reaction via an S_N1 mechanism:

A) Ph-C(Cl)(CH₃)-CH₃   B) H-C(Cl)H-H   C) H₃C-C(Cl)(CH₃)-CH₃   D) Ph-C(Cl)(CH₃)-Ph

5.8. Suggest the product(s) of carbocationic rearrangements in the reactions below. Provide a mechanistic explanation for the formation of each product.

a) PhCH₂CH(Br)CH₃ + H₂O →

b) cyclopentyl-CH(I)CH₃ + CH₃OH →

c) 1-bromo-2-methylcyclohexane + H₂O →

5.3. STEREOCHEMISTRY OF S_N2 AND S_N1 REACTIONS

Understanding the mechanism of a nucleophilic substitution reaction is essential for predicting the stereochemistry of the reaction products. In the bimolecular S_N2 reactions, initial approach of the negatively charged nucleophile to the substrate is possible only from the side that is opposite the leaving group (see Figure 5.5 and the related text). Because of this mechanism, the reaction product has a new C–Nu bond on the opposite side from the initial C–X bond, and therefore the configuration of the carbon atom is changed. In the case of a chiral substrate, the S_N2 substitution at a stereocenter changes configuration to the opposite, from R to S or from S to R. A process resulting in the change of stereochemical configuration of an atom to the opposite configuration is termed **inversion of configuration**. Stereochemistry of S_N2 substitution is important in the reactions involving single stereoisomers as the substrates (enantiomers or diastereomers). For example, the reaction of enantiomerically pure (S)-2-bromobutane with sodium iodide in acetone gives (R)-2-iodobutane as the only product (Figure 5.10). Likewise, the reaction of *cis*-1-iodo-4-isopropylcyclohexane with sodium azide in methanol produces only *trans*-1-azido-4-isopropylcyclohexane. Reactions that selectively produce a single stereoisomer as the product are called **stereoselective reactions**.

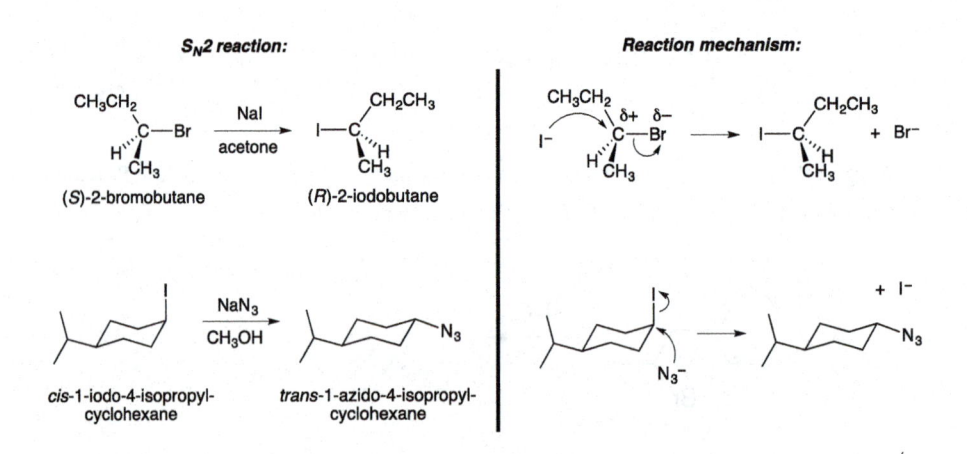

FIGURE 5.10. Stereoselective S_N2 reactions resulting in inversion of configuration.

In the unimolecular S_N1 reactions, the leaving group departs first forming a trigonal planar carbocationic intermediate which then combines with the nucleophile in the second, fast step of reaction (see Figure 5.7 and the related text). Since the carbocation is planar, the nucleophile can attack it from either side of the plane with equal probability, resulting in the formation of a pair of enantiomers as products. A process resulting in the formation of a pair of enantiomers (racemate) in the reaction of a chiral, nonracemic substrate is termed as **racemization**. Figure 5.11 shows an example of such an S_N1 reaction (the hydrolysis of 3-bromo-3-methylhexane).

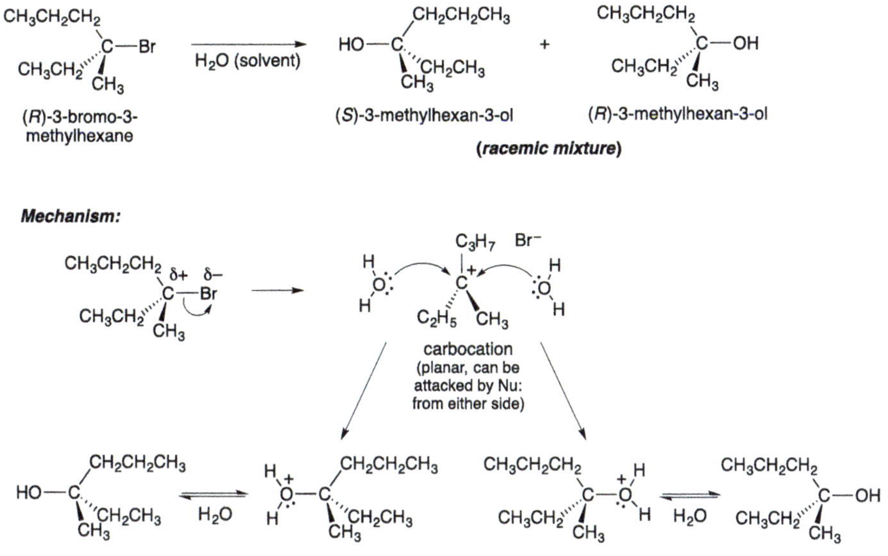

FIGURE 5.11. Stereochemistry of an S_N1 reaction (hydrolysis) of a 3° bromoalkane.

In reality, however, complete racemization (the formation of a 1:1 mixture of enantiomers starting from a single enantiomer) is rarely achieved because of ion-pairing. The leaving group X⁻ is electrostatically attracted to the carbocation R⁺, forming an ion-pair that can exist for some time in the solution. The anionic part of the ion-pair usually stays on the same side of the carbocation from which it initially departs, and thus blocks this side from nucleophilic attack. As a result, a partial inversion of configuration (an excess of the enantiomer with inverted configuration) may be observed in some S_N1 reactions.

Problems

5.9 Draw three-dimensional structures of product(s) for the reactions below. Assign (R) or (S) configuration to each stereocenter. Provide a mechanistic explanation for the formation of each product.

a)

b) [structure: CH2=CH-CH(H)(Br)-CH3 with H and Br shown] → KF / DMSO

c) [structure: cyclohexane with CH3, CH3, Br substituents] → H₂O / water

d) [structure: (CH3)2CH-C(Br)(H)-CH3 with Br and H shown] → NaSCH₃ / DMSO

e) [structure: cyclohexane with t-Bu and Cl] → NaN₃ / DMSO

5.4. E2 AND E1 MECHANISMS OF β-ELIMINATION REACTIONS

Similar to nucleophilic substitution, elimination reactions can proceed by a bimolecular mechanism (abbreviated as **E2**) or a unimolecular mechanism (abbreviated as **E1**). The bimolecular mechanism occurs in one step as a simultaneous transfer of electrons from the base (Base:⁻) to the leaving group (X), resulting in removal of the proton at the β-carbon, formation of the double bond, and departure of X:⁻ (Figure 5.12). Since two reactants (the base and the substrate) are involved in the transition state, the overall rate of E2 reaction depends on the concentrations of both participating reactants, and is mathematically expressed by the second order kinetics equation: Rate = k [Substrate][Base:⁻]. The energy diagram of an E2 reaction is similar to the S_N2 reaction. The presence of a strong base capable of deprotonating the hydrogen at the β-carbon is essential for an E2 reaction. Typical bases include alkoxides in solution of the related alcohol (CH_3ONa in CH_3OH, EtONa in EtOH, or t-BuOK in t-BuOH), or sodium amide ($NaNH_2$) in liquid NH_3. A halogen atom, Cl, Br, or I usually represents the leaving group X.

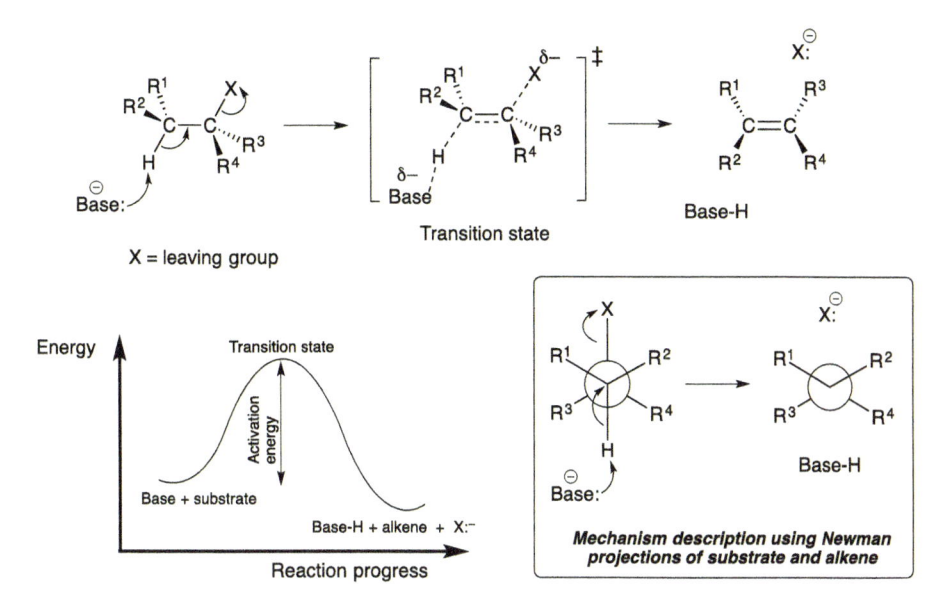

FIGURE 5.12. E2 mechanism and energy diagram.

An appropriate alignment of reactants is essential for E2-elimination: the initial system of atoms H–C–C–X must be in the same plane (referred to as a **coplanar** arrangement) with H and X in anti-conformation, as shown using the Newman projections in Figure 5.12. This anti-coplanar arrangement is required in order to achieve a better interaction of the molecular orbitals involved in concerted transfer of electrons in the course of a reaction. Because of this mechanism, the substituents R^1, R^2, R^3, and R^4 keep the same arrangement in the resulting alkene as in the initial substrate, with R^1 and R^3 on one side of the double bond and R^2 and R^4 on the other side, as shown in Figure 5.12. Therefore, the product of the reaction is a single stereoisomer and the E2 elimination can be defined as a stereoselective reaction.

Figure 5.13 shows several examples of stereoselective E2 reactions. Specifically, the E2 reactions of two diastereomeric 1-bromo-1,2-diphenylpropanes produce the appropriate stereoisomeric alkenes (see Chapter 6 for the E/Z-nomenclature of stereoisomeric alkenes). It should be emphasized that the molecule of substrate must be in the appropriate conformation in order to make the anti-coplanar elimination possible. In particular, the required anti-coplanar arrangement of the H–C–C–Br system in *trans*-1-bromo-2-methylcyclohexane can be achieved only when the leaving group Br and the hydrogen at the β-carbon are in the axial positions depicted in Figure 5.13. As a result, 3-methylcyclohexene is formed in this reaction as a single product.

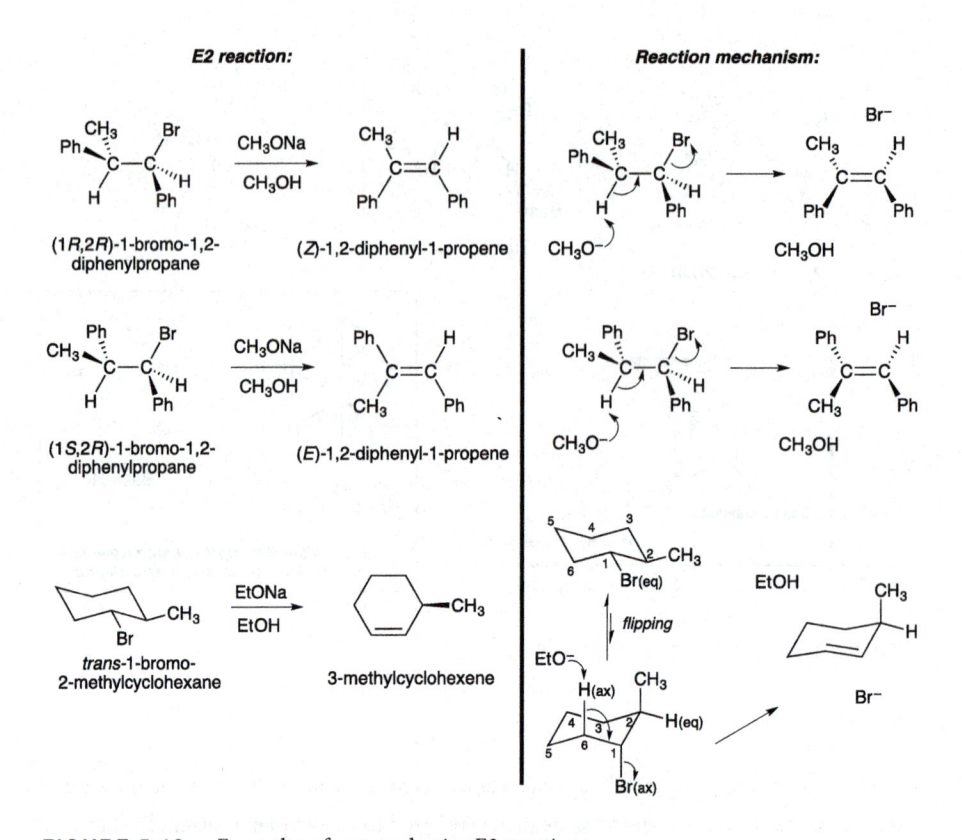

FIGURE 5.13. Examples of stereoselective E2 reactions.

E2 elimination reactions of substrates which have two different β-carbons bearing hydrogens yield two isomeric alkenes. Several examples of such reactions are shown in Figure 5.14. The products in each reaction differ by the position of the double bond in the chain, and therefore, products are positional isomers or regioisomers. As a rule, the major product in the β-elimination reaction of a haloalkane contains the double bond with a higher number of alkyl substituents. In other words, the main product is the more highly substituted alkene. This observation about the preferred regioselectivity of β-elimination reactions was published in 1875 by the Russian chemist Aleksander Zaitsev and is called **Zaitsev's rule**. The predominant formation of the more highly substituted alkenes in elimination reactions is generally explained by the higher thermodynamic stability of a double bond bearing a larger number of alkyl substituents. Zaitsev's rule usually gives a good prediction of the regioselectivity in the reactions of non-cyclic substrates bearing a good leaving group such as bromine or iodine. There are exceptions to this rule. For example, an E2 reaction of *trans*-1-bromo-2-methylcyclohexane (Figure 5.13) does not follow Zaitsev's rule and yields the less substituted 3-methylcyclohexene instead of the more substituted 1-methylcyclohexene because of conformational restrictions. In contrast, a similar E2 reaction of *cis*-1-bromo-2-methylcyclohexane (Figure 5.14) proceeds according to Zaitsev's rule since the anti-coplanar elimination is possible in this case.

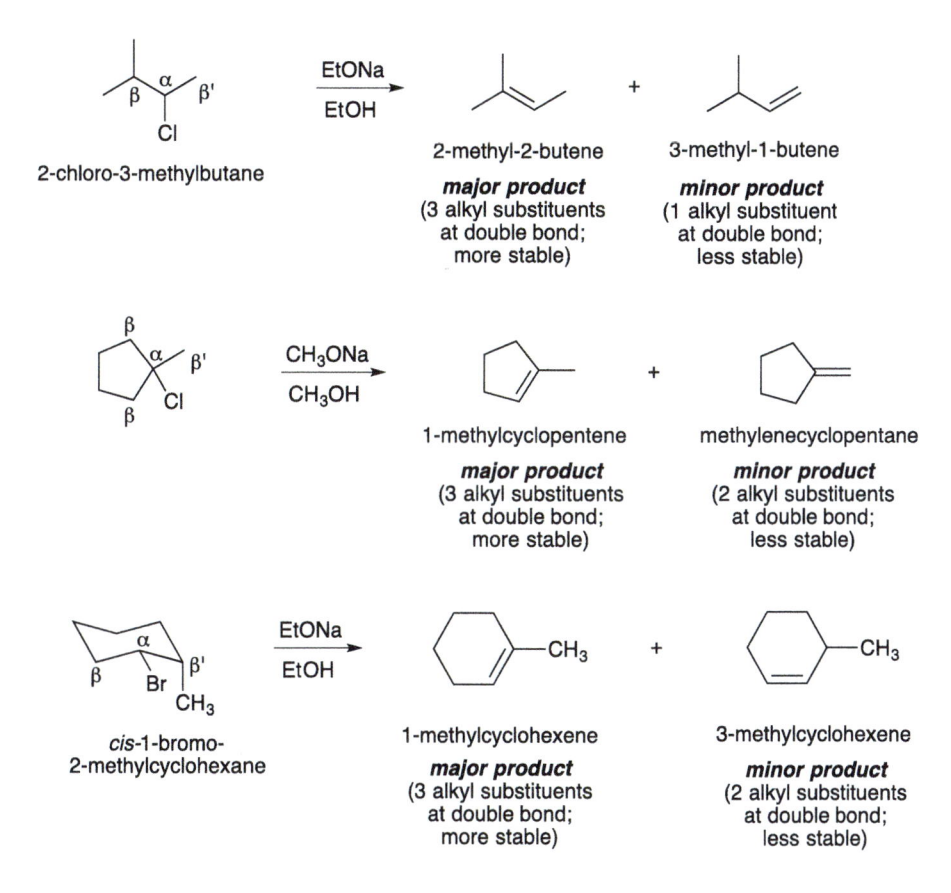

FIGURE 5.14. Regioselective E2 reactions following Zaitsev's rule.

The unimolecular E1 β-elimination commonly occurs as a competing process with S_N1 nucleophilic substitution producing an alkene as a by-product along with the product of substitution. An example of such a competing E1-S_N1 reaction is shown in Figure 5.15. A mechanistic description of E1 elimination reactions includes two steps, and the first is identical to the first step of the S_N1 mechanism. During the course of the first, rate-determining step, the C–X bond breaks, forming a carbocationic intermediate R⁺, which then participates in two competing fast processes: 1) reaction of R⁺ with a solvent molecule (or nucleophile), producing an S_N1 reaction (see Figure 5.7 for detailed S_N1 mechanism), or 2) proton transfer from the β-carbon of the carbocation to the molecule of solvent, producing alkene as the β-elimination product. Similar to E2 reactions, the E1 reaction also proceeds according to Zaitsev's rule promoting a more stable, more highly substituted alkene as the major product.

The overall rate of the E1 reaction is determined by the first, slow step of C–X bond breaking, requiring higher activation energy. The nature of the base involved in the next, fast step does not affect the overall reaction rate. This is why E1 reactions do not require the presence of strong bases, and solvent molecules serve as the proton acceptors. Since the rate of the E1 reaction is determined by the rate of the first step, which involves only substrate molecules, it is mathematically expressed by the first order kinetics equation: Rate = k [substrate].

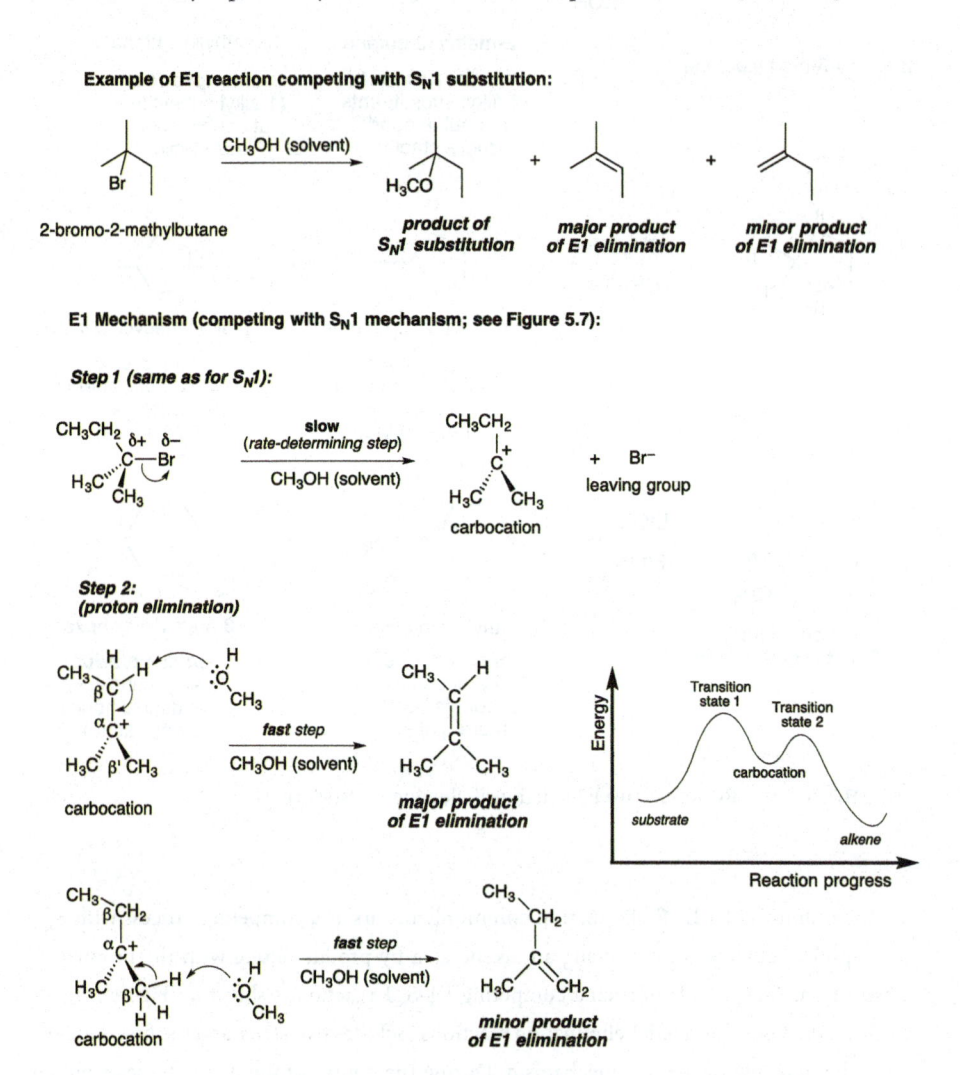

FIGURE 5.15. E1 mechanism and energy diagram.

104 ORGANIC CHEMISTRY

Problems

5.10 Suggest a mechanism explaining formation of the product in the following reaction:

CH₃-[cyclohexane with Br and CH₃]-CH₃ $\xrightarrow[C_2H_5OH]{C_2H_5ONa}$ CH₃···[cyclohexene]···CH₃

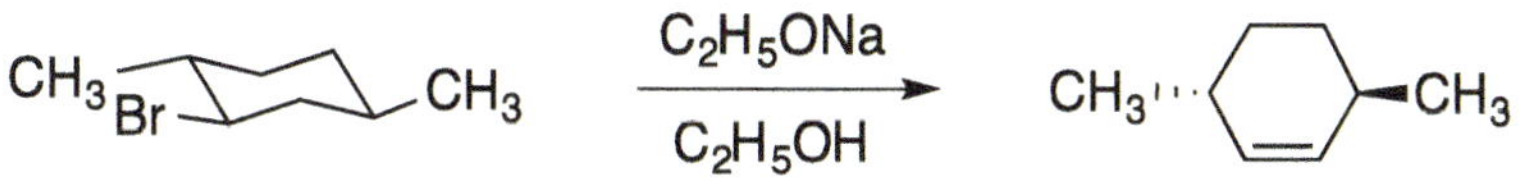

5.11 Provide an explanation for the outcome of the following reactions:

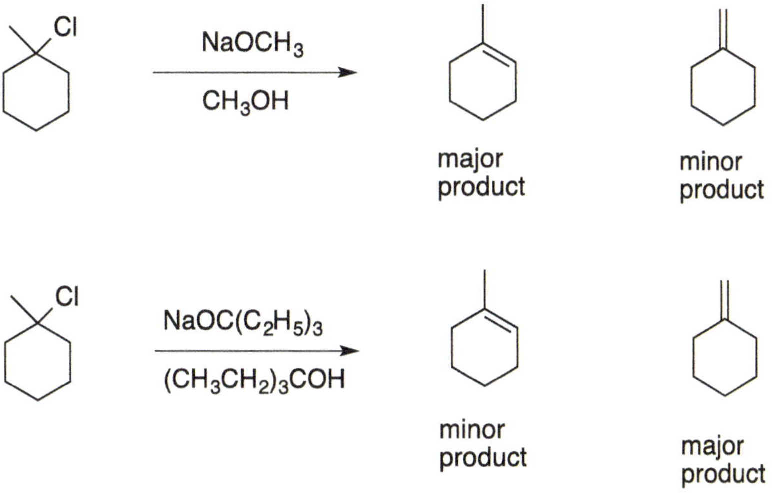

[1-chloro-1-methylcyclohexane] $\xrightarrow[CH_3OH]{NaOCH_3}$ [1-methylcyclohexene, major product] + [methylenecyclohexane, minor product]

[1-chloro-1-methylcyclohexane] $\xrightarrow[(CH_3CH_2)_3COH]{NaOC(C_2H_5)_3}$ [1-methylcyclohexene, minor product] + [methylenecyclohexane, major product]

5.12 Assume an E2-elimination mechanism. Which of the following will react faster: *trans*-1-*tert*-butyl-4-chlorocyclohexane or *cis*-1-*tert*-butyl-4-chlorocyclohexane?

5.13 For each of reactions provide the major product:

a) 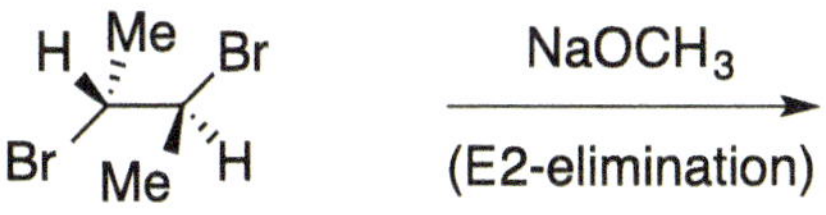 [H, Me, Br / Br, Me, H stereochemistry] $\xrightarrow[\text{(E2-elimination)}]{NaOCH_3}$

b) 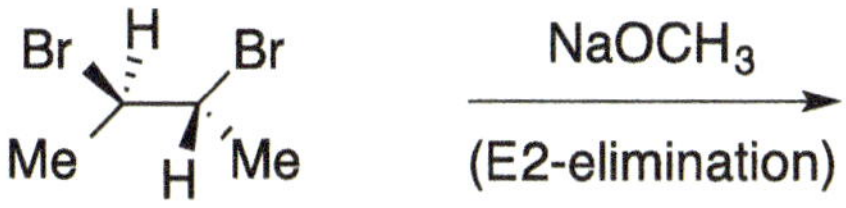 [Br, H, Br / Me, H, Me stereochemistry] $\xrightarrow[\text{(E2-elimination)}]{NaOCH_3}$

5.14. Regardless of mechanism, which of the following would only give a single alkene as the product of an elimination reaction?

A) 3-chloro-3-ethylpentane

B) 2-chloro-2-methylpentane

C) 3-chloro-3-methylpentane

D) 2-chloro-4-methylpentane

CHAPTER 6

Alkenes

Alkenes are generally defined as hydrocarbons with a carbon-carbon double bond in their structure. An older, common name for alkenes is olefins. Some complex alkenes and cycloalkenes (referred to as terpenes) can be found in natural sources such as the essential oils of plants, especially conifers. Simple alkenes such as ethene (common name ethylene) and propene (common name propylene) are important industrial products. In fact, worldwide production of ethylene (over 100 million tons per year) exceeds that of any other organic compound. Ethylene is mainly used for production of polyethylene, a common plastic containing a polymeric chain of ethylene units. Alkenes have a very rich chemistry and serve as precursors to a variety of important organic compounds. This chapter provides an overview of important structural features, nomenclature, and reactions of alkenes. Polymerization of ethylene and other alkenes is discussed at the end of the chapter.

6.1. STRUCTURE AND NOMENCLATURE OF ALKENES

Alkenes are unsaturated hydrocarbons (see Section 1.3 in Ch. 1) with a double bond in their structure and a general molecular formula of C_nH_{2n}. Cyclic hydrocarbons with a double bond in the ring are called cycloalkenes. Hydrocarbons with two double bonds are called dienes, and the general name of hydrocarbons with several double bonds is polyenes. Cycloalkenes, dienes and polyenes are usually considered subclasses of alkenes.

The double bond in alkene is formed by two sp² hybridized carbon atoms and consists of one σ-bond and one π-bond between the atoms (see Figure 1.20 and the related text in Chapter 1). The π-bond has two lobes, one above and the second below the planar σ-bond framework formed by two trigonal planar carbon atoms. In contrast to a C–C single bond, rotation about a carbon-carbon double bond is impossible without breaking the π-bond, which requires significant amount of energy. Due to the restricted rotation, alkenes with two different substituents at each carbon can form two different stereoisomers across the double bond. The older **cis-trans nomenclature** for stereoisomeric alkenes is similar to the nomenclature of disubstituted cycloalkanes: the unbranched alkenes with two substituents on the same side of the double bond are *cis* isomers, and the alkenes with substituents on the opposite sides of the double bond are *trans* isomers. Examples of *cis* and *trans* alkenes (*cis*- and *trans*-2-butene) are shown in Figure 6.1. It should be emphasized that two different substituents at each carbon of the double bond are required for this type of stereoisomerism. For example, the molecule of 1-butene, which is related to 2-butene as a regioisomer (see Section 5.4), does not have *cis* and *trans* stereoisomers because the substituents on one of the carbons are identical (two hydrogens on the same carbon).

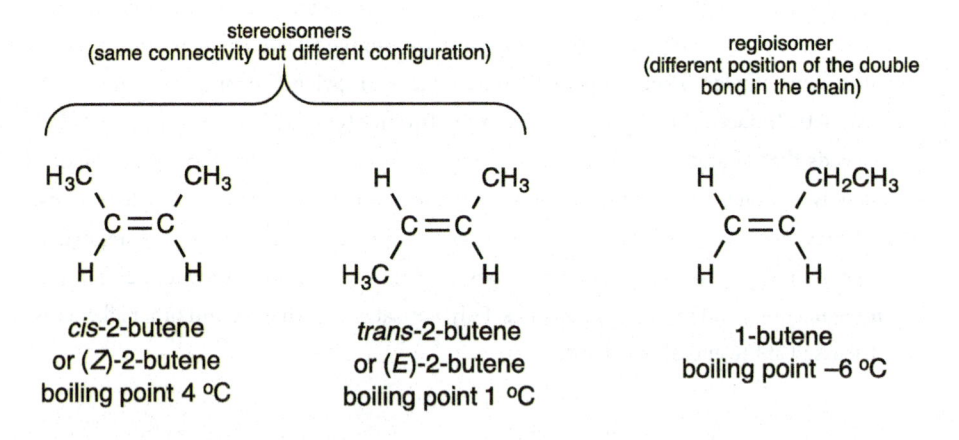

FIGURE 6.1. Isomers of butene.

The *cis-trans* nomenclature is limited mainly to unbranched disubstituted alkenes. In the case of tri- or tetrasubstituted double bonds, this nomenclature has limited use. According to IUPAC, a more general **E/Z nomenclature** is the preferred method of describing the absolute stereochemistry of double bonds. The E/Z convention uses the same rules as the R/S system (Cahn-Ingold-Prelog Priority Rules, see Section 4.2) to assign priority to the two substituents connected to each carbon of the double bond. If the high priority substituents at both carbons are on the same side of the double bond, the alkene has a Z configuration (from the German word *Zusammen*, which means "together"). When the high priority substituents at both carbons are on the opposite sides of the double bond, the alkene has an E configuration (from the German word *Entgegen*, which means "opposite"). Examples of E/Z nomenclature are shown and explained in Figure 6.2.

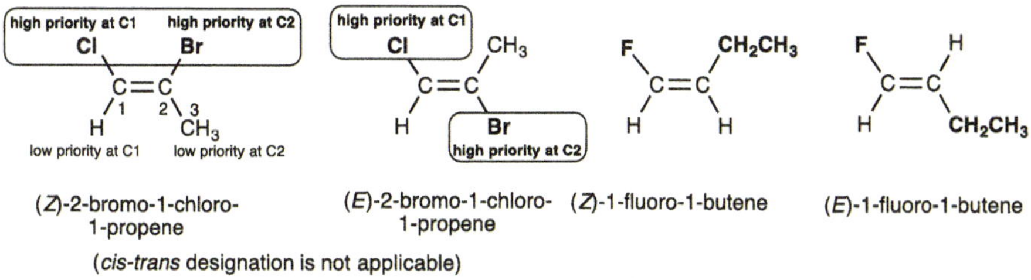

FIGURE 6.2. Examples of E/Z nomenclature.

According to the general IUPAC rules, the presence of a double bond in the carbon chain of alkene is indicated by changing the infix -an- in the related alkane to the infix -en- (Section 1.3).

The following rules are used to assign the IUPAC name for a substituted alkene:

1. Identify the longest chain (the parent chain) of carbon atoms that includes both carbons of the double bond and name it by changing infix -an- in the related alkane to the infix -en-. If there are two or more chains of the same length, choose the chain with the greater number of substituents. This is the root name of the alkene.
2. Put numbers on the carbons of the parent chain starting from the end nearest to the double bond. Smaller numbers on the carbons indicate the location of the double bond.
3. Use the general IUPAC rules to indicate each substituent and functional group in the parent chain (see Sections 1.3 and 3.2).
4. If needed, place the stereochemical descriptor (E) or (Z) in front of the name. The older designation *cis-trans* can also be used to indicate the configuration of the parent unbranched alkene. Note: In many cases, Z configuration corresponds to cis-alkenes and E configuration corresponds to trans-alkenes; however, exceptions are possible.

Figure 6.3 shows several examples of alkene nomenclature. These examples illustrate the process of selecting and numbering the parent chain correctly and assigning the configuration of the double bond. Specifically, the parent chain of the molecule of 2-ethyl-3-methyl-1-pentene has 5 carbons including both carbons of the double bond. A longer chain of 6 carbons is present in this molecule, but it includes only one carbon of the double bond. The molecule has no stereoisomers and so the E/Z descriptor is not needed. The second molecule, (Z)-4-(*tert*-butyl)-3-ethyl-2-methyl-3-heptene, has two stereoisomers and requires the stereochemical descriptor in front of the name. The parent chain has the configuration of *trans*-3-heptene; however, the name has a (Z) descriptor because the high priority substituents (isopropyl at C3 and *tert*-butyl at C4) are on the same side of the double bond. In (E)-3-methyl-2-hexene, the *trans*-configuration of the parent chain is in agreement with the (E) descriptor. Dienes and polyenes are named using the same general rules and indicating the stereochemistry of each double bond in front of the name (Figure 6.3).

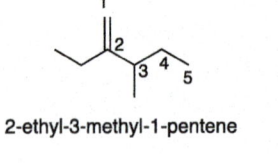

FIGURE 6.3. Examples of IUPAC nomenclature of alkenes.

Cycloalkenes are named according to the IUPAC rules for cycloalkanes by replacing the infix -an- to the infix -en- and assigning numbers 1 and 2 to the carbons of the double bond in the ring. Only *cis*-isomers are possible in the smaller rings, therefore, there is no need to show the stereochemical descriptor in front of the name. Several examples of cycloalkenes are shown in Figure 6.4.

FIGURE 6.4. Examples of IUPAC nomenclature of cycloalkenes.

110   ORGANIC CHEMISTRY

Naming cycloalkanes with a substituent that has a double bond requires using the names of the following alkenyl groups: $H_2C=$ methylene, $H_2C=CH-$ vinyl, and $H_2C=CH-CH_2-$ allyl. Several examples of such names are shown in Figure 6.5.

methylenecyclopropane   1-ethyl-3-vinylcyclopentane   1-allyl-3-methylenecyclohexane

FIGURE 6.5. Naming alkenes with an exocyclic double bond.

The names of alkenyl groups are often used in the common names of different compounds such as methylene chloride (CH_2Cl_2), vinyl chloride $(CH_2=CHCl)$, and allyl alcohol $(CH_2=CHCH_2OH)$.

Problems

6.1. Provide the IUPAC name for each of the following compounds:

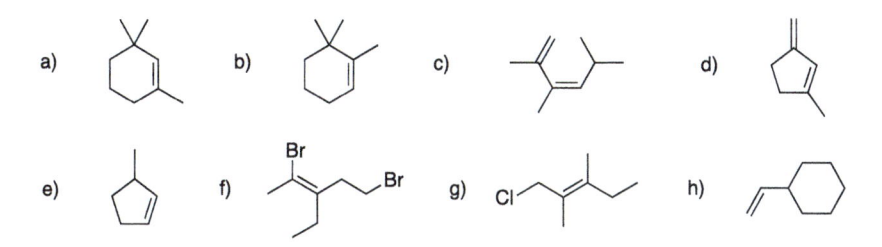

a) b) c) d) e) f) g) h)

6.2. Draw structural formulas for the following alkenes:

a) 1-methyl-1-vinyl-cyclohexane
b) vinyl chloride
c) allyl bromide
d) methylenecyclohexane

6.3. Draw the structures for the following alkenes and determine how many stereoisomers are possible (ignore any conformers):

a) 1-bromo-3-methyl-2-butene
b) 4-chloro-2-pentene
c) 2-iodo-3-methyl-2-pentene

ALKENES   111

6.4. Determine the configuration of the double bond in the following alkenes:

a)
```
   H        B(CH₂CH₃)₂
    \      /
     C == C
    /      \
  HO        CH(CH₃)₂
```

b)
```
   H₃C       CH₂OH
      \    /
       C==C
      /    \
  H₂NH₂C    CH₂Br
```

c)
```
   H      CH₃
    \    /
     C==C
    /    \
   Br    CH₂CH₃
```

d)
```
  F~~\        /~~Br
       C == C
      /      \
           CH(CH₃)₂ (iPr)
```

6.5. Determine the index of hydrogen deficiency (see Section 1.3) for the following molecules:

a) C_6H_8 b) $H_2C=CHCH_2NO_2$

c) cyclopentene-Cl   d) pyrrole (N-H)   e) bicyclic alkene

6.2. ELECTROPHILIC ADDITION: REACTIONS OF ALKENES WITH STRONG ACIDS

An electrophilic addition is defined an addition reaction of an electrophilic reagent to a double bond or a triple bond. An electrophilic reagent or electrophile is a molecule or atom that acts as an acceptor of electrons in a chemical reaction (see Section 2.1). An electrophilic atom is usually characterized by the presence of a partial or full positive charge. Strong Brønsted acids (HCl, HBr, HI, and H_2SO_4) are typical electrophiles in reactions with alkenes. Electrophilic addition is the most characteristic reaction of alkenes, widely used in chemical labs and industry for preparation of many important products.

Addition of hydrogen halides (HCl, HBr, or HI) to alkenes yields haloalkanes as illustrated in Figure 6.6. For an unsymmetrical alkene such as the 1-propene in Figure 6.6, this is a regioselective addition resulting in the formation of a single regioisomer, 2-bromopropane. No 1-bromopropane is formed in this addition. The first observation of regioselectivity of electrophilic addition reactions was published by the Russian chemist Vladimir Markovnikov in 1870 and is called **Markovnikov's rule**. According to Markovnikov's rule, the addition of an acid HX to an unsymmetrical alkene yields a product in which hydrogen is added to the carbon with more hydrogen atoms, while the halogen atom X is added to the carbon with fewer hydrogen atoms. Two additional examples of regioselective addition reactions (HCl and HBr) are shown in Figure 6.6.

$H_2C=CH_2$ + HX $\longrightarrow$ CH_3CH_2X

X = Cl, Br, or I

1-propene + H–Br $\longrightarrow$ 2-bromopropane ($BrCH_2CH_2CH_3$ 1-bromopropane is not observed)

FIGURE 6.6. Electrophilic addition of hydrogen halides to alkenes.

The observed regioselectivity of electrophilic addition is explained by a two-step reaction mechanism involving carbocationic intermediates (Figure 6.7). During the first, slow reaction step, electrons from the π-bond of alkene move to the proton of the acid HX to form a new C–H bond. In principle, this step may lead to any of the two intermediates: 1° carbocation or 2° carbocation. However, 1° carbocations have lower stability compared to the 2° carbocations (see Section 5.2.2), and the formation of 1° carbocations requires higher activation energy. This is why the more stable 2° carbocationic intermediate is formed in the first

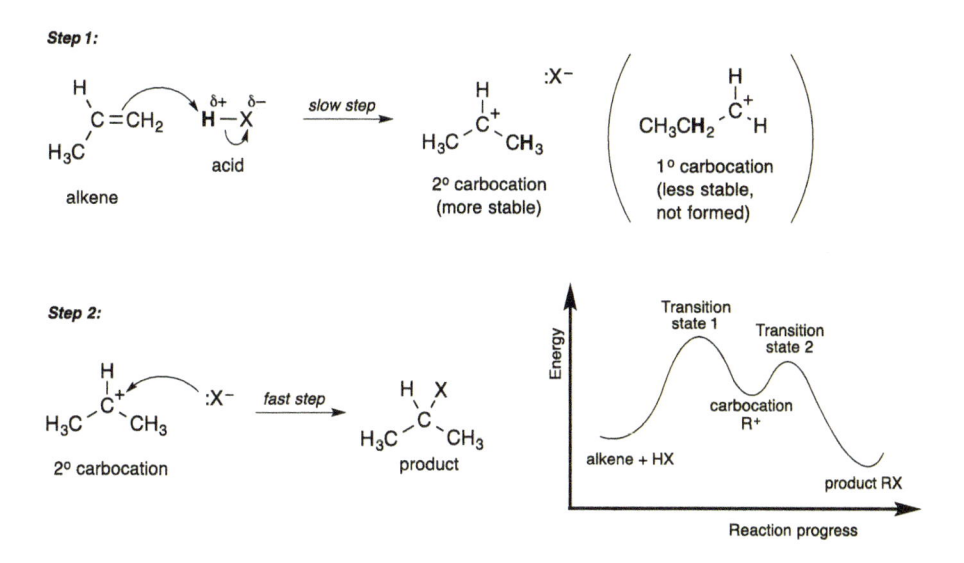

Step 1:

alkene + acid (H–X) $\xrightarrow{\text{slow step}}$ 2° carbocation (more stable) + :X⁻ (1° carbocation (less stable, not formed))

Step 2:

2° carbocation + :X⁻ $\xrightarrow{\text{fast step}}$ product

FIGURE 6.7. Mechanism of electrophilic addition reaction and the energy diagram.

step. In the second step, the 2° carbocation quickly combines with anion X⁻ as a nucleophile to form the final product. An energy diagram for the process of electrophilic addition is similar to the S$_N$1 reaction. It includes two transition states and a local energy minimum for the carbocationic intermediate.

Reactions of alkenes with strong acids (usually H_2SO_4) in water produce alcohols as main products. Several examples of acid-catalyzed hydration reactions are shown in Figure 6.8. Formation of alcohols in these reactions occurs because carbocationic intermediates combine with water as a nucleophile, similar to the second step of an S$_N$1 reaction (see Figure 5.7 in Chapter 5). Addition of water to an unsymmetrical alkene proceeds in accordance with Markovnikov's rule, via intermediate formation of the most stable carbocation. Reactions of alkenes with strong acids may result in rearrangement of a less stable carbocation to a more stable carbocation by 1,2-hydride shift or 1,2-alkyl shift, as explained in Section 5.2.2 (see Figure 5.9).

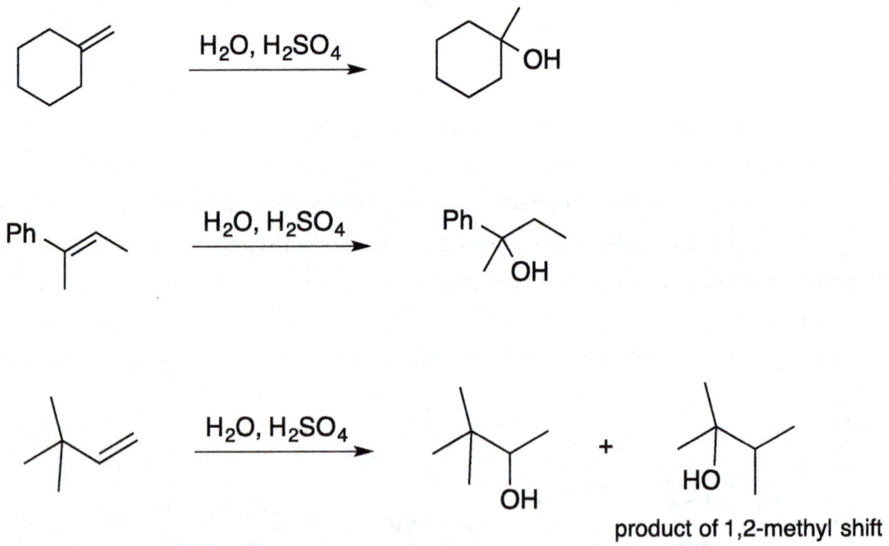

FIGURE 6.8. Acid-catalyzed hydration of alkenes.

Problems

6.6. Using curved arrows and showing reaction intermediates, write detailed mechanisms for the reactions shown in Figure 6.8.

6.7. Provide the products of electrophilic addition of HCl or HBr. In addition, give structures of the rearranged products when a 1,2-hydride shift or 1,2-alkyl shift is possible.

a) 1,3-dimethylcyclopentene + HCl →

b) 3,3-dimethyl-1-butene (2,3-dimethyl-1-butene) + HBr →

c) 1,1,3,3-tetramethylcyclopentene + HCl →

6.8. Provide the products of acid-catalyzed hydration in the reactions below. In addition, give structures of the rearranged products when a 1,2-hydride shift or 1,2-alkyl shift is possible.

a) ethylcyclohexene + H$_2$O, H$_2$SO$_4$ →

b) 1-vinylcyclohexane + H$_2$O, H$_2$SO$_4$ →

c) 4-methyl-1,3-pentadiene/4-methyl-2-pentene + H$_2$O, H$_2$SO$_4$ →

6.3. ELECTROPHILIC ADDITION OF CL$_2$ OR BR$_2$

Chlorine and bromine react with alkenes at room temperature, forming the corresponding dichloro- or dibromoalkanes (Figure 6.9). Only the reactions of Cl$_2$ and Br$_2$ are practically useful; fluorine reacts with alkenes violently, while iodine is not sufficiently reactive. The addition of halogens to alkenes is a stereoselective reaction, as illustrated by the bromination of cyclopentene producing *trans*-1,2-dibromocyclopentane as a single diastereomer (as a racemic mixture of two enantiomers, only one of which is shown in Figure 6.9). No *cis*-1,2-dibromocyclopentane is formed in this reaction. Likewise, the bromination of (E)-2-butene yields exclusively *meso*-2,3-dibromobutane, but not the other diastereomer. The stereoselective addition producing *trans* product is called **anti addition** and such reactions are referred to as **anti stereoselective addition reactions**. It should be noted that the reactions of achiral alkenes with halogens leading to chiral products of addition always give a racemic mixture of enantiomers (see Section 4.4).

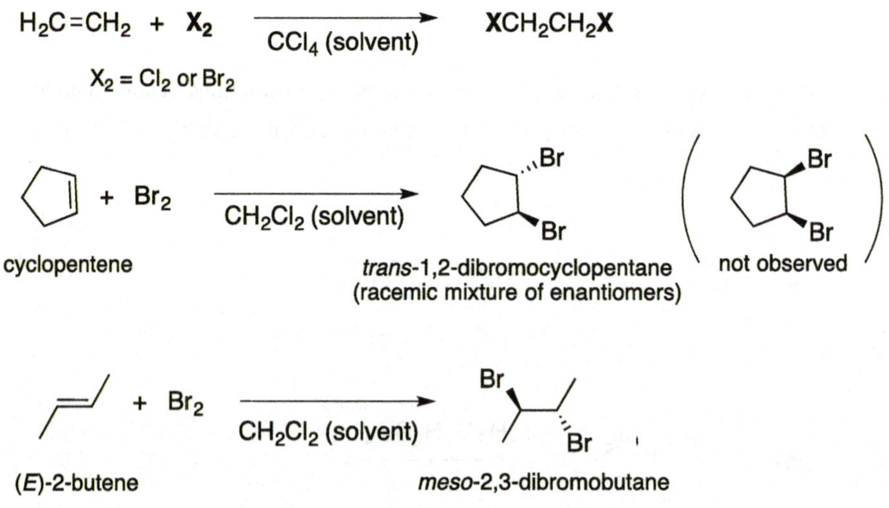

FIGURE 6.9. Addition of halogens to alkenes.

The *anti* stereoselective addition reactions of halogens to alkenes are explained by a mechanism involving a cyclic bromonium (or chloronium) ion as the reactive intermediate. A mechanism for the reaction of bromine with cyclopentene is shown in Figure 6.10. In the initial reaction step, the molecule of bromine acts as an electrophile, attracting π-electrons from the double bond of cyclopentene. As a result, a carbocationic intermediate is formed and one bromine atom is released as its anion. The carbocation formed in this step has a bromine atom attached to the carbon next to the positive charge. The lone electron pair from the bromine atom can move to the carbocation, creating a cyclic resonance form, referred to as a bromonium ion. The real structure of a bromonium ion intermediate (i.e., the resonance hybrid) is a mixture of all three possible resonance contributors (see Section 1.6), and therefore has partial positive charges on the carbon atoms and on the bromine.

In the second step, the bromine anion (a nucleophile) attacks carbon from the side opposite the bromonium atom (the leaving group), according to an S_N2 mechanism (Section 5.2.1). This is a stereoselective substitution resulting in inversion of configuration (Section 5.3), forming the final *trans* product.

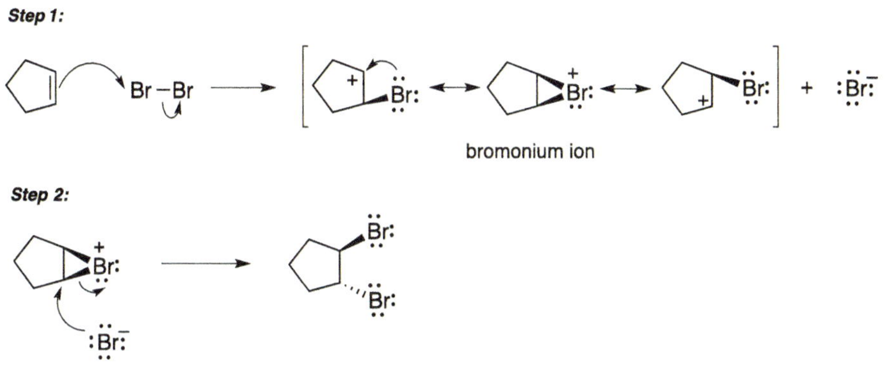

Step 1:

Step 2:

FIGURE 6.10. Mechanism of electrophilic bromination.

Reactions of alkenes with halogens in aqueous solution form halogenated alcohols (referred to as **halohydrins**) as main products. Figure 6.11 shows several examples of alkene halohydration reactions. Formation of halohydrins in these reactions results from participation of water as a nucleophile in the second step of the reaction mechanism. These are *anti* stereoselective addition reactions that result in the formation of products with *trans* configuration of the added groups. Addition of halogen and water to an unsymmetrical alkene proceeds via the formation of the most stable (more highly substituted) carbocationic resonance contributor as the reaction intermediate. Water (the nucleophile) adds to the more substituted carbon, while halogen (the electrophile) initially adds to the less substituted carbon of the double bond.

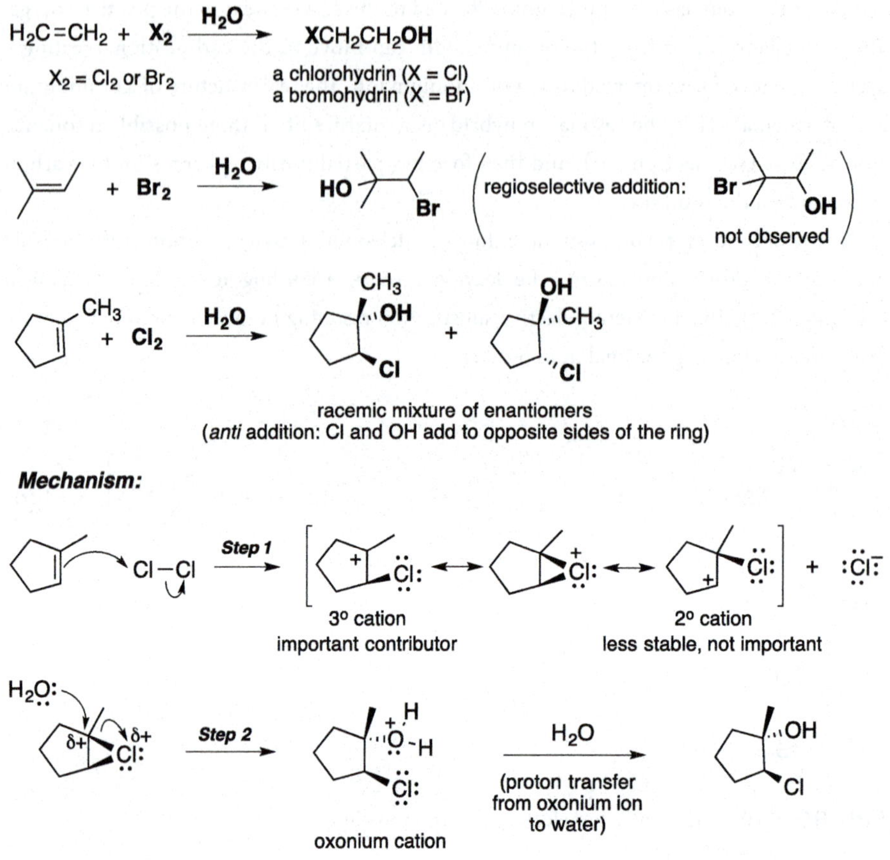

FIGURE 6.11. Electrophilic halogenation in the presence of water (halohydrin formation).

Problems

6.9. Draw three-dimensional structures of the products from the following stereoselective reactions:

a) [structure: (CH3)2C=C(CH3)CH2CH3] Cl₂ →

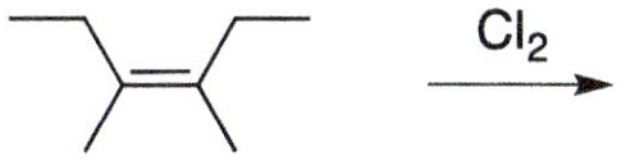

b) [structure] Cl₂ →

c) [structure] Br₂/H₂O →

d) [structure] Br₂/H₂O →

e) 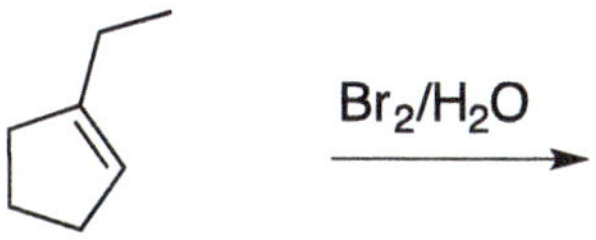 [ethylcyclopentene structure] Br₂/H₂O →

6.4. HYDROBORATION OF ALKENES

Hydroboration reaction involves the addition of borane, BH_3, to alkenes. The electronegativity value of boron (2.0) is slightly lower than that of hydrogen (2.1) and therefore, BH_3 cannot act as a Brønsted acid (see Section 2.1). Moreover, a hydrogen atom in BH_3 can act as a nucleophile because of partial negative charge. The boron atom in BH_3 has only 6 valence electrons, which makes this molecule a Lewis acid and an electrophile capable of accepting a pair of π-electrons from the double bond. Borane is a highly reactive and potentially dangerous compound; however, complexes of BH_3 with ethers (diethyl ether or a cyclic ether such as tetrahydrofuran, THF) are stable and commercially available solutions.

A hydroboration reaction is usually performed by mixing an alkene and BH_3 in ether solution. The reaction proceeds as a regioselective addition in accordance with the electrophilic reactivity of boron and the nucleophilic character of hydrogen in BH_3. The overall reaction proceeds as three consecutive additions of B–H bonds to three molecules of alkene, resulting in the formation of trialkylborane, R$_3$B, as the final product of hydroboration (Figure 6.12). Addition of the B–H fragment to a double bond occurs stereoselectively, with both boron and hydrogen atoms adding on the same side of the double bond, producing a product with *cis* configuration of the added atoms or groups. The stereoselective addition producing the *cis* product is called a ***syn* addition**, and such reactions are referred to as ***syn* stereoselective addition reactions**. The *syn* stereoselective addition reaction is illustrated by the hydroboration of a cyclic molecule, such as 1-methylcyclohexene in Figure 6.12.

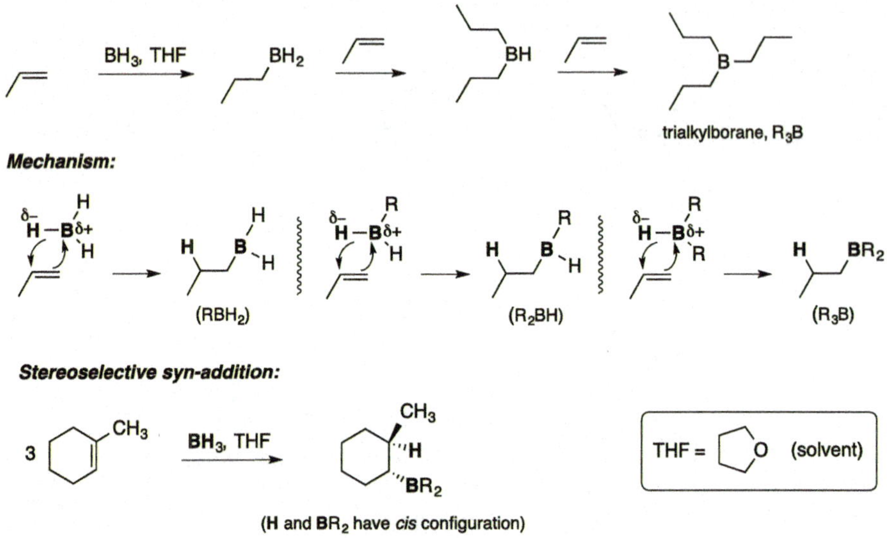

FIGURE 6.12. Hydroboration of alkenes.

The products of hydroboration, trialkylboranes, R$_3$B, are unstable, pyrophoric compounds, which are usually converted to alcohols by oxidative treatment with hydrogen peroxide (H$_2$O$_2$) in aqueous basic solution without isolation. The sequence of reactions converting alkene to alcohol via trialkylborane is called hydroboration-oxidation reaction (Figure 6.13). This reaction has important applications in organic chemistry and technology because it affords the conversion of unsymmetrical alkenes to alcohols with non-Markovnikov regioselectivity, opposite to that of acid-catalyzed hydration (see Figure 6.8). The overall regioselectivity and stereoselectivity of hydroboration-oxidation is the same as that for the hydroboration reaction, because the –BR$_2$ group is replaced by the –OH group, with retained configuration during oxidation. Herbert C. Brown, a professor of Purdue University, originally developed hydroboration-oxidation reactions. He was awarded a Nobel Prize in Chemistry in 1979 for his work with organoboranes.

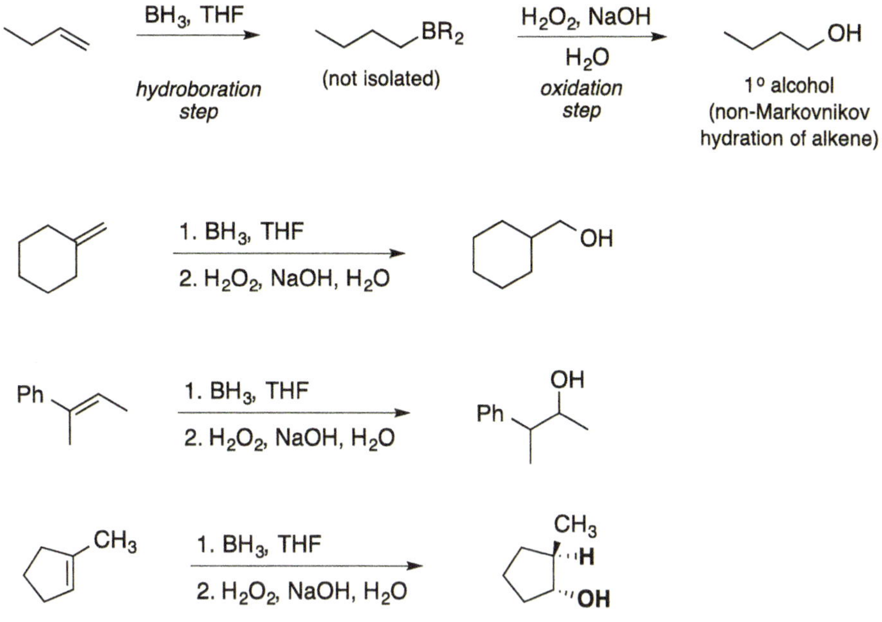

BH₃, THF → ~~~BR₂ → H₂O₂, NaOH, H₂O → ~~~OH

hydroboration step | (not isolated) | oxidation step | 1° alcohol (non-Markovnikov hydration of alkene)

methylenecyclohexane 1. BH₃, THF; 2. H₂O₂, NaOH, H₂O → cyclohexyl-CH₂OH

Ph_/=_ 1. BH₃, THF; 2. H₂O₂, NaOH, H₂O → Ph-CH(-)-CH(OH)-CH₃

1-methylcyclopentene 1. BH₃, THF; 2. H₂O₂, NaOH, H₂O → trans-2-methylcyclopentanol (CH₃ up with H, OH down)

FIGURE 6.13. A hydroboration-oxidation reaction.

Problems

6.10 Draw three-dimensional structures of the products in the following stereoselective reactions:

a) 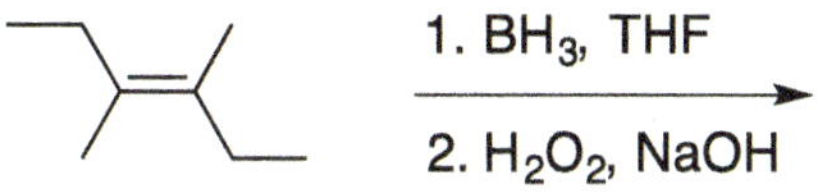 (E or Z)-3-methyl-3-hexene (tetrasubstituted-like alkene) — 1. BH₃, THF; 2. H₂O₂, NaOH

b) 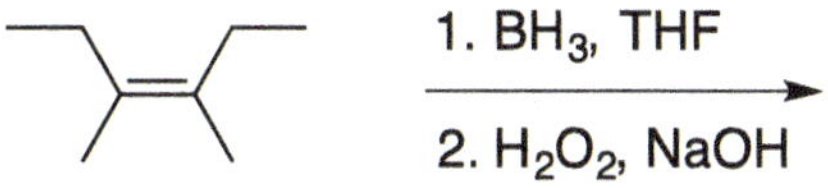 2,4-dimethyl-3-hexene — 1. BH₃, THF; 2. H₂O₂, NaOH

c) 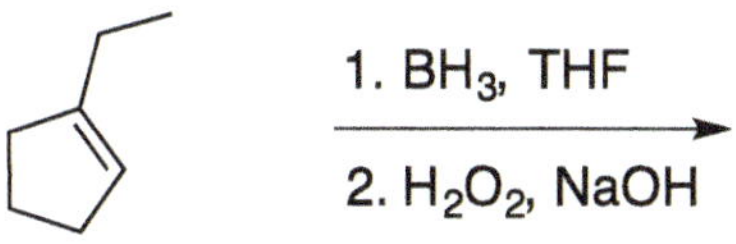 1-ethylcyclopentene — 1. BH₃, THF; 2. H₂O₂, NaOH

ALKENES 121

6.5. OXIDATION AND REDUCTION OF ALKENES

Oxidation of an organic compound is defined as addition of an oxygen atom to a carbon atom or removal of a hydrogen atom from a carbon atom. **Reduction** is defined analogously as addition of a hydrogen atom or removal of an oxygen atom. In a more general definition, oxidation results in the formation of a bond to an element that is more electronegative than carbon, and therefore carbon is losing electron density. Reduction leads to the increase of electron density on carbon. For example, the addition of a halogen molecule (Cl_2 or Br_2) to alkene can be considered an oxidation reaction because halogens are more electronegative than carbon. In contrast, the hydroboration reaction (addition of BH_3) is, in general sense, a reduction reaction, since boron is less electronegative than carbon. It should be noted that the addition of water to alkenes (acid-catalyzed hydration or hydroboration-oxidation of alkenes) is neither oxidation nor reduction since this reaction leads to addition of one new C–H and one C–O bond to the organic molecule.

Oxidation reactions require the use of oxidizing reagents (or **oxidants**), which can transfer oxygen atoms to organic molecules. The common oxidizing reagents include compounds with a metal-oxygen bond (compounds of osmium, chromium and manganese in the highest oxidation state), an oxygen-oxygen bond (peroxides and ozone), or an oxygen-halogen bond (hypochlorites and compounds of iodine in high oxidation states). Since oxidation involves movement of electrons away from carbon, common oxidants can also be considered as electrophilic reagents.

Reduction reactions utilize reagents that can transfer hydrogen with a pair of electrons (hydride anion, $H:^-$) to carbon atom. Common reducing reagents include molecular hydrogen (H_2) in the presence of transition metal catalysts (usually Pt, Pd, Ni, Rh), and metal hydrides.

The following subsections give an overview of several important oxidation reactions of alkenes and catalytic reduction of alkenes.

6.5.1. Dihydroxylation of alkenes

A **dihydroxylation** reaction (introduction of two hydroxyl groups, OH) converts alkene to a diol (also called a **glycol**). This reaction requires osmium tetroxide (OsO_4) or potassium permanganate ($KMnO_4$). Overall, a dihydroxylation reaction proceeds with *syn*-addition because of the initial formation of a cyclic intermediate, such as the cyclic osmate, which is further hydrolyzed by treatment with aqueous sodium bisulfite to the *cis*-1,2-diol, and some reduced osmium byproducts (Figure 6.14). Osmium is an expensive metal and in a more practical modification of this reaction, OsO_4 is used as a catalyst with stoichiometric amounts of hydrogen peroxide. The purpose of hydrogen peroxide is to reoxidize the reduced form of osmium back to OsO_4 in a catalytic cycle.

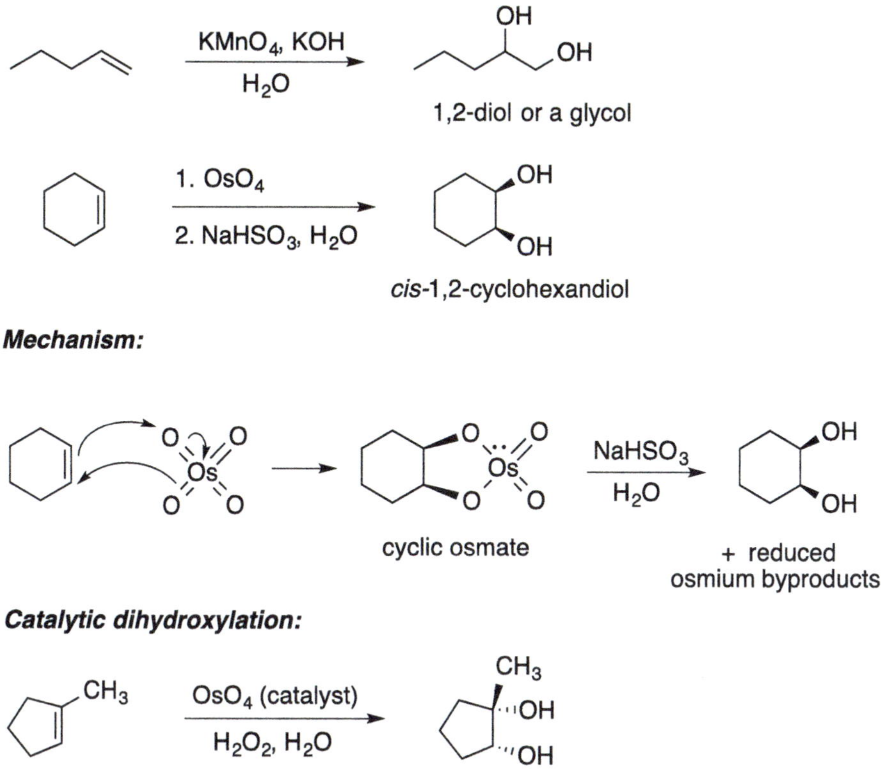

FIGURE 6.14. Dihydroxylation of alkenes.

6.5.2. Oxidation of alkenes with peroxycarboxylic acids: Preparation of epoxides

The oxidation of alkenes with a peroxycarboxylic acid (RCO$_3$H) produces the corresponding epoxides (or **oxiranes**, according to IUPAC nomenclature). The mechanism of this reaction (called **epoxidation**) involves electrophilic addition of the peroxy oxygen to the double bond. This is a stereoselective *syn*-addition, so all substituents in the oxirane ring keep the same arrangement as in the initial alkene. For example, the reaction of (Z)-2-butene yields *cis*-2,3-dimethyloxirane, while (E)-2-butene yields *trans*-2,3-dimethyloxirane (Figure 6.15). Note that IUPAC nomenclature of epoxides uses the parent name *oxirane* for the three-membered ring, with oxygen numbered as the first atom. Epoxides produced from cycloalkenes use a different naming system: cycloalkane is used as the parent name and the position of the oxygen is indicated by the prefix 1,2-epoxy.

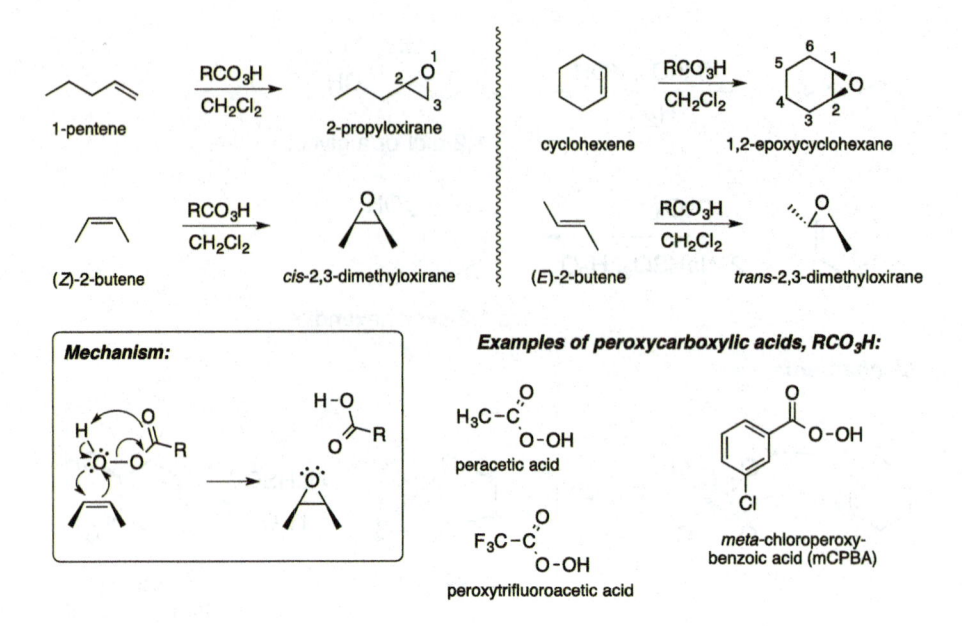

FIGURE 6.15. Epoxidation of alkenes with peroxycarboxylic acids.

Epoxides are important compounds with practical application for industrial production of glycols and as essential components of epoxy glues. The three-membered oxirane ring is easily opened by reaction with various nucleophilic reagents, as illustrated in Figure 6.16. The mechanism of these ring-opening reactions involves nucleophilic substitution of the oxygen atom at one carbon of the oxirane ring, and may require the presence of a strong acid as catalyst. The reaction of unsymmetrical epoxides with strong or moderately strong nucleophiles, such as NH_3, is regioselective with nucleophilic substitution occurring at the less sterically hindered electrophilic carbon. The oxirane ring opening is an *anti*-stereoselective reaction because the nucleophile attacks the carbon from the side opposite to oxygen (similar to the opening of bromonium ion, see Figure 6.10). Addition of water to 1,2-epoxycyclohexane yields *trans*-1,2-cyclohexanediol, in contrast to *cis*-1,2-cyclohexanediol obtained by direct dihydroxylation of cyclohexene using OsO_4 (Figure 6.14).

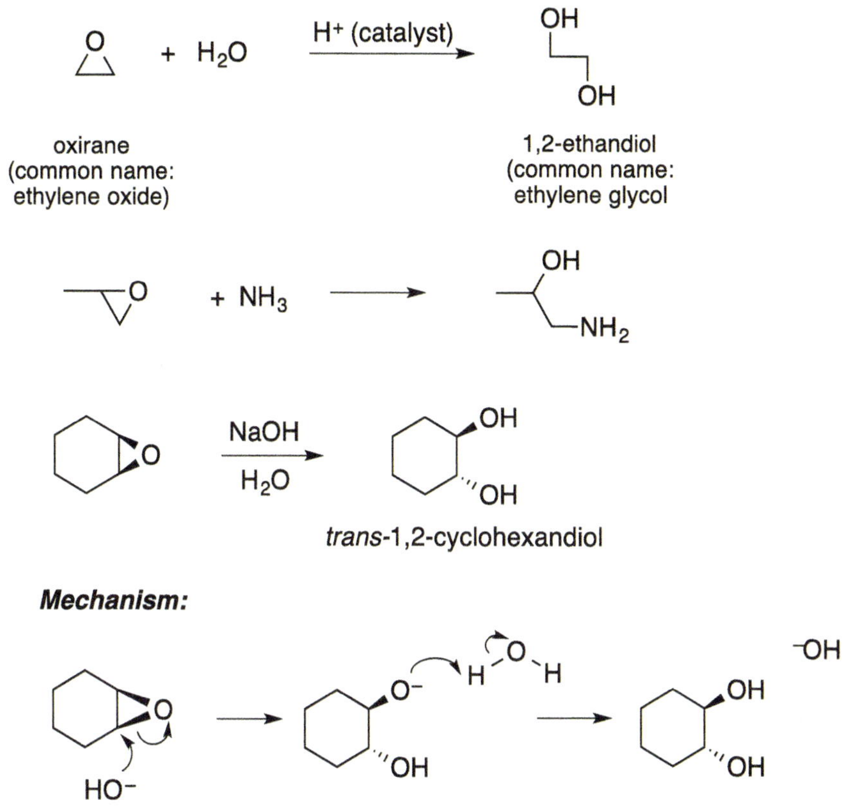

FIGURE 6.16. Epoxide Reactions.

6.5.3. Oxidative cleavage of double bond: Ozonolysis of alkenes

Oxidative cleavage breaks C=C double bonds into two carbonyl groups. This cleavage can be achieved by treating alkene with ozone (O_3), followed by reductive work-up with zinc metal (Zn) or dimethyl sulfide (CH_3SCH_3). This reaction sequence is called **ozonolysis**. The mechanism of ozonolysis involves initial addition of O_3 to a double bond, forming an intermediate molozonide, which rearranges to an ozonide. Ozonides are generally unstable and explosive compounds, which are converted to the final carbonyl products by treatment with a reducing reagent to remove the extra oxygen atom (Figure 6.17).

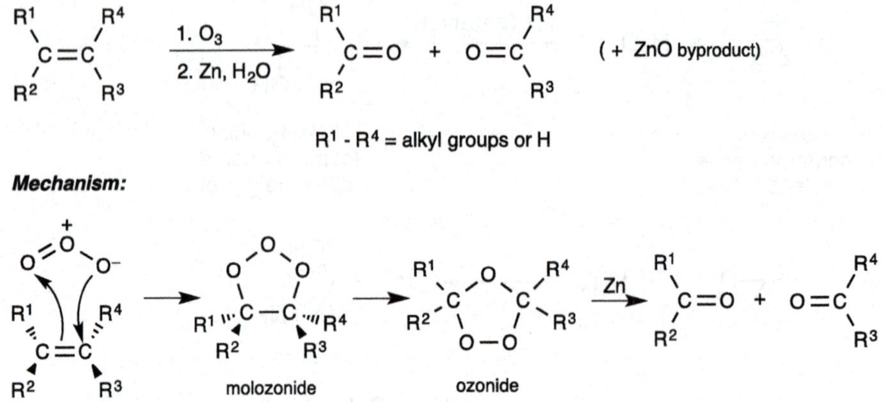

FIGURE 6.17. General scheme and simplified mechanism of an ozonolysis reaction.

Figure 6.18 shows examples of ozonolysis of alkenes and of a diene. Ozonolysis provides a useful tool for structural analysis of complex alkenes. Breaking a complex unknown molecule into several smaller, known aldehydes or ketones helps to locate the position of the double bond in the parent chain and identify the substituents at the carbon atoms.

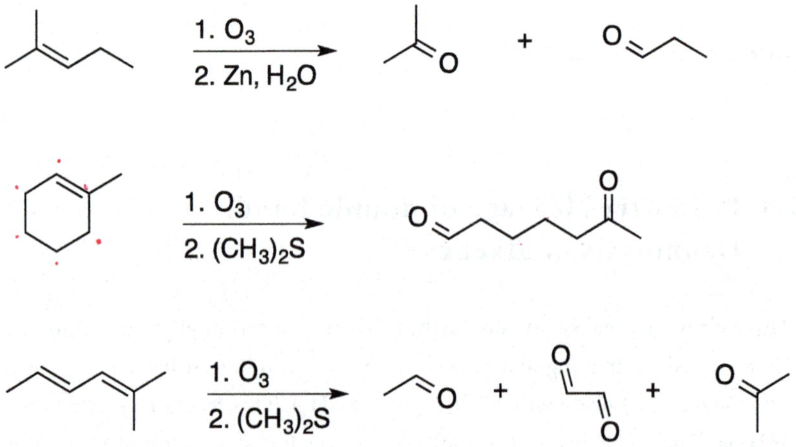

FIGURE 6.18. Examples of ozonolysis reactions.

126 ORGANIC CHEMISTRY

6.5.4. Reduction of alkenes: Catalytic hydrogenation

Alkenes can be reduced to the corresponding alkanes by addition of molecular hydrogen in the presence of a transition metal catalyst (a finely powdered Ni, Pd, or Pt, usually supported on some inert material). This process is called **catalytic hydrogenation**. This reaction proceeds as *syn*-addition involving molecules of hydrogen and alkene adsorbed on the surface of the metal catalyst. Examples of catalytic hydrogenation are shown in Figure 6.19. Hydrogenation of alkenes is an exothermic reaction. Heat of hydrogenation is used as a measure of stability of alkenes. For example, the hydrogenation of *cis*-2-butene and *trans*-2-butene yields the same alkane, butane. However, the reaction of *cis*-2-butene produces more heat, which means that the *cis* double bond is initially higher in energy (less stable) than the *trans* double bond. According to heats of hydrogenation, a double bond with a greater number of alkyl substituents has higher thermodynamic stability compared to the less substituted double bond.

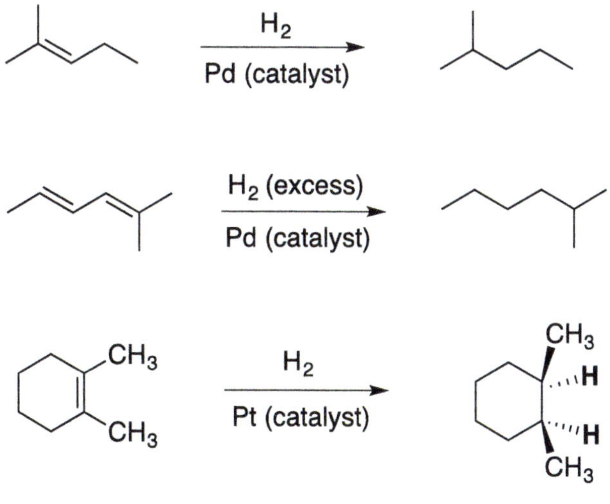

FIGURE 6.19. Catalytic hydrogenation of alkenes.

Problems

6.11. Determine which of the following transformations are oxidation reactions, reductions reactions, or neither oxidation nor reductions:

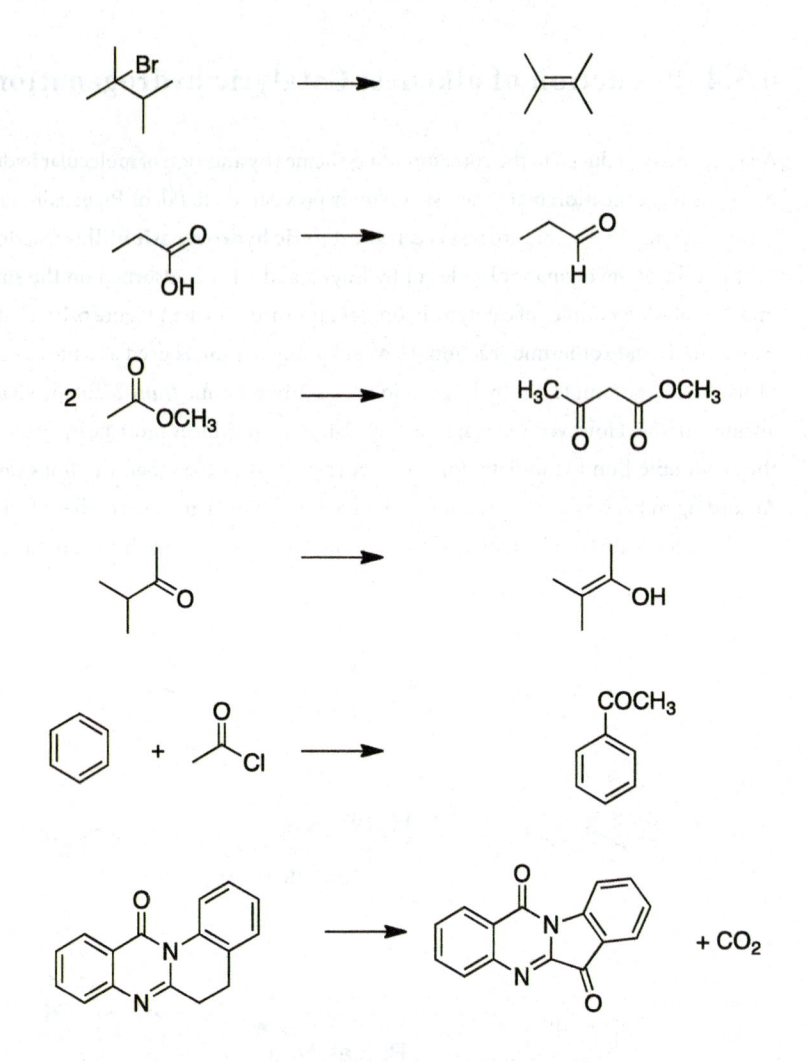

6.12. Rank the following compounds by relative stability/heat of hydrogenation:

A) 1-hexene
B) cis-3-hexene
C) trans-3-hexene
D) 2,3-dimethyl-2-butene
E) E-3-methyl-2-pentene

6.13. Draw three-dimensional structures of the products from the following stereoselective reactions:

a) ![alkene] $\xrightarrow{\text{1. OsO}_4}{\text{2. NaHSO}_3, \text{H}_2\text{O}}$

b) ![alkene] $\xrightarrow{\text{1. OsO}_4}{\text{2. NaHSO}_3, \text{H}_2\text{O}}$

c) ![ethylcyclopentene] $\xrightarrow{\text{1. OsO}_4}{\text{2. NaHSO}_3, \text{H}_2\text{O}}$

d) ![alkene] $\xrightarrow{\text{H}_2}{\text{Pd (catalyst)}}$

e) ![alkene] $\xrightarrow{\text{H}_2}{\text{Pd (catalyst)}}$

6.14. Draw the structure of the product in the following ozonolysis reaction:

![norbornene] $\xrightarrow{\text{1. O}_3}{\text{2. (CH}_3)_2\text{S}}$

6.6. RADICAL REACTIONS OF ALKENES

Similar to alkanes, alkenes can also participate in reactions involving radical intermediates (see Section 3.6). The radical reactions of alkenes can proceed as substitution of hydrogen at the sp³ hybridized carbon next to the sp² hybridized carbon (the so called allylic carbon), or as addition of radical intermediates to a double bond. Radical addition reactions are used in the industrial preparation of polymers by radical polymerization of alkenes.

6.6.1. Allylic halogenation of alkenes

Halogens (Cl_2 or Br_2) react with alkenes at room temperature in the absence of direct sunlight to yield the corresponding dichloro- or dibromoalkanes (Figure 6.9) by an electrophilic addition mechanism (Figure 6.10). However, in the presence of sunlight or at high temperature, different products are formed by radical substitution of hydrogen with halogen at the allylic position. Several examples of **allylic halogenation** reactions are shown in Figure 6.20. These reactions have the same radical substitution chain mechanism as shown for alkanes in Figure 3.13 and involve the formation of allylic radicals as reaction intermediates. Allylic radicals are more stable than 3° or 2° alkyl radicals (see Figure 3.15) because of the resonance delocalization of a single electron, as shown in Figure 6.20.

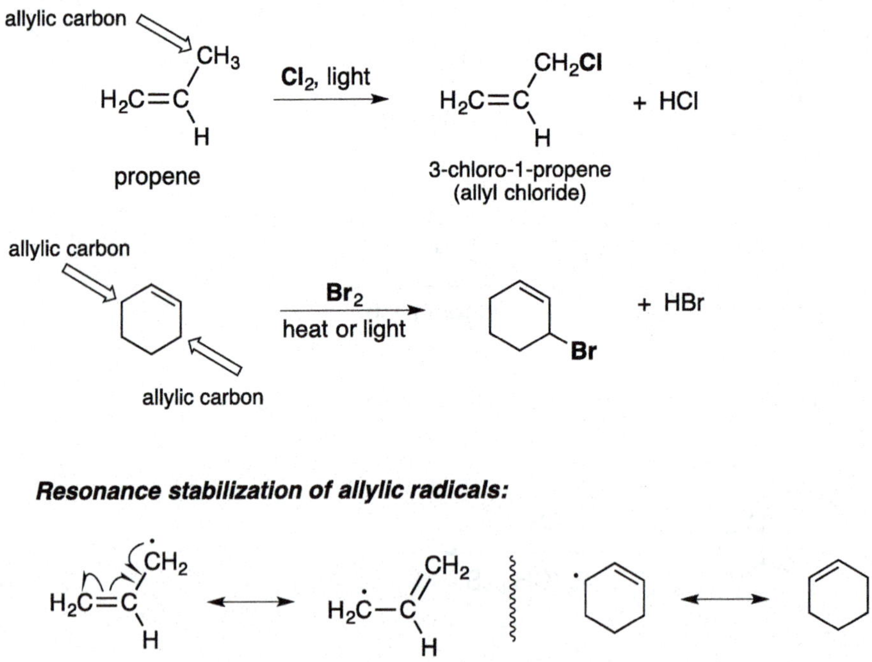

FIGURE 6.20. Allylic halogenation of alkenes.

Radical bromination with Br_2 is a practically inconvenient procedure because of the low yield of products and the toxicity of elemental bromine. Instead of bromine, a commercial, solid, and stable reagent, *N*-bromosuccinimide (NBS), is commonly used for allylic brominations. In the presence of light or other radical initiators, NBS serves as a source of bromine radical (Br•); molecular Br_2 in low concentration is required for the efficient radical allylic bromination (Figure 6.21).

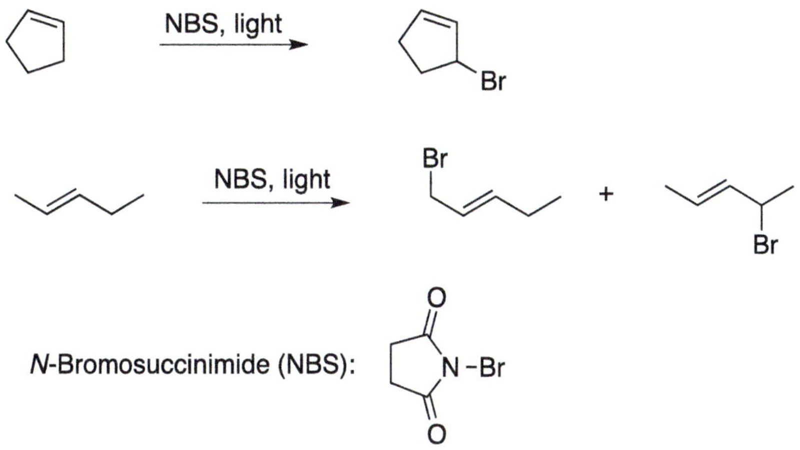

FIGURE 6.21. Allylic brominations using NBS.

6.6.2. Radical addition to double bond

Hydrogen bromide (HBr) reacts with alkenes regioselectively, producing the corresponding bromoalkanes according to Markovnikov's rule (Figure 6.6) by electrophilic addition mechanism via the more stable carbocationic intermediates (Figure 6.7). However, in the presence of special additives called peroxides (ROOR) the mechanism of reaction changes to **radical addition**, resulting in a change in regioselectivity. Under radical conditions, HBr adds to the double bond with a non-Markovnikov (also called anti-Markovnikov) regioselectivity, producing a product in which hydrogen is added to the carbon with the fewest number of hydrogen atoms. Several examples of the non-Markovnikov addition of HBr to alkenes are shown in Figure 6.22. Peroxides function as radical initiators and therefore, these reactions proceed according to a radical chain mechanism (Figure 6.22). Initially, the peroxide molecule breaks homolytically (see Section 3.6), producing two alkoxy radicals RO•, which react with HBr, releasing bromine radical Br•. The addition of the Br• radical to a double bond proceeds with regioselective formation of the more stable, more highly substituted alkyl radical. In the next step, the alkyl radical abstracts hydrogen from another molecule of HBr, propagating the chain of radical additions. The chain reaction is terminated when any two radicals present in the reaction mixture combine (see Section 3.6). It should be noted that in the presence of HBr, reactions of Br• with alkenes do not proceed as allylic substitution (Section 6.6.1) because the presence of molecular halogen (Br_2 or Cl_2) is required for the propagation of radical substitution reactions (see Figure 3.13).

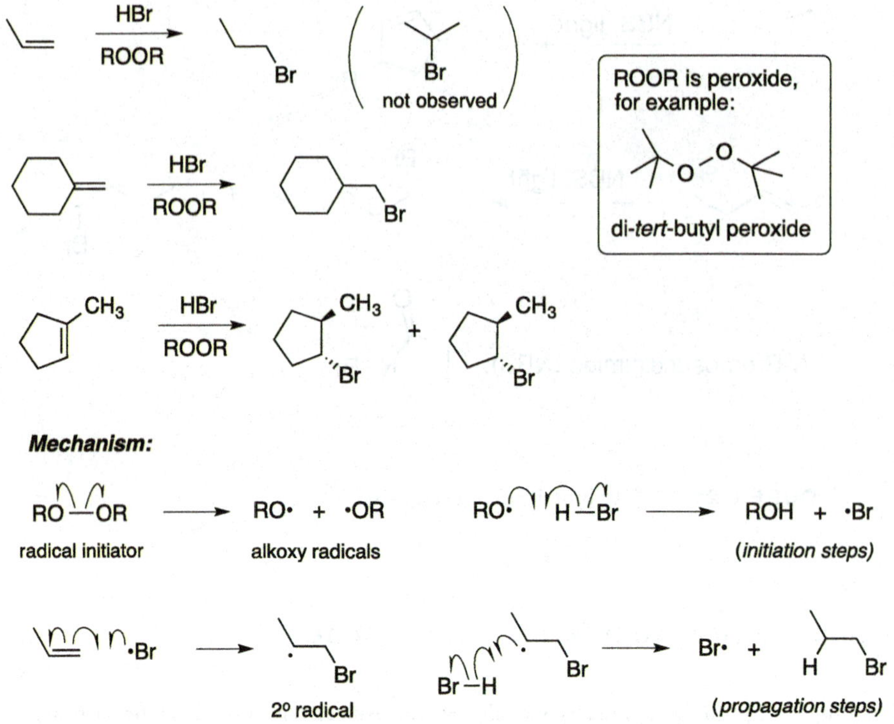

FIGURE 6.22. Radical addition of HBr to alkenes.

6.6.3. Radical polymerization of alkenes: Polymers

The mechanism of radical addition to alkenes is essential for understanding the industrial process of radical polymerization. **Polymers** are large molecules consisting of repeating units of small molecules called **monomers**. **Polymerization** is the process of building the polymer from monomeric molecules. Polymers are characterized by the average degree of polymerization and by their average molecular weight. Figure 6.23 shows examples of common polymers, such as polyethylene, polypropylene, polystyrene, PVC, and polybutadiene, and the corresponding monomers.

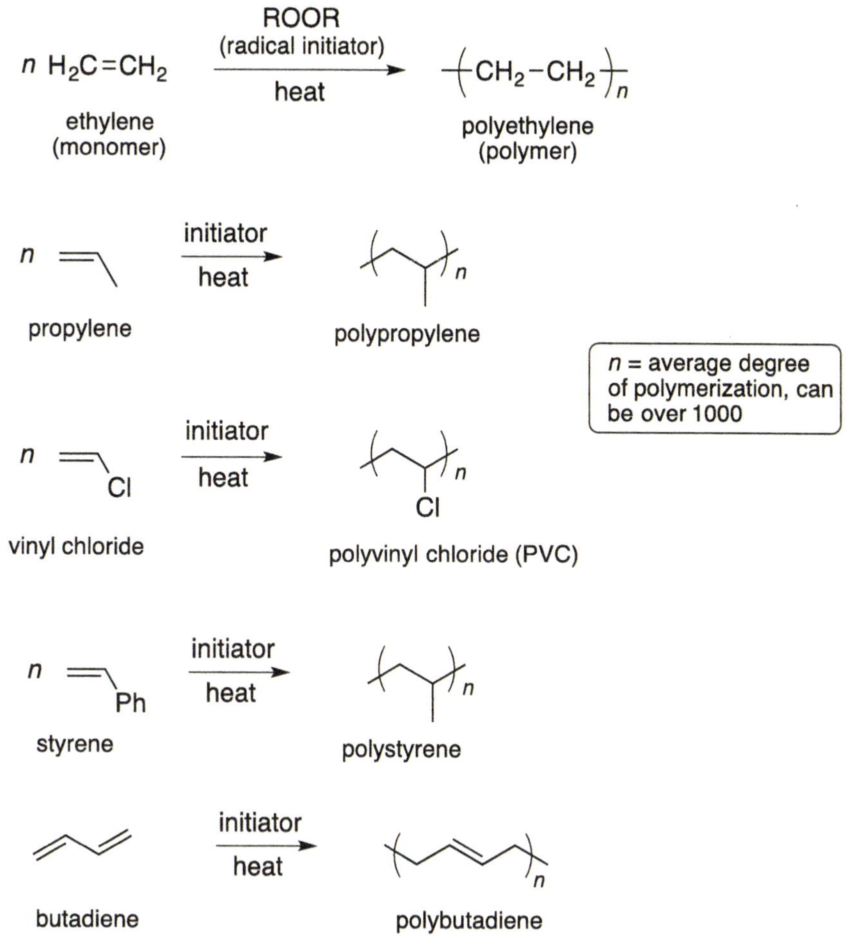

FIGURE 6.23. Examples of important polymers.

The process of radical polymerization involves peroxides as radical initiators, and consists of repeating radical addition steps (Figure 6.24). Propagation steps can repeat thousands of times until the growing chain is terminated by combination with any radical present in the reaction mixture. Polymerization of conjugated dienes, such as butadiene, involves conjugate addition as the propagation step (see Chapter 13 for additional information about non-cyclic conjugated systems). Modern industrial processes for polymerization employ many different types of reactions and initiators, and the chemistry of polymers has evolved into a large and important area of multidisciplinary science.

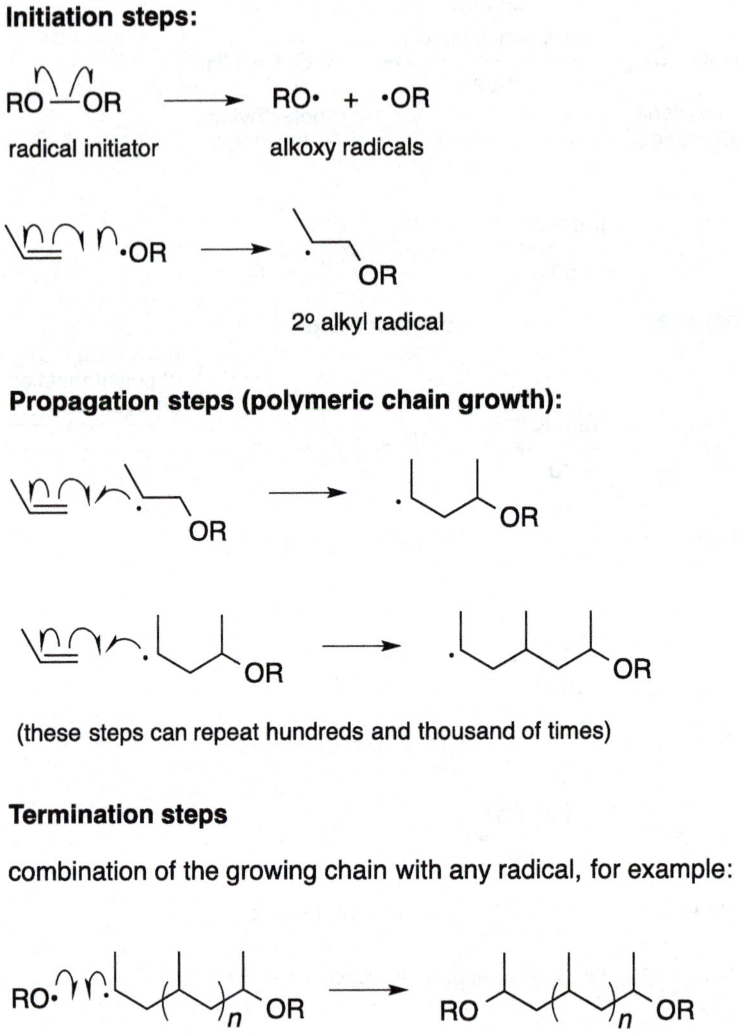

Initiation steps:

RO—OR ⟶ RO• + •OR

radical initiator   alkoxy radicals

CH₂=CH— + •OR ⟶ •CH—CH₂—OR

2° alkyl radical

Propagation steps (polymeric chain growth):

(these steps can repeat hundreds and thousand of times)

Termination steps

combination of the growing chain with any radical, for example:

RO• + •(—CH₂—CH—)ₙ—OR ⟶ RO—(—CH₂—CH—)ₙ—OR

FIGURE 6.24. Mechanism of radical polymerization.

Problems

6.15. Provide the products in the following reactions:

a) (2-methylpropene) + NBS, hv / CCl₄ →

b) (2-methyl-2-butene) + HBr / Peroxides →

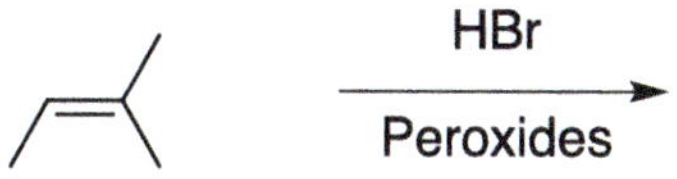

c) (methylcyclohexane) + Br₂, hv → then NaOCH₃ / CH₃OH →

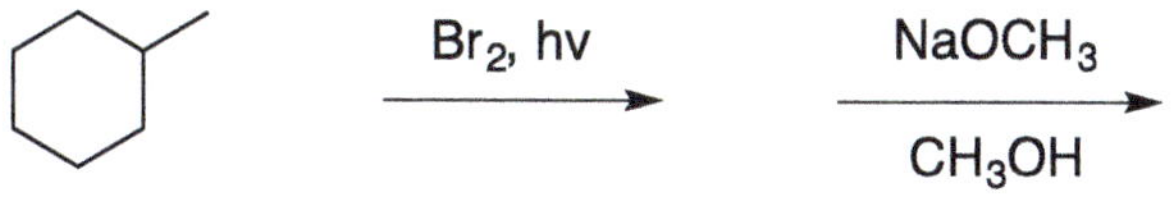

CHAPTER 7

Alkynes

Alkynes are organic compounds with a carbon-carbon triple bond in their structure. A common name for alkynes is acetylenes. Alkynes are not common in nature; however, some diynes and other organic molecules with conjugated carbon-carbon triple bonds have been isolated from various natural sources such as plants, fungi, bacteria, marine sponges, and corals. Some derivatives of alkynes are biologically active compounds and find application as pharmaceutical drugs. The simplest alkyne, ethyne (common name acetylene), is an important industrial product. In particular, acetylene is used in torches to generate temperatures of around 3,000 °C to weld metals. Alkynes have a very rich chemistry and serve as precursors in the synthesis of various organic products. This chapter deals with the structure, nomenclature, and reactions of alkynes. The chapter ends with an overview of the use of alkynes in organic synthesis.

7.1. STRUCTURE AND NOMENCLATURE OF ALKYNES

Alkynes are generally defined as unsaturated hydrocarbons with a triple bond and a general molecular formula of C_nH_{2n-2}, with two units of unsaturation (see Section 1.3). Hydrocarbons with two triple bonds are called diynes and the general name of hydrocarbons with several triple bonds is polyynes. The triple bond in alkynes consists of two sp hybridized carbon atoms with one σ-bond and two π-bonds between them (see Figure 1.23 and the related text in Chapter 1). The two sp orbitals are arranged in a straight line with carbon nuclei in center, similar to the linear geometry of acetylene. Because of the linear geometry of the alkyne unit, cyclic compounds (up to seven-membered rings) with a triple bond in the ring cannot exist as stable compounds.

According to the general IUPAC rules, changing the infix -an- in an alkane to the infix -yn- indicates the presence of a triple bond in the carbon chain of an alkyne (Section 1.3). **The following rules are used to assign IUPAC names for substituted alkynes:**

1. Identify the longest chain (the parent chain) of carbon atoms that includes the triple bond and name it by changing the infix -an- in the related alkane to the infix -yn-. If there are two or more chains of the same length, choose the chain with the greater number of substituents. This is the root name of the alkyne.
2. Number the carbons of the parent chain starting from the carbon nearest to the triple bond end. Smaller numbers on the carbons indicate the location of the triple bond.
3. For hydrocarbons that have both double and triple bonds, the location of the double bond takes precedence in numbering the parent chain.
4. Use the general IUPAC rules to indicate each substituent and functional group in the parent chain (see Sections 1.3 and 3.2).
5. Naming cycloalkanes with a substituent that has a triple bond employs the name **ethynyl** for the substituent.

Several examples of IUPAC names and common names of alkynes are shown in Figure 7.1. Alkynes that have a triple bond at the end of the chain (with a hydrogen atom attached to sp hybridized carbon atom) are commonly referred to as **terminal alkynes**, while alkynes with a triple bond within the parent chain are called **internal alkynes**.

IUPAC name: 3,3-dimethyl-1-butyne
Common name: *tert*-butylacetylene
a terminal alkyne

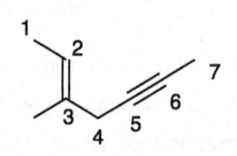

IUPAC name: 2-butyne
Common name: dimethylacetylene
an internal alkyne

(*E*)-3-methyl-2-hept-5-yne
an internal alkyne

ethynylcyclohexane
a terminal alkyne

FIGURE 7.1. Nomenclature of alkynes.

Problems

7.1. Provide the IUPAC name for each of the following compounds:

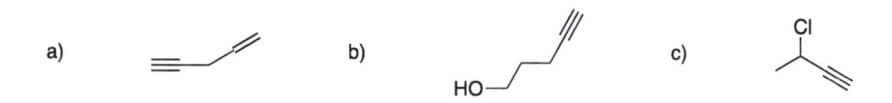

a)   b)   c)

7.2. Draw structural formulas for the following alkynes:
- a) 2,5-dimethyl-3-heptyne
- b) 3-ethynyl-1-cyclopentene
- c) 3-butyn-2-ol

7.2. PREPARATION AND REACTIONS OF ACETYLIDES

Because a hydrogen atom is attached to the sp hybridized carbon, terminal alkynes are characterized by a relatively high acidity with a pK_a value of approximately 25 (see Chapter 2, Section 2.3). In fact, terminal alkynes are more acidic than NH_3 ($pK_a = 38$) and H_2 ($pK_a = 35$) and can be deprotonated by conjugate bases of these weak acids. Figure 7.2 shows examples of terminal alkyne proton transfer reactions. The conjugate bases of alkynes are generally called alkynyl anions or acetylide anions and the metal salts of alkynes are called

$pK_a = 50$ (not acidic) $pK_a = 25$ (acidic) base $pK_a = 35$ conjugate acid

$(H_3C)_3C-C\equiv C-H$ + $Na^+H:^-$ $\longrightarrow$ $(H_3C)_3C-C\equiv C:^- Na^+$ + H_2

sodium hydride   sodium *tert*-butyl acetylide

$HC\equiv CH$ $\xrightarrow{NaNH_2}$ $HC\equiv CNa$ $\xrightarrow{NaNH_2 \text{ (excess)}}$ $NaC\equiv CNa$

sodium acetylide (monosodium acetylide)   disodium acetylide

Ph–C≡CH $\xrightarrow[\text{hexane}]{BuLi}$ Ph–C≡CLi

ethynylbenzene (common name: phenylacetylene)   lithium phenylacetylide

FIGURE 7.2. Preparation of acetylides from terminal alkynes.

metal **acetylides**. Internal alkynes, such as 2-butyne, do not have the acidic sp hybridized C–H bond and have pK_a values comparable to those of alkanes (about 50-60).

Alkynyl anions are strong bases and strong nucleophiles. Metal acetylides react with water as the acid (pK_a = 15.7), releasing the less acidic terminal alkynes (pK_a = 25). The reactions of acetylides as nucleophiles in S_N2 reactions with unhindered haloalkanes, epoxides (see Section 6.5.2), and other electrophilic substrates are especially valuable because these reactions form new carbon-carbon bonds and thus can be used for the synthesis of bigger molecules from small building blocks.

Figure 7.3 shows several examples of nucleophilic substitution reactions of acetylides. The products of these reactions are alkynes with a longer carbon chain. The reactions of monosodium acetylide with unhindered haloalkanes allow the preparation of various terminal alkynes, which can be further converted to the unsymmetrical internal alkynes.

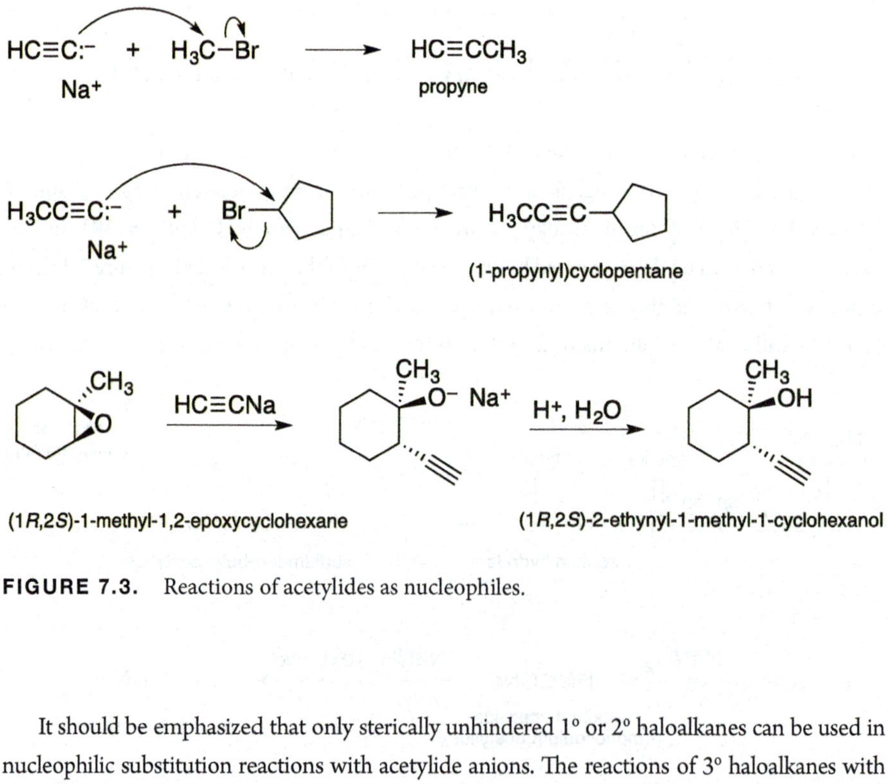

FIGURE 7.3. Reactions of acetylides as nucleophiles.

It should be emphasized that only sterically unhindered 1° or 2° haloalkanes can be used in nucleophilic substitution reactions with acetylide anions. The reactions of 3° haloalkanes with acetylides by the S_N2 mechanism are impossible, due to steric hindrance (see Section 5.2). S_N1 reactions require the use of protic solvents (ROH), which immediately convert alkynyl anions to the non-nucleophilic terminal alkynes by proton transfer. A different method via a β-elimination reaction of dibromoalkanes with strong bases represents an alternative general approach to the synthesis of alkynes bearing bulky substituents. The starting dibromoalkanes can be conveniently prepared by electrophilic bromination of the corresponding alkenes. The reaction of dibromoalkanes with a strong base, such as NaNH$_2$, results in the β-elimination of two HBr units by E2 mechanism (see Section 5.4). The preparation of alkynes from alkenes by the bromination-dehydrobromination reaction sequence is illustrated in Figure 7.4. Note that terminal alkynes are initially formed as sodium acetylides due to the presence of a strong base in the reaction mixture. The initially formed acetylides can be converted to alkynes by treatment with water.

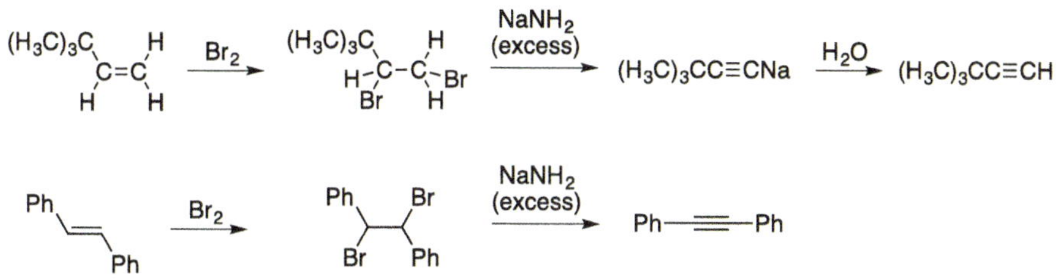

FIGURE 7.4. Preparation of alkynes from alkenes by the bromination-dehydrobromination reaction sequence.

Problems

7.3. Write detailed mechanisms for the reactions shown in Figures 7.3 and 7.4.

7.4. Explain the following observation:

however:

7.5. Provide the products of the following transformations:

a) 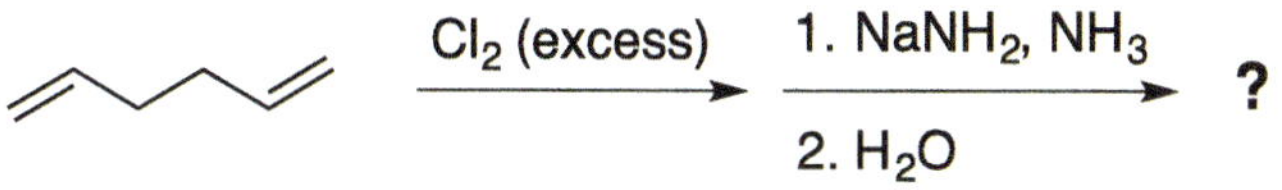 [pent-1-en-4-yne] $\xrightarrow{\text{Cl}_2 \text{ (excess)}} \xrightarrow[\text{2. H}_2\text{O}]{\text{1. NaNH}_2, \text{NH}_3}$?

b) 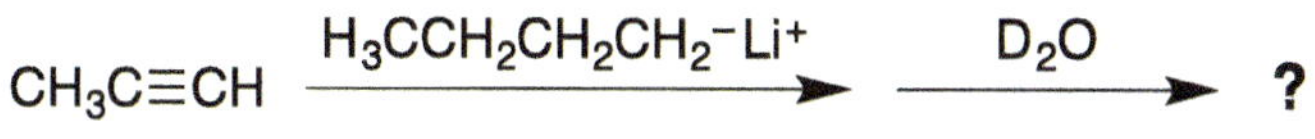 $\text{CH}_3\text{C}\equiv\text{CH} \xrightarrow{\text{H}_3\text{CCH}_2\text{CH}_2\text{CH}_2^-\text{Li}^+} \xrightarrow{\text{D}_2\text{O}}$?

7.6. Why does the attempt to synthesize 2,2-dimethyl-3-hexyne via the sequence of reactions shown below fail? Explain why and provide an alternative approach to synthesize this compound.

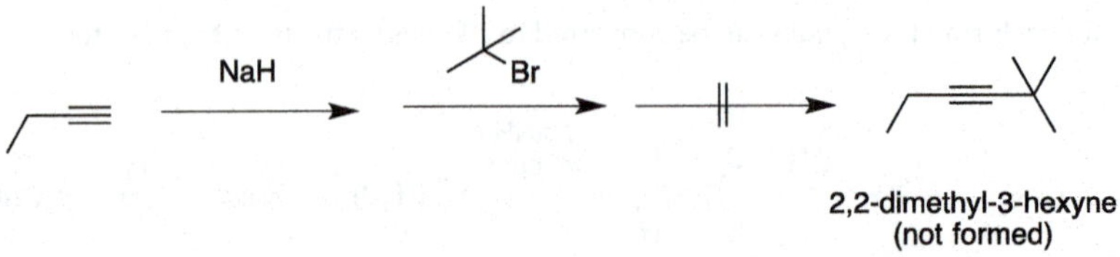

2,2-dimethyl-3-hexyne
(not formed)

7.3. ELECTROPHILIC ADDITION REACTIONS OF ALKYNES

Reactions of electrophilic reagents with alkynes proceed by the same mechanism as electrophilic additions to a double bond, and also involve carbocationic intermediates (see Chapter 6). Either one or both π-bonds of the triple bond can participate in the reaction. Typical electrophilic reagents include halogens (Cl_2 and Br_2), hydrogen halides (HCl, HBr, or HI) and other strong acids, and borane (BH_3).

7.3.1. Electrophilic addition of halogens

A triple bond in alkyne can add one or two molecules of halogen (Cl_2 or Br_2), producing the corresponding dihaloalkenes or tetrahaloalkanes respectively. For example, the reaction of 2-butyne with one molar equivalent of bromine yields (E)-2,3-dibromo-2-butene as the main product (Figure 7.5). If two or more molar equivalents of bromine are present in the reaction mixture, the second molecule of Br_2 adds to the initially formed dibromoalkene, yielding 2,2,3,3-tetrabromobutane.

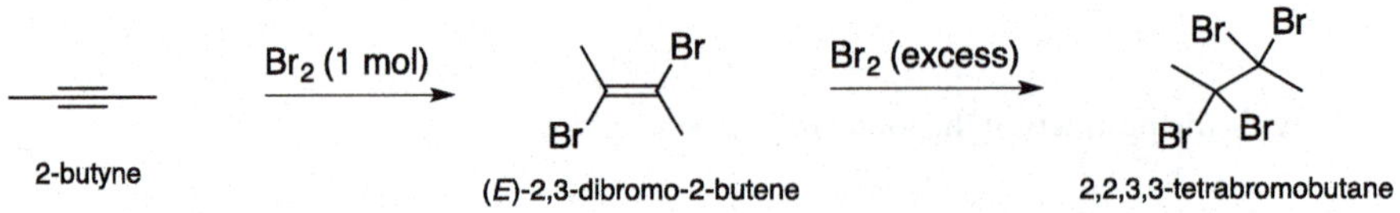

FIGURE 7.5. Bromination of 2-butyne.

Analogous to alkenes (see Section 6.3), addition of halogens to alkynes is an anti stereoselective reaction. The mechanism of this reaction involves the initial formation of a cyclic bromonium ion as the reactive intermediate. In the second step, the bromine anion (a nucleophile) attacks the carbon from the side opposite to the bromonium atom (the leaving group) according to the S_N2 mechanism, and forming the final *trans* product (Figure 7.6).

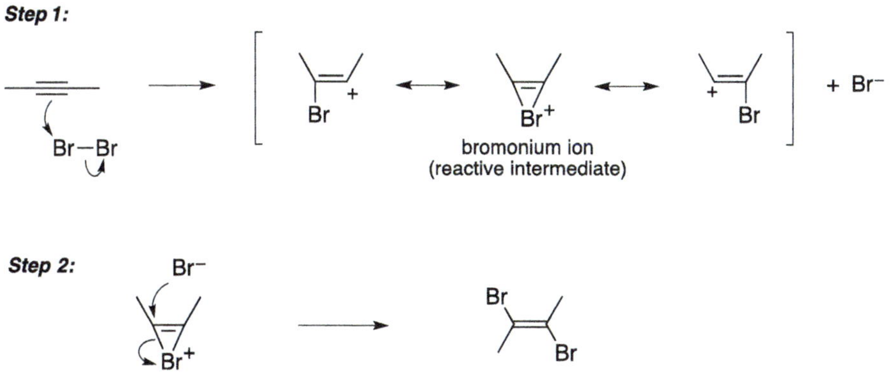

Step 1:

Step 2:

FIGURE 7.6. Mechanism of bromination of 2-butyne.

7.3.2. Addition of hydrogen halides

Analogous to the reaction of halogens, either one or two molecules of hydrogen halide (HCl, HBr, or HI) can add to the triple bond of alkyne. The addition of one molecule of HCl to acetylene yields vinyl chloride (Figure 7.7), which is an important monomer in industrial production of polyvinyl chloride (see Section 6.6.3). Addition of hydrogen halide HX to terminal alkynes is a regioselective reaction with hydrogen being added to the terminal carbon atom in agreement with Markovnikov's rule (see Section 6.2).

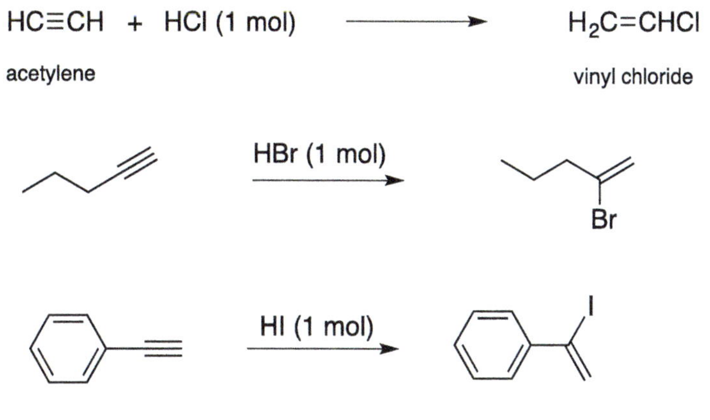

FIGURE 7.7. Addition of hydrogen halides to alkynes.

The second molecule of hydrogen halide adds to the initially formed haloalkene in agreement with greater stability of the intermediate carbocation. Thus, the addition of HCl to vinyl chloride yields 1,1-dichloroethane, but not the regioisomeric 1,2-dichloroethane (Figure 7.8). The regioselectivity of this reaction is explained by the resonance stabilization of the carbocationic intermediate with a chlorine atom attached to the positively charged carbon atom as shown in Figure 7.8.

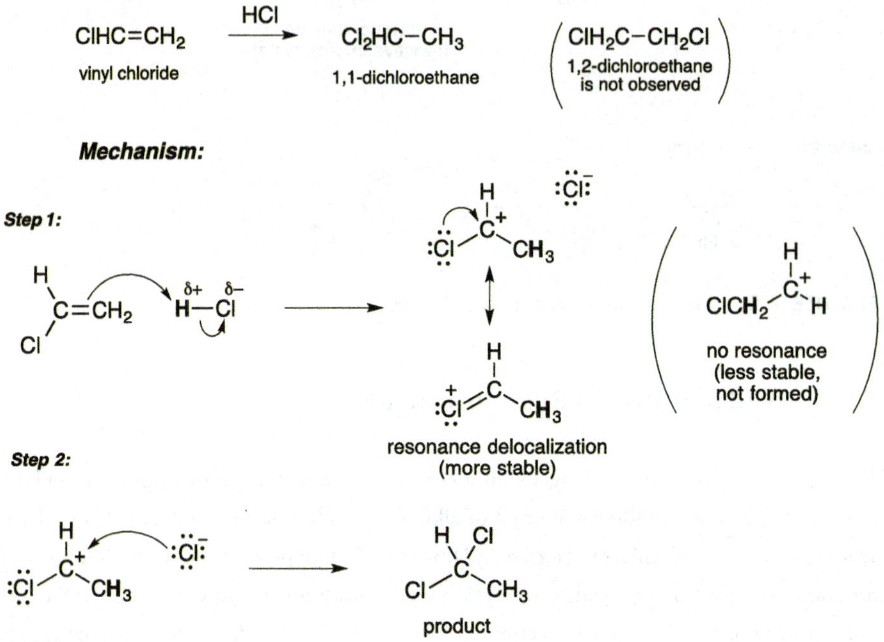

FIGURE 7.8. Regioselective addition of HCl to vinyl chloride.

7.3.3. Acid-catalyzed hydration of alkynes

Reactions of alkynes with strong acids (usually H_2SO_4) in water initially give alkenyl alcohols (the so-called enols) as the products of acid-catalyzed hydration of the π-bond (see Section 6.2). This reaction may require the presence of a mercury salt (usually $HgSO_4$) as catalyst and proceeds with Markovnikov's regioselectivity. The initially formed enols are unstable and exist in equilibrium with the isomeric carbonyl compound (ketone). The conversion of an enol to a ketone is achieved by the movement of a hydrogen atom from oxygen to carbon, which can be catalyzed either by acid or by base. Two constitutional isomers that differ by the position of a hydrogen atom and a double bond are called **tautomers** and the equilibrium between tautomers is called **tautomeric equilibrium** (or **tautomerization**). Figure 7.9 shows the mechanism of acid-catalyzed tautomerization. Ketone is the predominant component (usually more than 99%) in keto-enol tautomeric equilibrium and is shown as the final product of alkyne hydration.

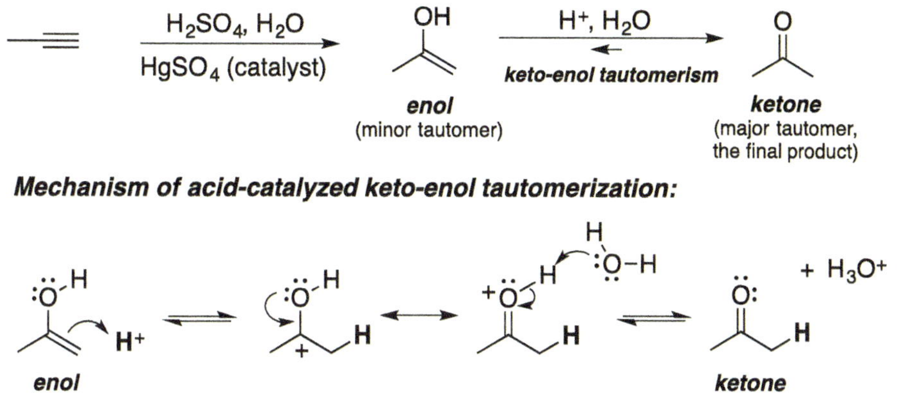

FIGURE 7.9. Acid-catalyzed hydration of alkynes.

Additional examples of acid-catalyzed hydration reactions of alkynes are shown in Figure 7.10. Note that the hydration of ethyne (acetylene) gives acetaldehyde, and the reactions of all other terminal or internal alkynes yield ketones.

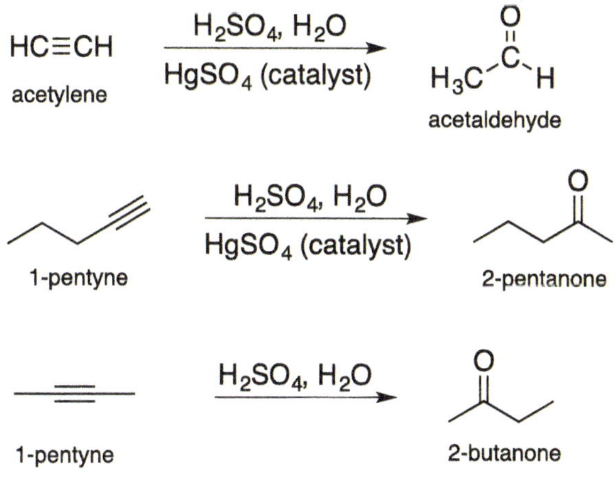

FIGURE 7.10. Acid-catalyzed hydration of alkynes.

7.3.4. Hydroboration-oxidation of alkynes

Similar to alkenes, hydroboration-oxidation of terminal alkynes occurs with non-Markovnikov regioselectivity, (see Section 6.4) forming enols as the initial products of hydration (Figure 7.11). Note that enols formed in this reaction are regioisomers of the enols

ALKYNES 145

formed by acid-catalyzed hydration of alkynes (compare with Figure 7.9). Base-catalyzed tautomerization of the initially formed enols yields the corresponding aldehydes as the final products.

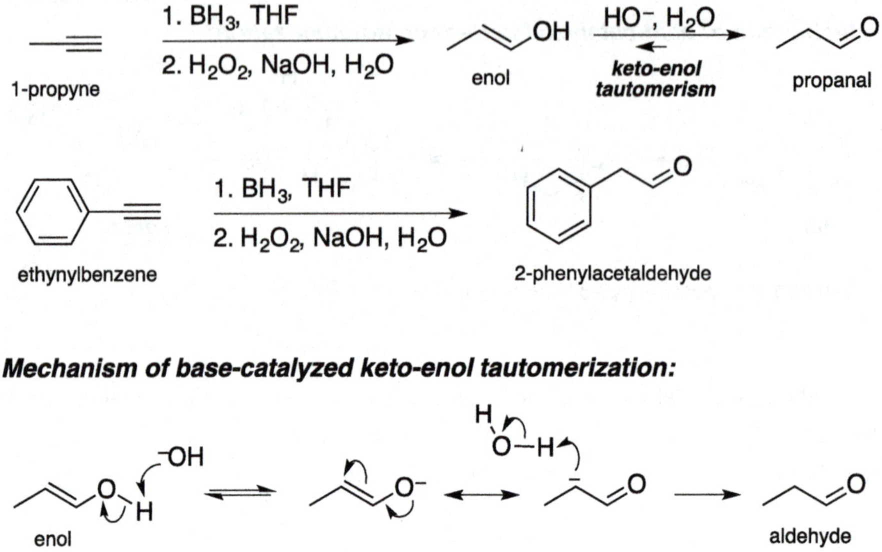

Mechanism of base-catalyzed keto-enol tautomerization:

FIGURE 7.11. Hydroboration-oxidation of alkynes.

Problems

7.7. Propose detailed mechanisms for the reactions shown in Figures 7.7, 7.10, and 7.11.

7.8. Draw the enol form of the following compounds:

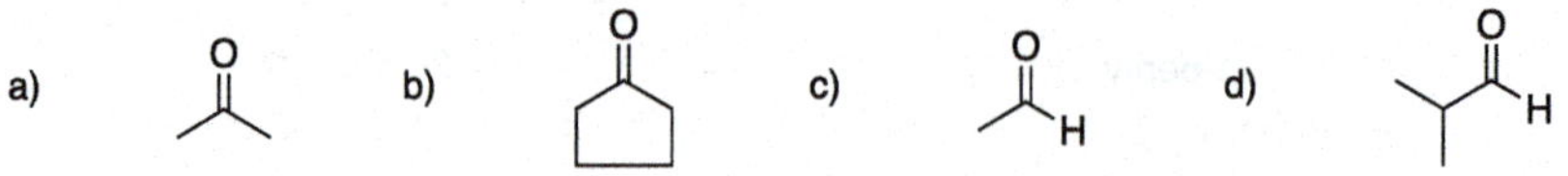

a) acetone   b) cyclopentanone   c) acetaldehyde   d) isobutyraldehyde

7.9. Explain why the following carbonyl compounds do not have an enol form.

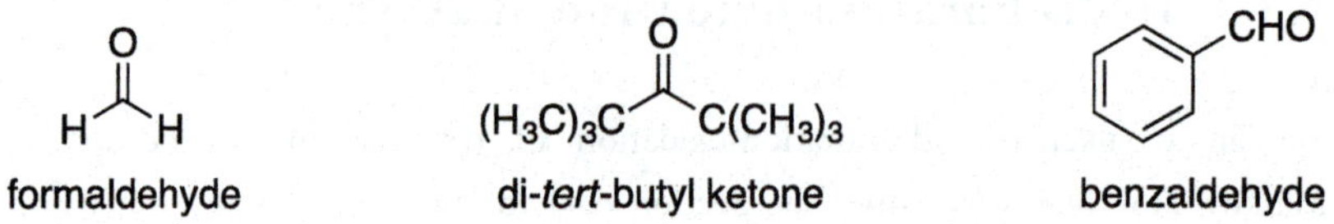

formaldehyde   di-*tert*-butyl ketone   benzaldehyde

7.10. Determine which reagents are required to obtain the indicated products in high yield.

a) CH≡CH →[?] →[1. OsO₄ / 2. NaHSO₃, H₂O] HO⧸⧹OH

b) CH≡CH →[?] →[Br₂] Br⧸⧹Br (with additional Br, Br)

c) CH≡CH →[?] →[1. RCOOH / 2. H₂O, H₃O⁺] HO⧸⧹OH

7.11. Provide the product of the following transformation:

CH₂=CHCH₂CH₃ →[NBS, light] →[HC≡C⁻Na⁺] →[1. BH₃ / 2. NaOH, H₂O₂] ?

7.4. REDUCTION OF ALKYNES

Analogous to alkenes, alkynes can be reduced to the corresponding alkanes by addition of molecular hydrogen (2 molar equivalents) in the presence of a transition metal catalyst (Ni, Pd, or Pt). The process of addition proceeds in two steps: 1) addition of one H_2 molecule forms an alkene; and 2) hydrogenation of the intermediate alkene produces the alkane. Both steps proceed as *syn*-additions involving molecules of hydrogen and alkyne or alkene adsorbed on the surface of the metal catalyst (see Section 6.5.4). It is difficult to selectively add one H_2 molecule to a triple bond when the highly active, pure metal catalysts are used. However, it is possible to reduce the catalytic activity of palladium powder by mixing it with calcium carbonate and some other additives. Such a poisoned catalyst is known as a **Lindlar** catalyst and it is used for stereoselective hydrogenation of alkynes to (Z)-alkenes. Examples of catalytic hydrogenation reactions of alkynes are shown in Figure 7.12.

It is also possible to reduce internal alkynes with anti-stereoselectivity to (E)-alkenes by using sodium or lithium metal in liquid ammonia at low temperature (NH_3 boils at –33 °C). This reaction has a complex mechanism involving a single electron transfer from the metal to the triple bond forming a radical anion intermediate in the initial step. In the next steps, the radical anion is protonated by NH_3 to yield an alkenyl radical, which is reduced to alkenyl

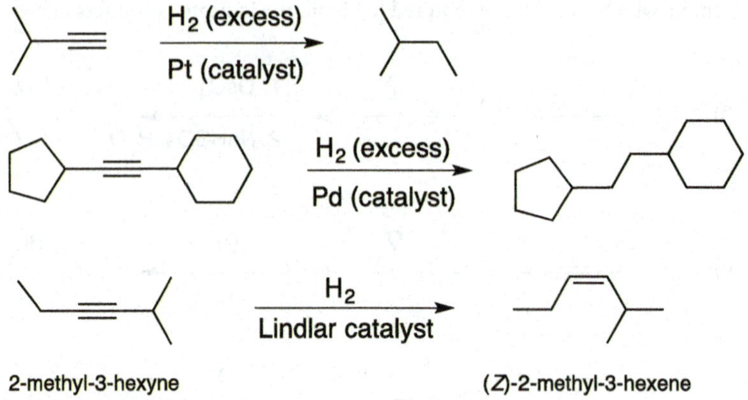

FIGURE 7.12. Catalytic hydrogenation of alkynes.

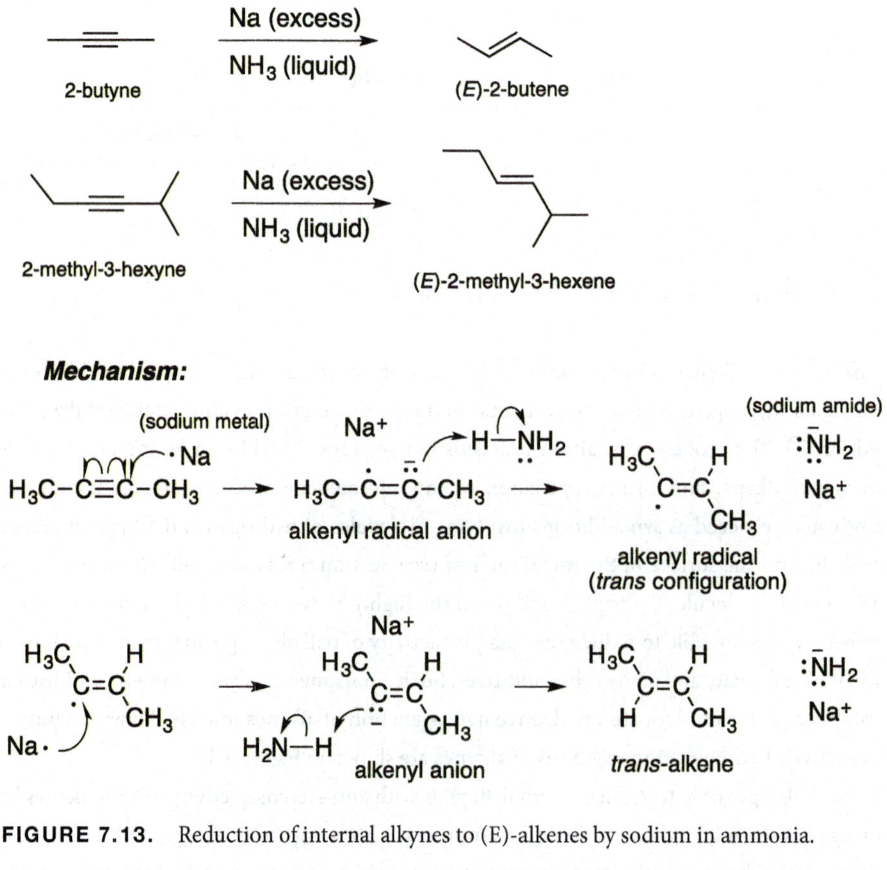

Mechanism:

FIGURE 7.13. Reduction of internal alkynes to (E)-alkenes by sodium in ammonia.

148 ORGANIC CHEMISTRY

anion by the second atom of sodium. The alkenyl anion formed in this step has a *trans* configuration because the *cis* product is less stable, due to steric hindrance (repulsion of the bulky CH_3 groups). Protonation of alkenyl anion gives the final *trans*-alkene (Figure 7.13). Note that two different types of arrows are used in this mechanism: fishhook arrows (see Figure 1.2 in Chapter 1) indicate the movement of a single electron, and the regular curved arrows show the movement of a pair of electrons from one position to another. Overall, two sodium atoms donate electrons to the triple bond, reducing it to the *trans* double bond, and two NH_3 molecules serve as the source of protons required for the hydrogenation. The by-product of this reaction is sodium amide $(NaNH_2)$, which can be easily separated from the nonpolar alkene by treatment with water.

Problems

7.12. Provide the products of the following transformations:

a) $H_3C-C\equiv CH$ → NaH → 1. (epoxide) 2. H_2O → H_2, Lindlar cat. → ?

b) (pent-4-en-1-yne) → NaH → CH_3I → Na, NH_3 → ?

c) (pent-4-yne) → NaH → D_2O → H_2, Pd → ?

7.5. INTRODUCTION TO ORGANIC SYNTHESIS

Organic synthesis is an important subdiscipline within the general field of organic chemistry, dealing with the construction of complex organic molecules that start with simple organic and inorganic compounds. Development of effective synthetic approaches to complex organic compounds (including naturally occurring products) is a critically important area of modern medicinal science and the pharmaceutical industry.

Development of a synthetic approach to a complex molecule (the target molecule) usually starts with careful evaluation of the molecule's structure, trying to find the best answers to the following two questions: 1) How to create the carbon skeleton from smaller building blocks? 2) How to introduce the required functional groups? This evaluation process is called **retrosynthetic analysis** and it goes backwards, step-by-step, starting from the target molecules, to the smaller building blocks. The final goal of retrosynthetic analysis is the development of the best synthetic route to the target molecule from simple, commercially available starting materials. Figure 7.14 shows an example of a retrosynthetic analysis of a molecule of butanone. Note that the retrosynthetic schematic involves the use of open arrows, indicating that the product is on the left and the starting material on the right side of the schematic.

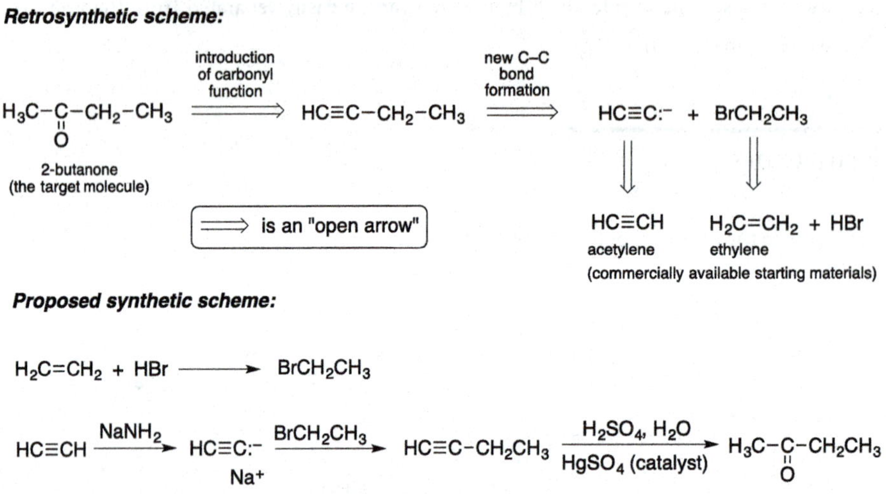

Retrosynthetic scheme:

$H_3C-C(=O)-CH_2-CH_3$ (2-butanone, the target molecule) ⇒ [introduction of carbonyl function] ⇒ $HC≡C-CH_2-CH_3$ ⇒ [new C–C bond formation] ⇒ $HC≡C:^-$ + $BrCH_2CH_3$

⇒ is an "open arrow"

↓ $HC≡CH$ (acetylene) ↓ $H_2C=CH_2$ + HBr (ethylene)

(commercially available starting materials)

Proposed synthetic scheme:

$H_2C=CH_2$ + HBr → $BrCH_2CH_3$

$HC≡CH$ —NaNH₂→ $HC≡C:^- Na^+$ —$BrCH_2CH_3$→ $HC≡C-CH_2CH_3$ —H_2SO_4, H_2O / $HgSO_4$ (catalyst)→ $H_3C-C(=O)-CH_2CH_3$

FIGURE 7.14. Retrosynthetic analysis of a molecule of butanone.

Problems

7.13. Provide the reagents and reaction conditions to accomplish the following syntheses in high yield:

a) 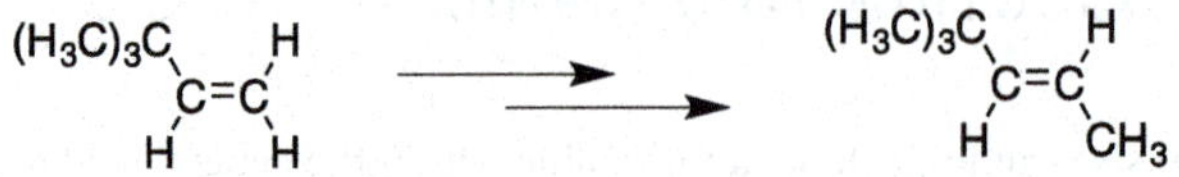 $(H_3C)_3C$–CH=CH–H (cis) → $(H_3C)_3C$–CH=CH–CH_3

b) 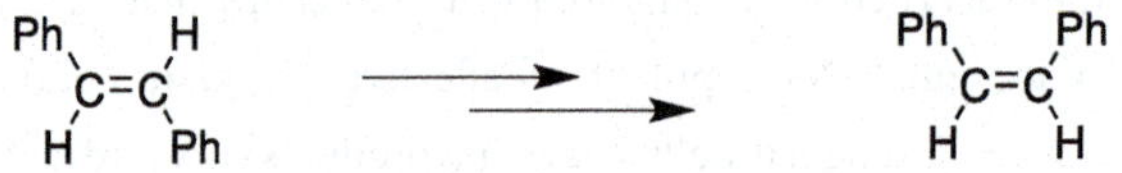 Ph–CH=CH–Ph (cis) → Ph–C(Ph)=CH–H (trans geometry with two Ph groups)

c) 2 $PhCH_2Br$ and $HC≡CH$ → 2 $PhCH_2CHO$

150 ORGANIC CHEMISTRY

CHAPTER 8

Alcohols

Alcohols are generally defined as organic compounds with a hydroxyl functional group (–OH) connected to sp^3 hybridized carbon. Organic molecules with a hydroxyl group are very common in nature. In particular, the hydroxyl groups are present in carbohydrates (such as glucose, sucrose, and cellulose) and many other multifunctional natural compounds. In a more specific definition, the class of alcohols is limited to alkyl alcohols in which the hydroxyl group is bound to an alkyl chain. Simple alkyl alcohols play important role in our lives. Methanol (CH_3OH), also known as methyl alcohol or wood alcohol, is an industrial product with numerous applications such as solvent, fuel, and antifreeze. Ethanol or ethyl alcohol (CH_3CH_2OH) is the principal component of alcoholic beverages naturally produced by fermentation of sugars. Ethanol is also widely used as fuel or a fuel additive. 1,2-ethanediol or ethylene glycol ($HOCH_2CH_2OH$) is used as engine coolant/antifreeze and also as the starting material in industrial production of polyesters. Alcohols serve as precursors in the synthesis of various important organic products such as ethers, esters, and carbonyl compounds. In this chapter the structure, nomenclature, properties, and reactions of alcohols are discussed.

8.1. CLASSIFICATION, NOMENCLATURE, AND PHYSICAL PROPERTIES OF ALCOHOLS

Alcohols are generally classified as **primary (1°)**, **secondary (2°)**, or **tertiary (3°) alcohols** depending on the number of carbon atoms attached to the carbon bearing the hydroxyl group (–OH). According to this classification, ethanol CH_3CH_2OH is a 1° alcohol, isopropanol $(CH_3)_2CHOH$ is a 2° alcohol, and *tert*-butanol $(CH_3)_3COH$ is a 3° alcohol.

According to the general rules of IUPAC nomenclature, the presence of the hydroxyl group in a molecule is indicated by the suffix -ol for all compounds where the hydroxyl group is the functional group with the highest priority (see Table 1.4 and the related text in Chapter 1). In the compounds where a higher priority group is present, the prefix hydroxy- is placed at the beginning of the name. Compounds with two hydroxyl groups are called **diols** (common name **glycols**). The carbons of the parent chain should be numbered in such a way that the hydroxyl group has the lowest possible number unless a higher priority group is present. In cyclic molecules, the carbon bearing hydroxyl group is counted as number 1. Figure 8.1 shows several examples of IUPAC names for alcohols.

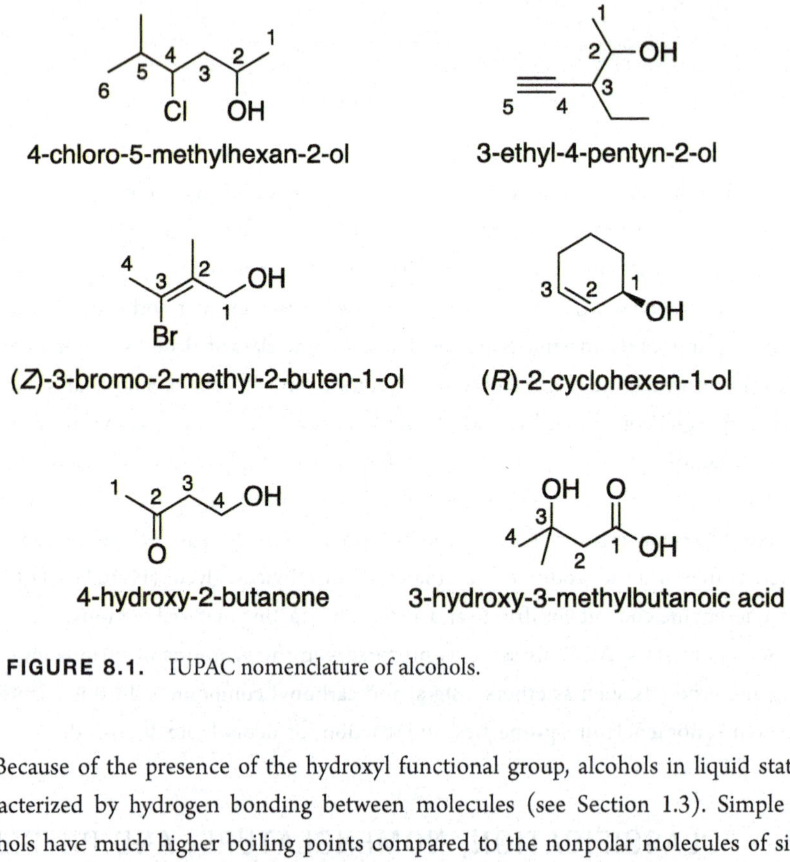

4-chloro-5-methylhexan-2-ol

3-ethyl-4-pentyn-2-ol

(Z)-3-bromo-2-methyl-2-buten-1-ol

(R)-2-cyclohexen-1-ol

4-hydroxy-2-butanone

3-hydroxy-3-methylbutanoic acid

FIGURE 8.1. IUPAC nomenclature of alcohols.

Because of the presence of the hydroxyl functional group, alcohols in liquid state are characterized by hydrogen bonding between molecules (see Section 1.3). Simple alkyl alcohols have much higher boiling points compared to the nonpolar molecules of similar size. For example, methanol CH_3OH, has a bp = +65 °C, which is much higher compared to ethane CH_3CH_3 (bp = −89 °C). Alcohols also have high solubility in water because they form hydrogen bonds with molecules of H_2O.

Problems

8.1. Which of the following is a 2° alcohol?

   Ethylene glycol, ethyl alcohol, cyclopentanol, 2-methyl-2-propanol

8.2. Sort the following according to increasing boiling point. List the molecule with the lowest boiling point first.

   1,2,3-Propantriol, hexane, 1-pentanol, 1,2-butandiol

8.3. For each reaction, provide the IUPAC name of the major product. If applicable, indicate stereochemistry.

a) $\equiv$ $\xrightarrow{\text{H}_2}{\text{Lindlar Catalyst}}$ $\xrightarrow{\text{1. BH}_3}{\text{2. NaOH, H}_2\text{O}_2, \text{H}_2\text{O}}$

b) [cyclohexane with tert-butyl and Br substituents, H shown] $\xrightarrow{\text{NaOH, H}_2\text{O}}$

c) [epoxide structure] $\xrightarrow{\text{1. HC}\equiv\text{C}^-\text{Na}^+}{\text{2. H}_2\text{O}}$

d) [1,2-dimethylcyclohexene] $\xrightarrow{\text{1. OsO}_4}{\text{2. NaHSO}_3, \text{H}_2\text{O}}$

8.2. ACIDITY AND BASICITY OF ALCOHOLS

Simple alkyl alcohols (ROH) in general have an acidity similar to water, with pK_a values around 16-18. The exact pK_a values of substituted alcohols are slightly affected by the inductive effect of a substituent in the alkyl chain (see Figure 2.6 and the related text in Chapter 2). Therefore, a 1° alcohol $CH_3CH_2CH_2OH$ (pK_a = 16) is slightly more acidic than a 2° alcohol $(CH_3)_2CHOH$ (pK_a = 17), and a 3° alcohol $(CH_3)_3COH$ (pK_a = 18) has the lowest acidity. Alcohols can be deprotonated by a strong inorganic base such as a hydride anion, (H)⁻ or by the reaction with active metals (sodium or potassium). The conjugate bases of alcohols are generally called **alkoxides**. The preparation of metal alkoxides (e.g., sodium ethoxide and potassium t-butoxide) is illustrated in Figure 8.2. Sodium ethoxide (as a solution in ethanol) and potassium t-butoxide (as a solution in t-butanol) are commonly used as bases in β-elimination reactions (see Section 5.4).

$$2CH_3CH_2OH + 2Na \longrightarrow 2CH_3CH_2O^-Na^+ + H_2$$
ethanol   sodium metal   sodium ethoxide

$$(CH_3)_3COH + K^+H:^- \longrightarrow (CH_3)_3CO^-K^+ + H_2$$
t-butanol   potassium hydride   potassium t-butoxide

FIGURE 8.2. Conversion of alcohols to alkoxides.

In the presence of a strong inorganic acid, alcohols act as acceptors of protons (see Figure 2.2 and the related text in Chapter 2). The conjugate acid of an alcohol (ROH_2^+) is called an oxonium ion and has a similar pK_a value as a hydronium ion (H_3O^+). Oxonium ions are formed as intermediate products in reactions of alcohols with strong acids (see Section 8.4).

Problems

8.4. a) Arrange the alcohols below according to increasing acidity. List the least acidic compounds first.

   Methanol   trifluoromethanol   tert-butanol

b) Sort the conjugated bases of these alcohols according to increasing basicity. List the least basic species first.

8.5. Sort the following ions according increasing basicity. List the least basic ion first.

CH_3NH^- CH_3^- CH_3O^- $HCOO^-$

8.6. a) Give the product and write the mechanism for the following transformation:

CH_3OH $\xrightarrow{\text{NaH}}$ [trans-1-chloro-2-methylcyclohexane] $\xrightarrow{\text{E2-Mechanism}}$

b) What would be the outcome of the above transformation if *trans*-2-methyl-cyclohexanol would be used instead of the haloalkane?

8.3. REACTIONS OF ALKOXIDES: PREPARATION OF ETHERS

In reactions with haloalkanes (RX), alkoxide anions can act either as strong bases, producing the products of E2 β-elimination (alkenes), or as nucleophiles, forming ethers as the products of an S_N2 reaction (Chapter 5). The β-elimination pathway is typically observed in reactions of alkoxides with 2° and 3° haloalkanes (Section 5.4), while the reaction of nonsterically hindered 1° haloalkanes yields ethers as major products. The reaction of 1° haloalkanes (RCH_2X) or halomethanes (CH_3X) with metal alkoxides is known as **Williamson ether synthesis**, named after the British chemist Alexander Williamson who published a study of this reaction in 1850. Several examples of Williamson ether syntheses are shown in Figure 8.3. In the IUPAC nomenclature system, ethers are named using the prefix *alkoxy*, such as methoxy (CH_3O) and ethoxy (CH_3CH_2O), for the corresponding substituents in the parent chain.

Ethers (R-O-R) belong to an important class of organic compounds with numerous practical applications such as solvents and drugs for anesthesia. The ether functional group is present in carbohydrates and many other natural products. Diethyl ether ($CH_3CH_2OCH_2CH_3$), also known as ethyl ether or simply ether, is a colorless, highly volatile, flammable liquid, used as a solvent and a general anesthetic for over a thousand years. Distilling a mixture of ethanol and sulfuric acid can produce diethyl ether.

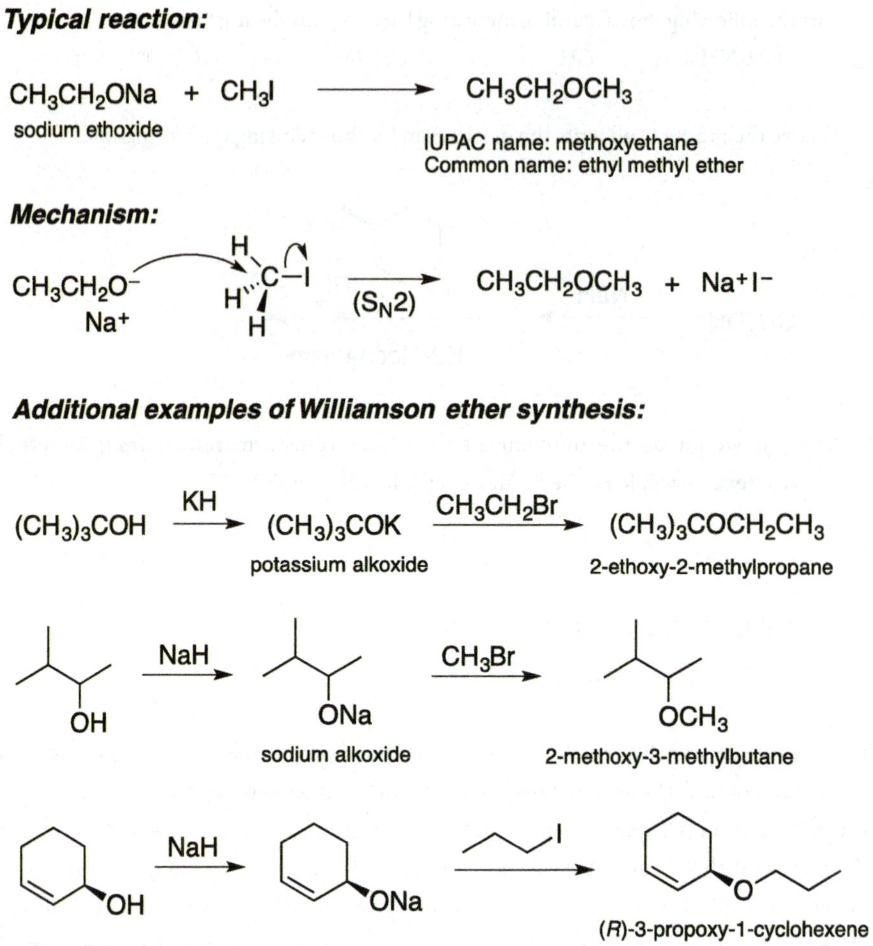

Typical reaction:

CH₃CH₂ONa + CH₃I ⟶ CH₃CH₂OCH₃
sodium ethoxide

IUPAC name: methoxyethane
Common name: ethyl methyl ether

Mechanism:

CH₃CH₂O⁻ Na⁺ + CH₃I ⟶ (S_N2) ⟶ CH₃CH₂OCH₃ + Na⁺I⁻

Additional examples of Williamson ether synthesis:

(CH₃)₃COH —KH→ (CH₃)₃COK —CH₃CH₂Br→ (CH₃)₃COCH₂CH₃
            potassium alkoxide   2-ethoxy-2-methylpropane

(3-methyl-2-butanol) —NaH→ sodium alkoxide —CH₃Br→ 2-methoxy-3-methylbutane

(cyclohexenol) —NaH→ (cyclohexenyl ONa) —propyl I→ (R)-3-propoxy-1-cyclohexene

FIGURE 8.3. Williamson ether synthesis.

Problems

8.7. Provide the major product in each of the following reactions:

156   ORGANIC CHEMISTRY

a) [cyclohexene] $\xrightarrow{Br_2, H_2O}$ $\xrightarrow{NaH}$

b) [CH₃CH₂CH(Cl)(H)CH₃ — chiral] $\xrightarrow[S_N2\text{-Mechanism}]{NaOH, H_2O}$ $\xrightarrow{NaH}$ $\xrightarrow{CH_3CH_2I}$

c) [cyclohexene] $\xrightarrow[\text{2. NaHSO}_3, H_2O]{\text{1. OsO}_4}$ $\xrightarrow[\text{2. CH}_3I \text{ (excess)}]{\text{1. NaH (excess)}}$

8.8 Propose a mechanism for the formation of diethyl ether from ethanol in the presence of sulfuric acid as catalyst.

8.4. CONVERSION OF ALCOHOLS TO ALKYL HALIDES

Alcohols can be converted to alkyl halides (RX) by treatment with the corresponding acid HX (HCl, HBr, or HI) in aqueous solution. The mechanism of this reaction involves the initial protonation of the hydroxyl group forming an oxonium ion. In contrast to the hydroxyl group, the H_2O^+ substituent in oxonium ion is an excellent leaving group, which can be readily replaced by the halide anion either by an S_N1 or S_N2 mechanism, depending on the structure of alkyl group. Examples of such reactions are shown in Figure 8.4.

Reactions of alcohols with HCl, HBr, or HI are not very selective and usually produce alkyl halides in a mixture with several by-products, such as ethers, alkenes, and rearranged haloalkanes formed by carbocationic rearrangements (see Figure 5.9 and the related text in Chapter 5). The more selective experimental procedures for converting alcohols to alkyl chlorides and alkyl bromides require the use of thionyl chloride ($SOCl_2$) and phosphorus tribromide (PBr_3) as reagents. Figure 8.5 shows examples of these reactions with the corresponding mechanisms. Both of these reagents initially react with alcohols, forming intermediate products with a better leaving group, which is then replaced by a halide anion according to S_N2 mechanism.

ALCOHOLS

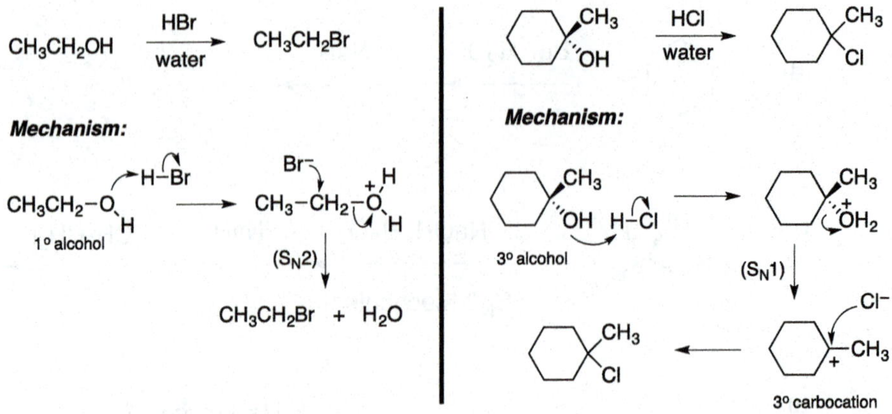

FIGURE 8.4. Conversion of alcohols to alkyl halides by treatment with HBr or HCl.

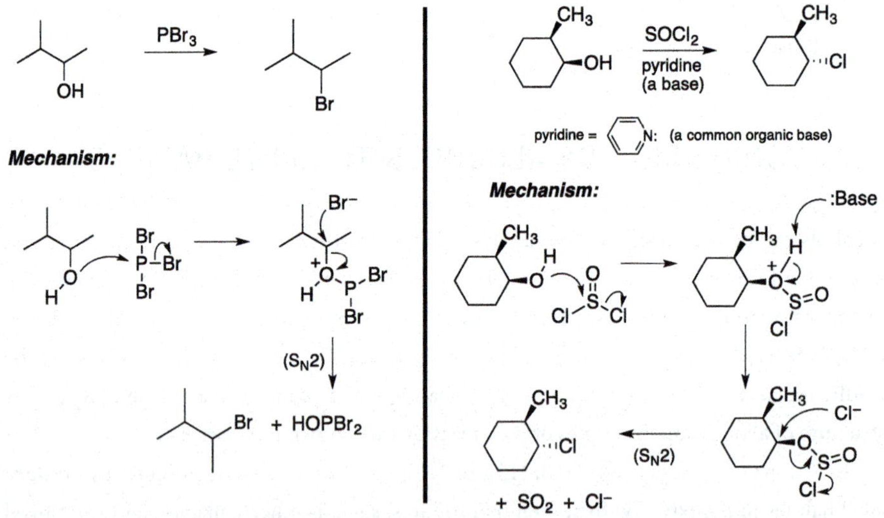

FIGURE 8.5. Conversion of alcohols to alkyl halides by reactions with PBr$_3$ or SOCL$_2$.

Problems

8.9. Suggest a mechanism that accounts for the formation of 2-bromo-2-methylbutane from 3-methyl-2-butanol and HBr.

8.10 Provide the major product in each of the following reactions:

a) PhCH₂OH $\xrightarrow{HI}$ $\xrightarrow{KCN}$

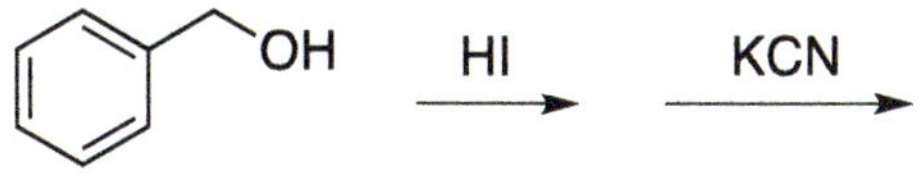

b) methylenecyclohexane $\xrightarrow[\text{2. H}_2\text{O}_2,\text{ NaOH, H}_2\text{O}]{\text{1. BH}_3}$ $\xrightarrow[\text{pyridine}]{SOCl_2}$

c) trans-2-methylcyclohexanol $\xrightarrow{PBr_3}$ $\xrightarrow{CH_3ONa}$

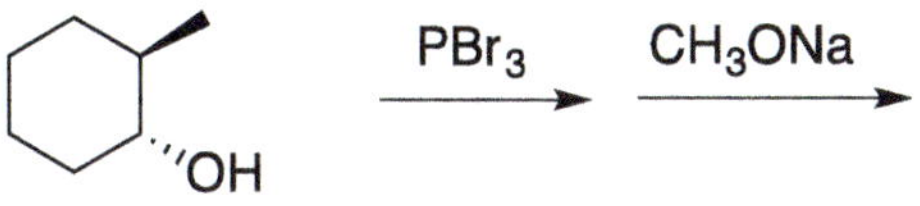

d) CH₃CH₂CH₂OH $\xrightarrow[\text{pyridine}]{SOCl_2}$ $\xrightarrow{KI}$ $\xrightarrow{HC\equiv C^-Na^+}$

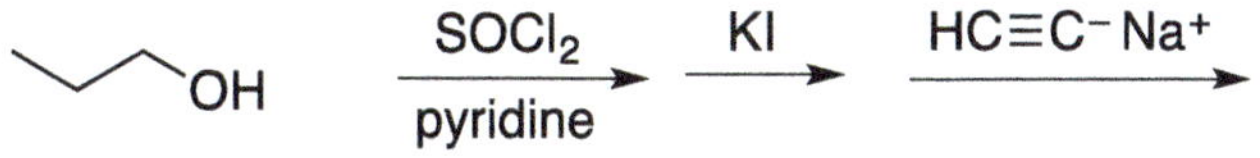

8.5. ACID-CATALYZED DEHYDRATION

Alcohols can be converted to alkenes by treatment with concentrated sulfuric acid (H_2SO_4) or phosphoric acid (H_3PO_4). The mechanism of this reaction involves initial protonation of the hydroxyl group, forming an oxonium ion, followed by elimination of water (dehydration). Elimination of water from 2° and 3° alcohols proceeds according to an E1 β-elimination mechanism, yielding a more stable, more highly substituted alkene as the major product, in agreement with Zaitsev's rule (see Section 5.4). Reactions of 1° alcohols, such as ethanol, with sulfuric acid normally produce ethers (Section 8.3); however, they may also form alkenes at high temperatures (about 180 °C). Examples of dehydration reactions are shown in Figure 8.6.

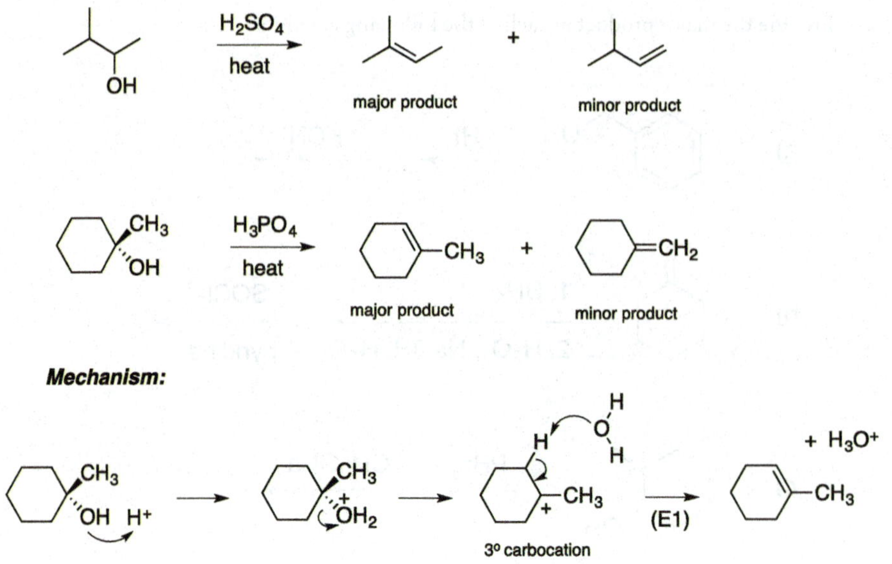

FIGURE 8.6. Dehydration of alcohols.

Dehydration of alcohols with sulfuric acid is often accompanied by carbocationic rearrangements via 1,2-hydride shift or 1,2-alkyl shift (see Chapter 5), resulting in the formation of various byproducts.

Problems

8.11. Provide major products in the following reactions:

a) 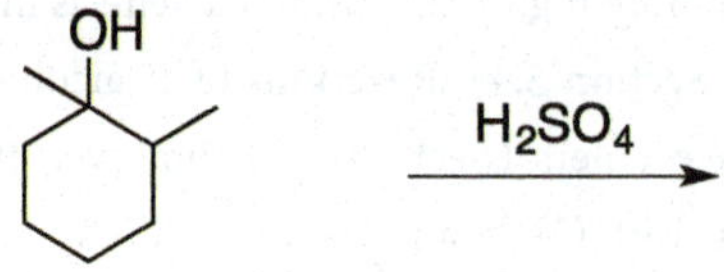 $\xrightarrow{H_2SO_4}$

b) 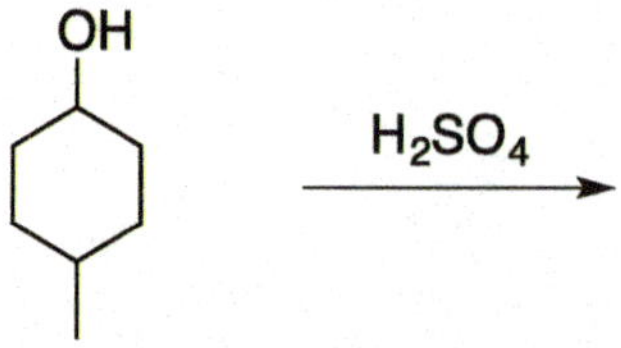 $\xrightarrow{H_2SO_4}$

8.12. For each reaction provide a mechanism that accounts for the rearranged product.

a) [cyclopentane with C(OH)(CH₃) substituent] + H₂SO₄ → [1,2-dimethylcyclohexene]

b) [(CH₃)₃C-CH(OH)-CH₃... tert-butyl group with CH(OH)] + H₂SO₄ → [2,3-dimethyl-2-butene]

8.6. PREPARATION AND REACTIONS OF ALKYL SULFONATES

The hydroxyl functional group is a poor leaving group and therefore cannot be directly replaced by nucleophiles in nucleophilic substitution reactions (see Chapter 5). A common method of improving its leaving group ability is based on the conversion of hydroxyl to a sulfonate group (RSO_2O-), which is an excellent leaving group. The primary reagents that are used for the conversion of a hydroxyl to a sulfonate group are sulfonic acids, the derivatives of hexavalent sulfur formed by oxidation of thiols.

8.6.1. Thiols and sulfonic acids

Sulfur belongs to the same group of the periodic table as oxygen and therefore has a similar valence shell electronic configuration. However, because of its position in the third period, sulfur can form more than 4 bonds in covalent molecules (see Section 1.1). **Thiols** (RSH) are the sulfur analogs of alcohols, with different physical and chemical properties from alcohols. Compared to alcohols, thiols have higher acidity (pK_a values of approximately 10) due to the larger size of the sulfur anion leading to a better delocalization of negative charge in the anion (see Section 2.3). Simple alkanethiols (e.g., methanethiol and ethanthiol), also known as **mercaptans**, are colorless gases or clear liquids with a very characteristic, strong odor. Thiols are very important in biochemistry. The thiol functional group is found in the amino acid cysteine, which is an essential building block in many proteins and enzymes. Thiols react with mild oxidants (e.g., atmospheric oxygen and elemental iodine) forming disulfides (RSSR), which is an important reaction in the chemistry of proteins. Simple alkylthiols can be oxidized to form alkanesulfonic acids by the action of strong oxidants, such as potassium permanganate or nitric acid (Figure 8.7).

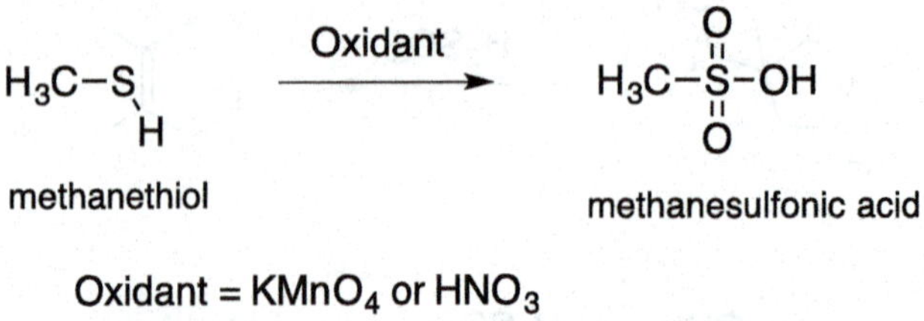

FIGURE 8.7. Formation of methanesulfonic acid by oxidation of methanethiol.

Sulfonic acids have high acidity ($pK_a = -3$), comparable to that of sulfuric acid, and have numerous applications in organic chemistry and chemical industry. Especially important is **π-toluenesulfonic acid**, also known as tosic acid or tosylic acid (Figure 8.8). The $CH_3C_6H_4SO_2-$ group is known as the tosyl group and is usually abbreviated as Ts. The anion of tosic acid ($CH_3C_6H_4SO_2O^-$) is called a tosylate anion and abbreviated as TsO⁻. An important derivative of tosic acid is tosyl chloride (TsCl). Alkyl esters of sulfonic acids are generally called alkyl sulfonates. The esters of tosic acid, known as alkyl tosylates (ROTs), are important reagents in organic chemistry. Alkyl esters of methanesulfonic acid (CH_3SO_2OR) are called alkyl mesylates and abbreviated as ROMs.

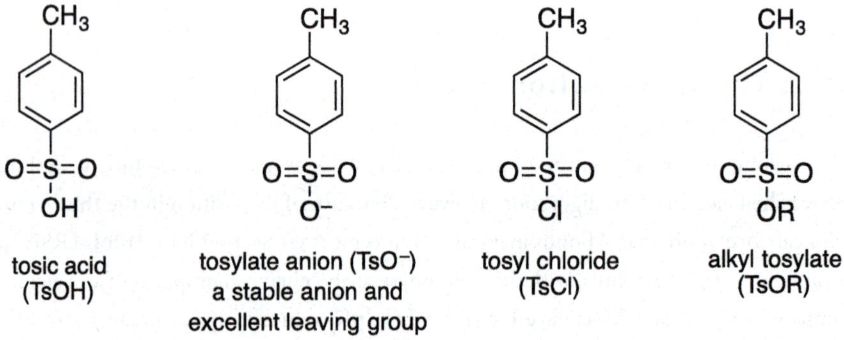

FIGURE 8.8. *p*-toluenesulfonic acid and related compounds.

8.6.2. Alkyl tosylates

Alcohols can be converted to **alkyl tosylates** by treatment with tosyl chloride in the presence of an organic base, usually pyridine or alkyl amines. The base is required to neutralize HCl, formed as a by-product in the reaction (Figure 8.9). It is important that the carbon atom in alkyl tosylate retains the same stereochemical configuration as alcohol.

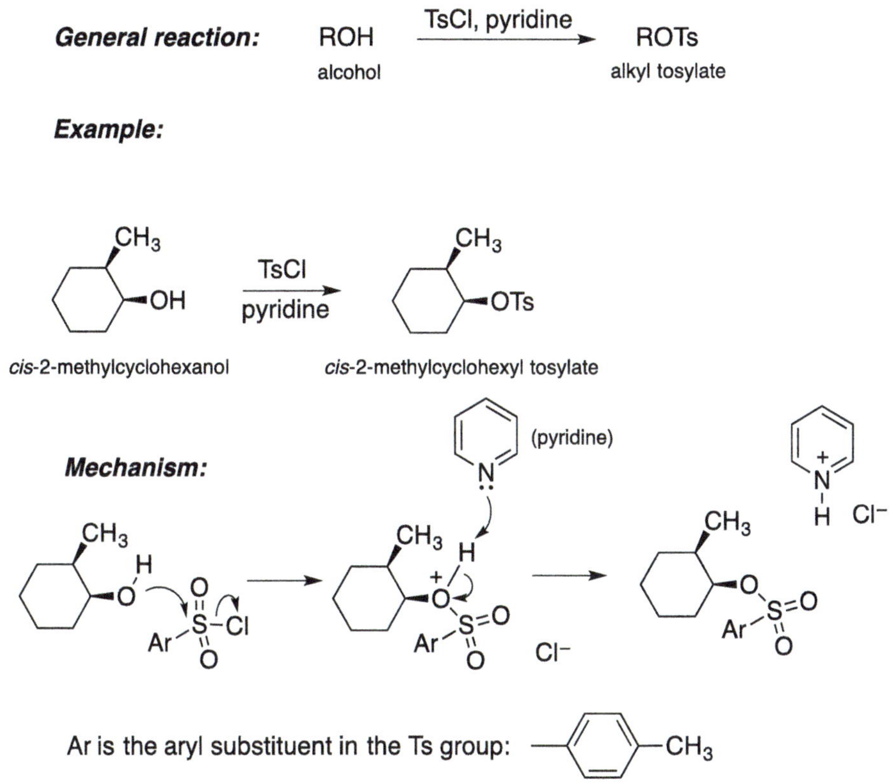

General reaction: ROH $\xrightarrow{\text{TsCl, pyridine}}$ ROTs
alcohol → alkyl tosylate

Example:

cis-2-methylcyclohexanol $\xrightarrow{\text{TsCl, pyridine}}$ cis-2-methylcyclohexyl tosylate

Mechanism:

Ar is the aryl substituent in the Ts group: —C₆H₄—CH₃

FIGURE 8.9. Conversion of alcohols to alkyl tosylates.

Alkyl tosylates are commonly used as highly reactive substrates in nucleophilic substitution reactions. Reactions of 1° and 2° alkyl tosylates with nucleophiles proceed according to the S_N2 mechanism, forming products with inverted configuration (see Chapter 5). Several examples of nucleophilic substitution reactions involving alkyl tosylates are shown in Figure 8.10.

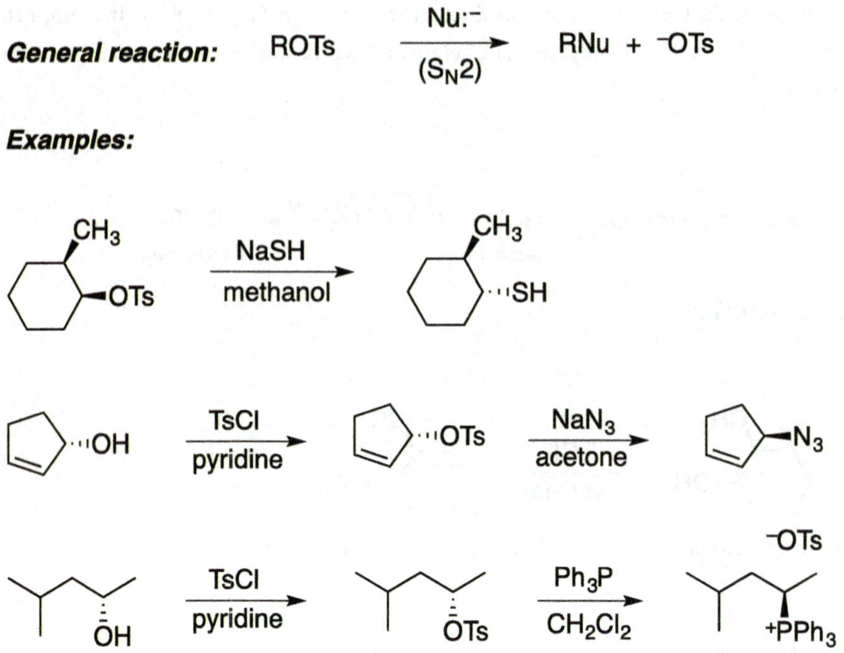

General reaction: ROTs $\xrightarrow[(S_N2)]{Nu:^-}$ RNu + ⁻OTs

Examples:

FIGURE 8.10. Reactions of alkyl tosylates with nucleophiles.

Problems

8.13. Provide major products in the following reactions:

a) 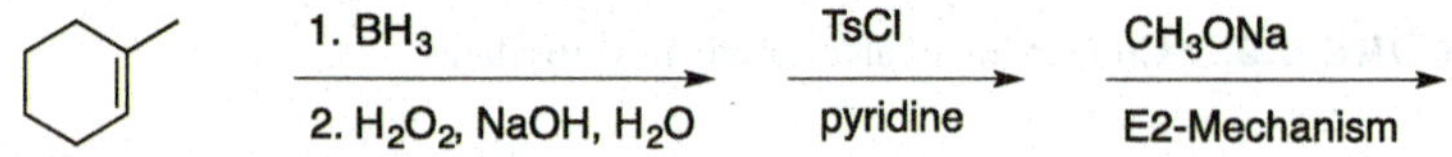 cyclohexene $\xrightarrow[\text{2. H}_2\text{O}_2,\text{ NaOH, H}_2\text{O}]{\text{1. BH}_3}$ $\xrightarrow[\text{pyridine}]{\text{TsCl}}$ $\xrightarrow[\text{E2-Mechanism}]{\text{CH}_3\text{ONa}}$

b) 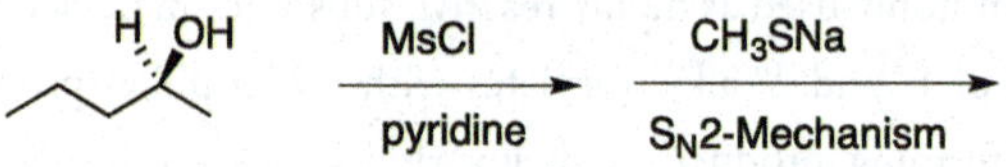 $\xrightarrow[\text{pyridine}]{\text{MsCl}}$ $\xrightarrow[\text{S}_N2\text{-Mechanism}]{\text{CH}_3\text{SNa}}$

8.7. OXIDATION OF ALCOHOLS

Oxidation of a 1° alcohol can lead to the creation of either one new C–O bond, forming an aldehyde, or two new C–O bonds, producing carboxylic acid (see Section 6.5 for a general description of oxidation reactions). Stronger oxidants (e.g., chromium trioxide, chromic acid, and potassium permanganate) oxidize 1° alcohols directly to carboxylic acids, while milder oxidants, e.g., [2-iodoxybenzoic acid (IBX) and pyridinium chlorochromate (PCC)], selectively produce aldehydes (Figure 8.11).

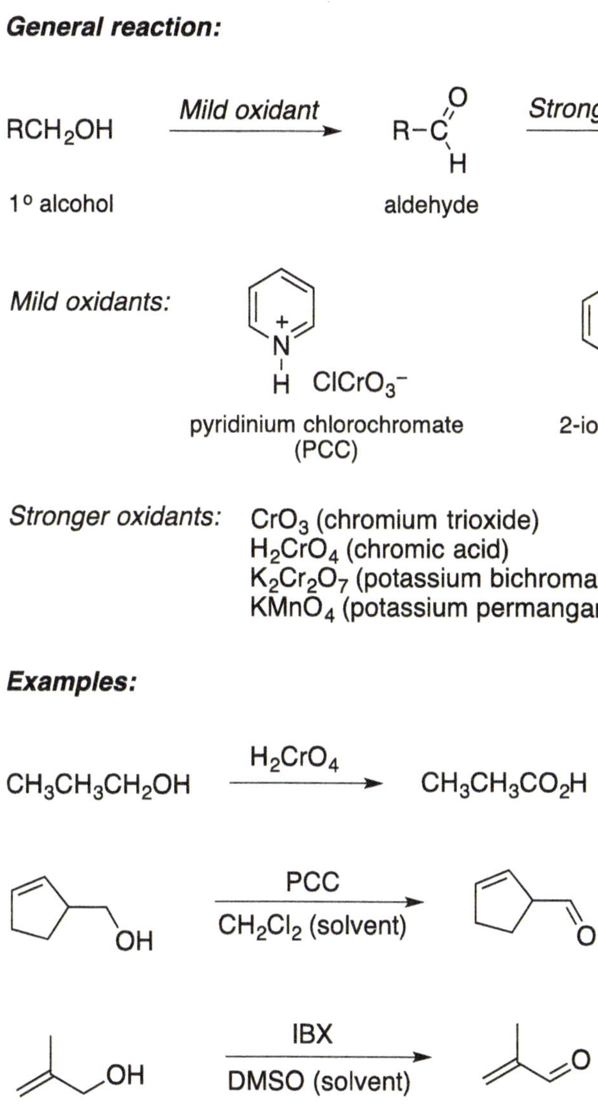

General reaction:

$$RCH_2OH \xrightarrow{\text{Mild oxidant}} R-\underset{H}{\overset{O}{\underset{\|}{C}}} \xrightarrow{\text{Stronger oxidant}} R-\underset{OH}{\overset{O}{\underset{\|}{C}}}$$

1° alcohol → aldehyde → carboxylic acid

Mild oxidants: pyridinium chlorochromate (PCC); 2-iodoxybenzoic acid (IBX)

Stronger oxidants: CrO_3 (chromium trioxide)
H_2CrO_4 (chromic acid)
$K_2Cr_2O_7$ (potassium bichromate) with sulfuric acid
$KMnO_4$ (potassium permanganate)

Examples:

$$CH_3CH_2CH_2OH \xrightarrow{H_2CrO_4} CH_3CH_2CO_2H$$

cyclopentenyl-CH₂OH $\xrightarrow[\text{CH}_2\text{Cl}_2 \text{ (solvent)}]{\text{PCC}}$ cyclopentenyl-CHO

CH₂=C(CH₃)CH₂OH $\xrightarrow[\text{DMSO (solvent)}]{\text{IBX}}$ CH₂=C(CH₃)CHO

FIGURE 8.11. Oxidation of 1° alcohols.

Oxidation of 2° alcohols with any common oxidant yields the corresponding ketones by replacing the C–H bond with the C–O bond (Figure 8.12). A typical reaction mechanism involves the initial formation of alkyl chromate (or alkyl iodate in the reaction of IBX), followed by the elimination of a reduced form of chromium (or iodine in the case of IBX), as shown in Figure 8.12. This mechanism includes deprotonation of a C–H bond at the carbon atom connected to the hydroxyl group in 1° and 2° alcohols. Tertiary alcohols cannot be oxidized under these conditions.

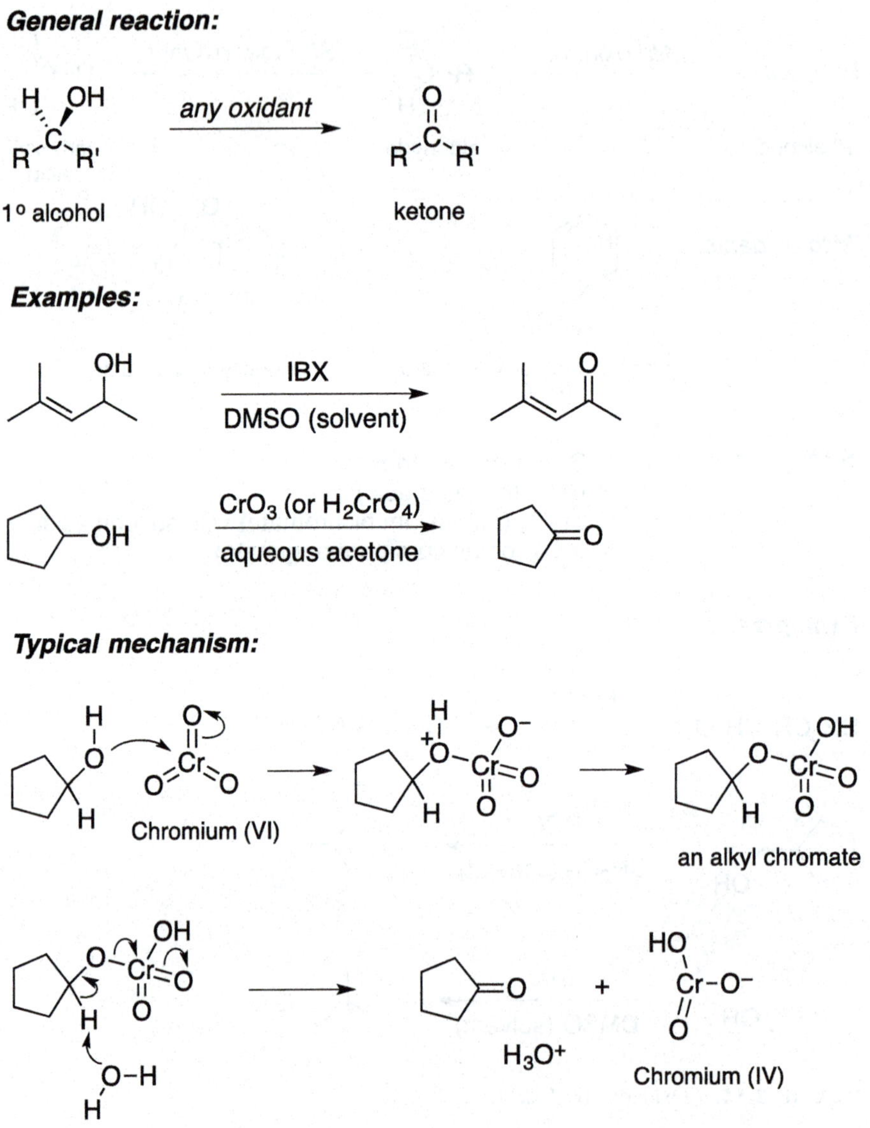

General reaction:

1° alcohol → ketone (any oxidant)

Examples:

4-methylpent-3-en-2-ol → 4-methylpent-3-en-2-one (IBX, DMSO solvent)

cyclopentanol → cyclopentanone (CrO$_3$ (or H$_2$CrO$_4$), aqueous acetone)

Typical mechanism:

Chromium (VI) → alkyl chromate → cyclopentanone + HO–Cr–O⁻ (Chromium (IV)), H$_3$O⁺

FIGURE 8.12. Oxidation of 2° alcohols and the general mechanism of oxidation.

Problems

8.14 Propose a mechanism for the oxidation of a 1° alcohol to aldehyde using IBX.

8.15 Provide major products in the following reactions:

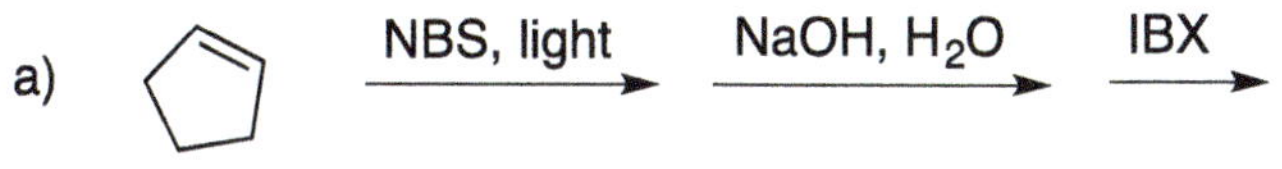

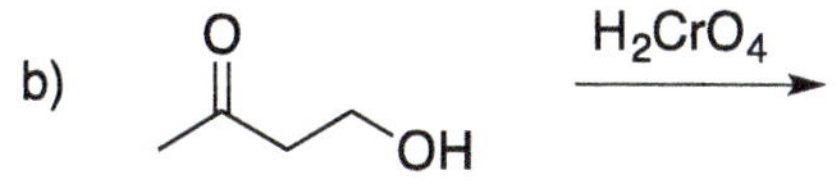

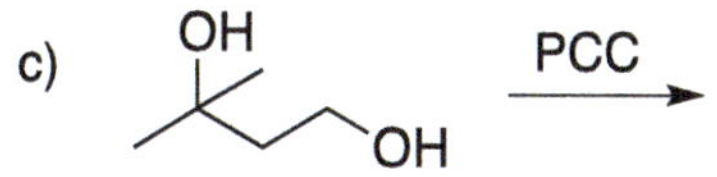

a) cyclopentene $\xrightarrow{\text{NBS, light}}$ $\xrightarrow{\text{NaOH, H}_2\text{O}}$ $\xrightarrow{\text{IBX}}$

b) 4-hydroxybutan-2-one $\xrightarrow{\text{H}_2\text{CrO}_4}$

c) 3-methyl-1,3-butanediol (OH on C3 tertiary, OH on C1 primary) $\xrightarrow{\text{PCC}}$

CHAPTER 9

Spectroscopy of Organic Compounds

Spectroscopic methods are a key technique for determining the structure of organic molecules. Spectroscopy centers on the measurement and interpretation of spectra that appear as a result of the interaction of electromagnetic radiation with chemical compounds. Scientists study the absorption, emission, or scattering of electromagnetic radiation by atoms or molecules using spectrometers. This chapter gives an overview of infrared (IR) spectroscopy, nuclear magnetic resonance (NMR) spectroscopy and mass spectrometry.

9.1. INFRARED SPECTROSCOPY

Infrared spectroscopy (**IR spectroscopy** or Vibrational Spectroscopy) utilizes the infrared region of the electromagnetic spectrum, which is light with lower frequencies and longer wavelengths than visible light.

Molecules are not rigid. There are many types of molecular vibration and each type is called a **vibrational mode**. For molecules with N number of atoms, linear molecules have $3N$-5 degrees of vibrational modes, whereas nonlinear molecules have $3N$-6 degrees of vibrational modes (also called **vibrational degrees of freedom**). These modes of vibration can be induced by infrared light. IR absorptions occur when the incoming light frequency matches the frequency of a vibrational mode.

The most useful region of IR for analysis of organic compounds has a wavelength range between 2,500 and 25,000 nm. Consequently, virtually all organic compounds will absorb infrared radiation within the same range. Infrared spectrometers permit chemists to obtain the unique absorption spectra of compounds, which reflect the molecular structure of compound.

The infrared spectrum of methanol may serve as an example (Figure 9.1). The scale at the bottom of the IR spectrum (the **wavenumbers**) shows frequency in units of reciprocal centimeters (cm^{-1}) to make the scale proportional to the energy of the absorption. In the IR spectrum of methanol, the strong and somewhat broad band around 3350 cm^{-1} is very characteristic of O–H bond stretches. Alkyl C–H stretching is observed around 2850–3000 cm^{-1}.

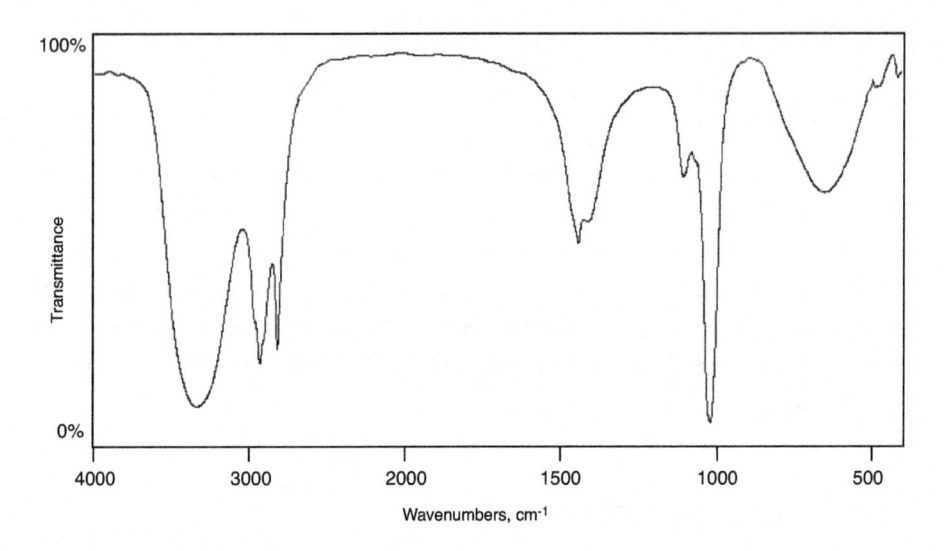

FIGURE 9.1. IR spectrum of methanol.

In particular, the region between 4000 cm^{-1} and 1450 cm^{-1} (Table 9.1) is very useful for identifying which functional groups are present within a molecule. Absorption bands in this region of the spectrum are usually due to bond stretching vibrations.

TABLE 9.1. Important IR bond stretching absorptions.

3200–3650 cm^{-1}	O-H stretch	Alcohols: strong intensity, usually broad; carboxylic acids may absorb slightly lower wavenumbers
3100–3550 cm^{-1}	N-H stretch	medium intensity
3000–3100 cm^{-1}	C(sp^2)-H (arene, alkene) stretch	usually several bands, weak to medium intensity
2850–3000 cm^{-1}	C(sp^3)-H (alkyl) stretch	medium intensity
2200–2250 cm^{-1}	C(sp)≡N (nitrile) stretch:	medium intensity
2100–2250 cm^{-1}	C(sp)≡C(sp) (alkyne) stretch	medium/weak intensity; terminal alkynes also show a second strong absorption around 3300 cm^{-1} (C(sp)-H stretch)
1630–1820 cm^{-1}	C(sp^2)=O stretch	strong intensity; different carbonyl functional groups can be distinguished by location of the C=O band
1450–1680 cm^{-1}	C(sp^2)=C(sp^2) (arene, alkene) stretch	weak to medium intensity – do not confuse with C=O band (=> strong intensity)
1550–1475 cm^{-1} and 1360–1290 cm^{-1}	NO$_2$ stretch	two strong set of bands

The complexity of infrared spectra in the 1450 to 600 cm^{-1} region makes it difficult to assign all the absorption bands conclusively. For example, in the spectrum of methanol (Figure 9.1), C-O stretching occurs at 1030 cm^{-1}, while C-H bending occurs just below 1450 cm^{-1}. Because of these unique patterns, this region is often called the **fingerprint region**. In general, the spectra in the fingerprint area tend to be very complex and not commonly used for functional group identification.

9.2. NMR SPECTROSCOPY

9.2.1. Overview of the underlying concepts of NMR spectroscopy

Nuclear magnetic resonance (NMR) spectroscopy exploits the magnetic properties of certain atomic nuclei. Nuclei of elements with an odd number of protons and/or neutrons possess a nuclear spin and can interact with external magnetic field acting as tiny magnets. For organic chemists, the nuclear magnetic properties of the ^{1}H isotope of hydrogen and the ^{13}C isotope of carbon are particularly important and can provide detailed information about the structure of organic molecules.

The nucleus of each isotope possesses two spin states that differ in energy within the magnetic field acting on the nucleus. The energy difference between these two spin states is proportional to the strength of the magnetic field. Per convention, these spin states are usually termed alpha and beta, with the alpha spin state being lower in energy. Interconversion between the alpha spin state and the beta spin state can be achieved by applying electromagnetic radiation. Within the realm of currently available technology for designing magnets, the frequency of the applied electromagnetic radiation is similar to radio and TV broadcasts (60-1000 MHz). Chemists say that a nucleus is in resonance if the frequency of the applied radiation equals the energy difference of the alpha spin state and beta spin state. At the resonance frequency the equilibrium between the two spin states is disturbed, resulting in a spin-flip from the alpha spin-state to the beta spin-state, and a signal can be recorded.

Note that there is a major difference between ^{1}H NMR and ^{13}C NMR spectroscopy. Within a given magnetic field the resonance frequency of a ^{13}C isotope is about a quarter of the resonance frequency of an ^{1}H isotope. In addition, also only 1% of all carbon atoms in a molecule are the ^{13}C isotope while more than 99.9% of the hydrogen atoms are NMR active ^{1}H isotopes. Thus, overall, ^{13}C NMR spectroscopy is less sensitive and may require a larger quantity of compound than ^{1}H NMR spectroscopy. The ^{12}C isotope – the major isotope of carbon – is NMR inactive, because it has an even number of protons and neutrons.

9.2.2. Basic information derived from NMR spectroscopy

While IR spectroscopy is a very powerful technique to determine which functional groups are present in a molecule, it struggles to distinguish between constitutional isomers. For example, from the IR spectra of 2-pentanone and 3-pentanone (Figure 9.2) we can easily determine that these compounds contain a carbonyl group (a strong band around 1700 cm^{-1}). While closer inspection reveals the presence of alkyl groups (e.g., alkyl-C-H stretching; several bands between 2900 and 3000 cm^{-1}) IR spectroscopy cannot determine the connectivity of the alkyl groups. NMR spectroscopy is by far more helpful in this case.

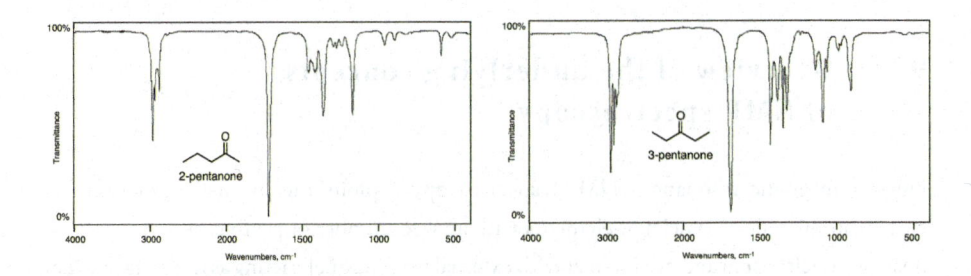

FIGURE 9.2.   IR spectra of 2-pentanone and 3-pentanone.

Figure 9.3 shows the ^{1}H NMR and ^{13}C NMR spectra of 2-pentanone and 3-pentanone. NMR spectra are recorded as **chemical shifts** (δ), whose units are **parts per million** (ppm). The measure of NMR experiments is a relative scale; that is, there is no absolute measure. Experimentally measured proton chemical shifts are referenced to the ^{1}H signal of tetramethylsilane (TMS, Me$_4$Si), while carbon chemical shift are referenced to ^{13}C signal of TMS. While ^{1}H NMR spectra range from 0-14 ppm, ^{13}C NMR spectra range from 0 – 220 ppm.

Several pieces of basic information can be derived from NMR spectra:

A. Number of signals

There is a very obvious difference between the ^{13}C NMR spectra of 2-pentanone and 3-pentanone. The spectrum of 3-pentanone only shows three signals, while the spectrum of 2-pentanone shows five signals (Figure 9.3). This results from the symmetry of the two molecules: 3-pentanone is a more symmetric molecule than 2-pentanone. In 3-pentanone, we can superpose the two methyl groups by reflection through the plane of the carbonyl C=O bond or by rotation around the C=O bond. We can do the same with the two methylene groups in this molecule. In NMR spectroscopy, atoms or groups that can be interconverted by a symmetrical operation are considered to be magnetically equivalent and emit the same signal. Non-equivalent atoms or groups give more than one signal. Thus in 3-pentanone we observe one signal for the methyl groups, one signal for the methylene groups and one signal for the carbonyl group. It is impossible to convert a methyl group into a methylene group or a carbonyl group via a symmetry operation. The same applies for a methylene group: a methylene group cannot be superposed onto a carbonyl group. All of these groups are non-equivalent, so we can measure one signal from each group, with a total of three signals for the whole molecule.

In contrast, we cannot superpose the two methyl groups by reflection or rotation for 2-pentanone. Therefore, the two methyl groups in this molecule are non-equivalent, resulting in one signal for each group. The same applies for the methylene groups in this molecule. Taken together – 4 signals for the alkyl groups and one signal for the carbonyl – we observe five signals in the ^{13}C NMR of 2-pentanone.

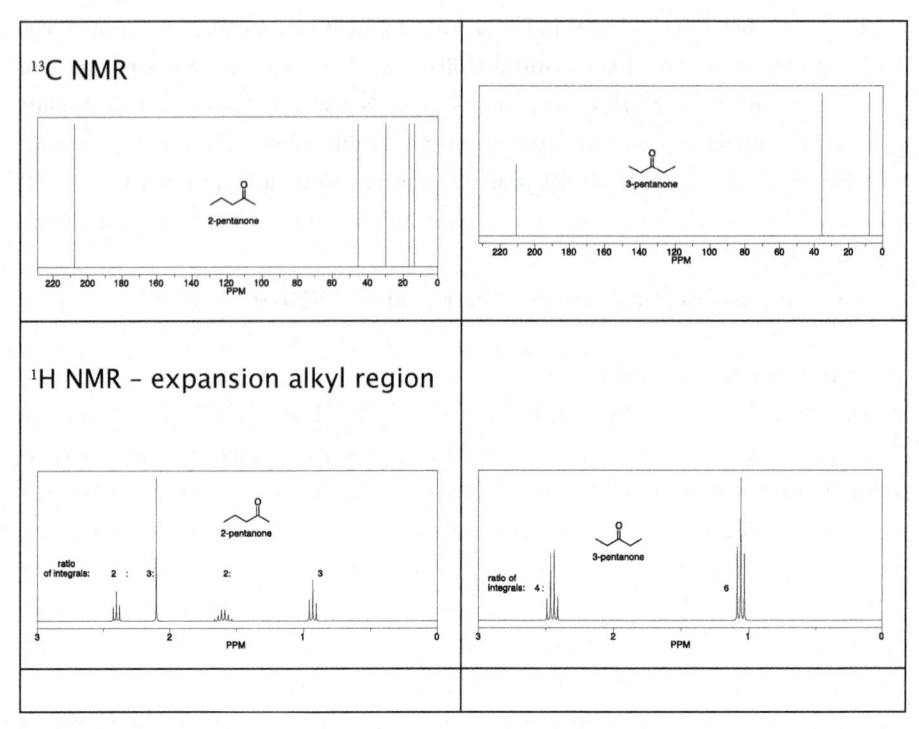

FIGURE 9.3. ^{1}H NMR and ^{13}C NMR spectra of 2- and 3-pentanone.

The ^{1}H NMR spectra of 2- and 3-pentanone are slightly more complex since most of the signals show a fine structure, with additional lines present within some of the signals. However, the ^{1}H NMR of 3-pentanone shows fewer signals than the ^{1}H NMR of 2-pentanone. The ^{1}H NMR of 3-pentanone has two signals and the ^{1}H NMR of 2-pentanone has four.

We can use reflection or rotation to superimpose the protons of the methyl groups within the 3-pentanone molecule. All six protons within those methyl groups are equivalent, resulting in one signal. We can do the same for the four protons of the two methylene groups, resulting in a second signal. From the discussion of ^{13}C NMR of 2-pentanone we already know that the four alkyl groups in this molecule are non-equivalent. Within each of the four alkyl groups in 2-pentanone, all protons are equivalent, resulting in a total of four signals.

An alternative approach to determine the number of signals in an NMR spectrum is a substitution test (Figure 9.4). For ^{1}H NMR, the substitution test is also used to label protons as homotopic, enantiotopic, diastereotopic, or unrelated. Homotopic, and generally enantiotopic protons as well, emit one signal, while a set of two diastereotopic or unrelated protons produce two different signals. The substitution test for protons works as follows:

1. Pick a proton within the molecule and draw a new structure by replacing the proton with a monovalent test group, X (e.g., X can be the hydrogen isotope deuterium).

2. Repeat this step by picking a second proton.
3. Compare the two structures created in steps 1 and 2. There are several possibilities:
- Two identical structures => the protons are homotopic – the protons are equivalent and give the same signal.
- The structures are enantiomers => the protons are enantiotopic and therefore in an achiral environment (which is generally the case), the protons are equivalent and give the same signal.
- The structures are diastereomers => the protons are diastereotopic. Remember that diastereomers usually have different physical properties. Thus, diastereotopic protons will produce two different signals.
- The structures are constitutional isomers => the protons are unrelated, yielding two different signals – one for each proton.

Note that NMR signals may overlap, which is quite often the case with diastereotopic protons.

As a rule of thumb, protons within a methyl group (CH_3) are always homotopic and unrelated to protons of methylene (CH_2) or methyne (CH) groups. Methylene groups are always unrelated to methyne groups. Great care should be taken with protons within methylene groups since methylene groups in molecules than contain chiral centers or alkyl rings may be diastereotopic. Terminal protons on mono-substituted alkenes ($RHC=CH_2$) are always diastereotopic.

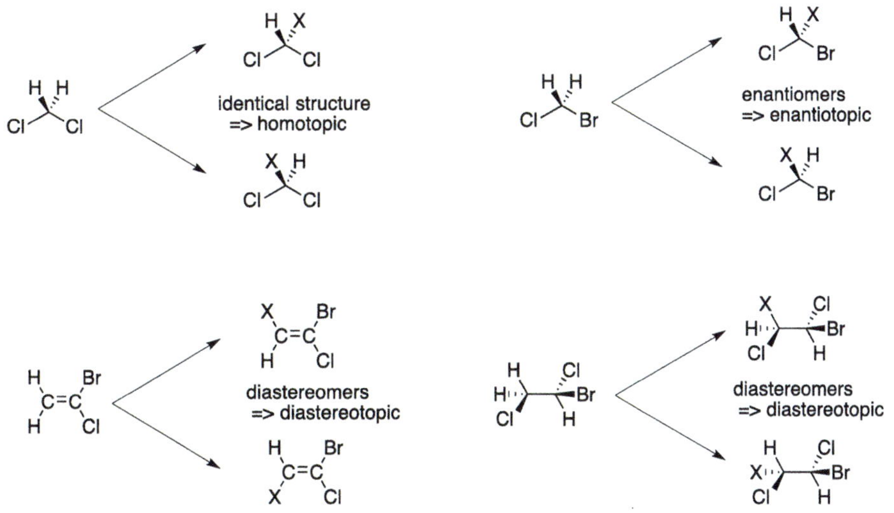

FIGURE 9.4. Substitution test.

Problems

9.1. How many signals do you expect in the ^{13}C NMR spectrum for each of the following compounds:

a) [methylenecyclohexane structure]   b) [2-methyl-2-pentene structure]   c) [branched alkene structure]   d) [methylenecyclohexane structure]

B. Integration

NMR is unique among common spectroscopic methods in that signal intensities are directly proportional to the number of nuclei causing the signal. In other words, all absorption coefficients for a given nucleus are identical. This is why proton NMR spectra are routinely integrated, whereas IR and UV spectra are not. It is important to point out that an **integral** only gives the relative number of protons. It is not possible to determine the absolute number of protons without additional information (such as a molecular formula).

There are 10 protons present in 3-pentanone according to its molecular formula $C_5H_{10}O$. We have recorded the ratio of integrals as 4 : 6 (Figure 9.3). Since 3-pentanone is a relatively simple molecule after having determined the integration of the signals we can actually assign the signals. There are two equivalent methyl groups in this molecule, containing 6 protons total. Thus the signal at δ = 1.06 ppm, for which an integration of 6 was recorded, is due to the resonance of the methyl protons. On the other hand, the signal at δ = 2.45 ppm (integration of 4) is due the methylene protons.

We can do the same with the spectrum of 2-pentanone (Figure 9.3). The signal at δ = 0.93 ppm and the signal at δ =2.10 ppm have an integration of 3. Therefore these signals are due to the methyl groups in the molecule, while the signals at δ = 2.40 and at δ = 1.60 ppm (both signals have an integration of 2) are due to the methylene groups. We will have to consider further information from the spectra to determine which signal is which.

In routine ^{13}C NMR spectra peak intensities are distorted. Integrations of such spectra will not give accurate ratios of peak areas. Under normal acquisition conditions, the signal heights are more indicative of the number of attached protons than the number of carbons. Thus in general, carbons of methyl (CH_3) and methylene (CH_2) groups produce tall signals, carbons of methyne groups (CH) produce medium signals, and quaternary carbons produce short signals.

Problems

9.2 For each compound below, determine the number of the signals and the integration.

    a) $CH_3CH_2COOCH_3$ b) $(CH_3)_2CHCOOH$

    c) $CH_3CH_2OCOCH_3$ d) $CH_3OCH_2COCH_3$

C. Signal splitting

Nuclei must have nonequivalent chemical shifts to show spin-spin interaction also called coupling to each other. While non-equivalent protons couple over up to 4 bonds, for the purpose of this chapter we will concentrate on vicinal or ^3J coupling that occurs over three bonds. In general, the number of lines into which a signal is split is given by the "*n* + 1" rule, where n is the number of protons on neighboring carbons. The intensities of these multiplets follow **Pascal's triangle** (Figure 9.5).

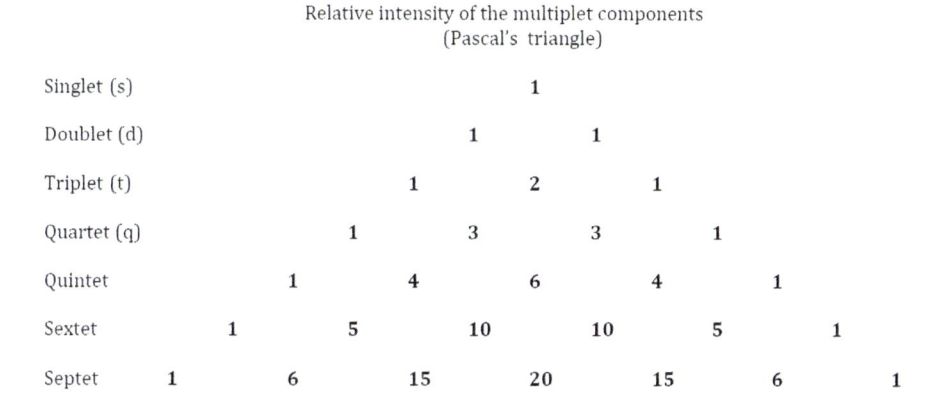

Relative intensity of the multiplet components
(Pascal's triangle)

Singlet (s)					1				
Doublet (d)				1		1			
Triplet (t)				1	2	1			
Quartet (q)			1	3		3	1		
Quintet			1	4	6	4	1		
Sextet		1	5	10		10	5	1	
Septet	1	6	15		20		15	6	1

Observed shape of signals:

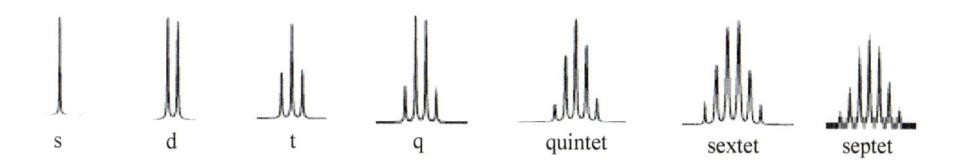

s   d   t   q   quintet   sextet   septet

FIGURE 9.5. Pascal's triangle and the observed shape of multiplets.

The distance between individual lines in a signal is called the **coupling constant** (or J value). Spin-spin splitting is not dependent on the external field, so we use energy units for coupling constants (Hz).

In the ^{1}H NMR spectrum of 2-pentanone, due to their integration of 3 we were able to assign the signals at δ = 0.93 ppm and δ = 2.10 ppm to the methyl groups. The signal at δ = 2.10 ppm is singlet, while the signal at δ = 0.93 appears as a triplet. When we count the distance between the protons of the each methyl group and its closest non-equivalent neighboring protons, we note that the protons bound to the methyl group adjacent to the carbonyl group have no nonequivalent protons within its three bonds. In contrast, the protons of the other methyl group are within three bonds of a methylene group. Thus, these methyl protons are able to "see" the spin state of each of the methylene protons and their signal is split into $n+1$ lines. With $n = 2$, those methyl protons "see" two non-equivalent protons and their signal is split into a triplet. Thus the signal at δ = 2.10 ppm – a singlet – is due the methyl protons adjacent to the carbonyl group, while the signal at δ = 0.93 ppm – a triplet – is due to the methyl protons further away from the carbonyl group. We can repeat this with the signals for the methylene groups in this molecule and determine that the signal at 2.40 ppm – a triplet with an integration of 2 – is due to the methylene group adjacent to the carbonyl group. Its protons are three bonds from the two nonequivalent protons of the other methylene group. Thus with $n = 2$ this signal is expected to be a triplet. The remaining methylene group signal at δ = 1.60 ppm is a sextet due its proximity to the methyl group, and the methylene group adjacent to the carbonyl group (Figure 9.6).

FIGURE 9.6. Signal splitting in the ^{1}H NMR spectra of isomeric pentanones.

The signal at δ = 2.45 ppm, which we already assigned as being the methylene group in 3-pentanone, appears as a quartet. The signal at δ = 1.06 ppm – the methyl protons – appears as a triplet. This triplet/quartet coupling pattern with a ratio of integration of 3:2 is very characteristic of ethyl groups.

Typical splitting patterns for common alkyl groups and substituted benzenes are summarized in Table 9.2 and Table 9.3, respectively.

TABLE 9.2. Splitting pattern of some alkyl groups R-X.

R-X	NUMBER OF SIGNALS	SPLITTING PATTERN/ INTEGRATION
Methyl: CH_3-X	1	One singlet
Ethyl: CH_3CH_2-X	2	One 2H quartet One 3H triplet
Propyl: $CH_3CH_2CH_2$-X	3	One 2H triplet One 3H triplet One 2H sextet
Isopropyl: $(CH_3)_2$CH-X	2	One 6H doublet One 1H septet
Butyl: $CH_3CH_2CH_2CH_2$-X	4	One 2H triplet One 3H triplet One 2H quintet One 2H sextet
tert-Butyl: $(CH_3)_3$C-X	1	One singlet

TABLE 9.3. Splitting pattern of benzene and substituted aromatic compounds.

	NUMBER OF SIGNALS	SPLITTING PATTERN/ INTEGRATION
benzene	1	One singlet
mono-X	3	One 2H doublet One 2H triplet One 1H triplet
ortho-X,X	2	One 2H doublet One 2H triplet
meta-X,X	3	One 1H singlet One 2H doublet One 1H triplet
para-X,X	1	One singlet
ortho-X,Y	4	Two 1H doublets Two 1H triplets

SPECTROSCOPY OF ORGANIC COMPOUNDS 179

	NUMBER OF SIGNALS	SPLITTING PATTERN/ INTEGRATION
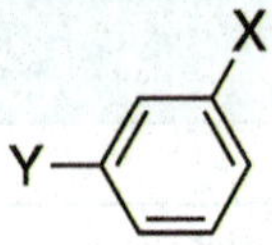 1,2,3-trisubstituted (X, Y on adjacent)	4	One 1H singlet Two 1H doublets One 1H triplet
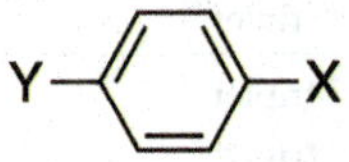 Y—◯—X (para)	2	Two 2H doublets

Note: Table assumes ^{1}H-^{1}H coupling over three bonds only and X ≠ Y.

Because of the naturally low quantity of ^{13}C nuclei, it is very unlikely to find two ^{13}C atoms near each other in the same molecule, and therefore we do not see spin-spin coupling between neighboring carbons in a ^{13}C NMR spectrum. There is, however, **heteronuclear coupling** between ^{13}C carbons and the hydrogens to which they are bound. Carbon-proton coupling constants can be very large, on the order of 100 – 250 Hz. For clarity, chemists generally use a technique called broadband decoupling, which essentially 'turns off' C-H coupling, resulting in a spectrum in which all carbon signals are singlets.

Problems

9.3. Identify the ^{1}H NMR spectrum for each of the following compounds:

I) CH_3CH_2Cl II) $(CH_3)_2CHCl$ III) $CH_3CH_2CH_2Cl$ IV) $CH_3CH_2CH_2CH_2Cl$

Spectrum A: Ratio of integrals: 2 : 2 : 2 : 3

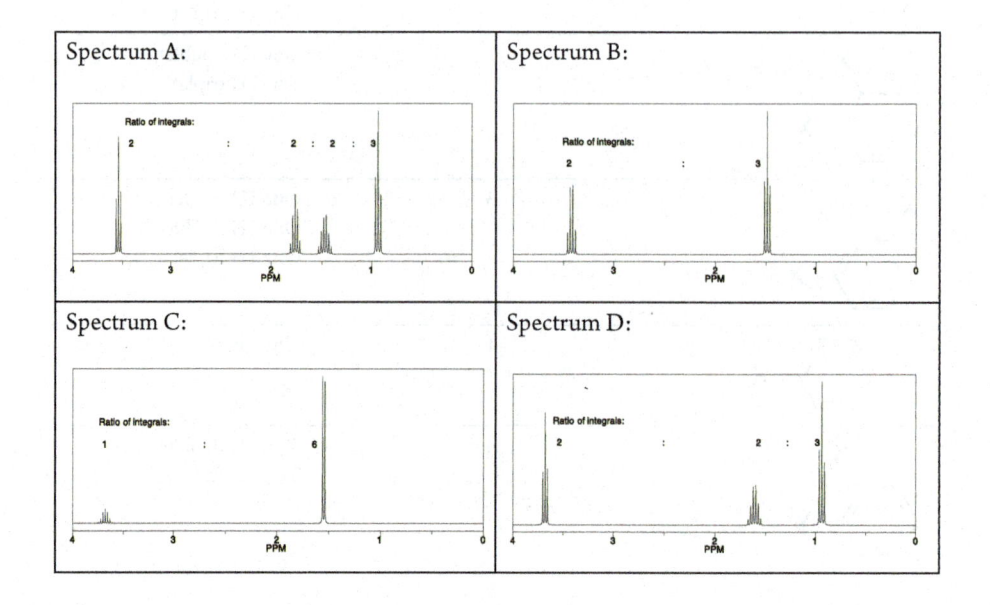

Spectrum B: Ratio of integrals: 2 : 3

Spectrum C: Ratio of integrals: 1 : 6

Spectrum D: Ratio of integrals: 2 : 2 : 3

94. Identify the ¹H NMR spectrum for each of the following compounds:

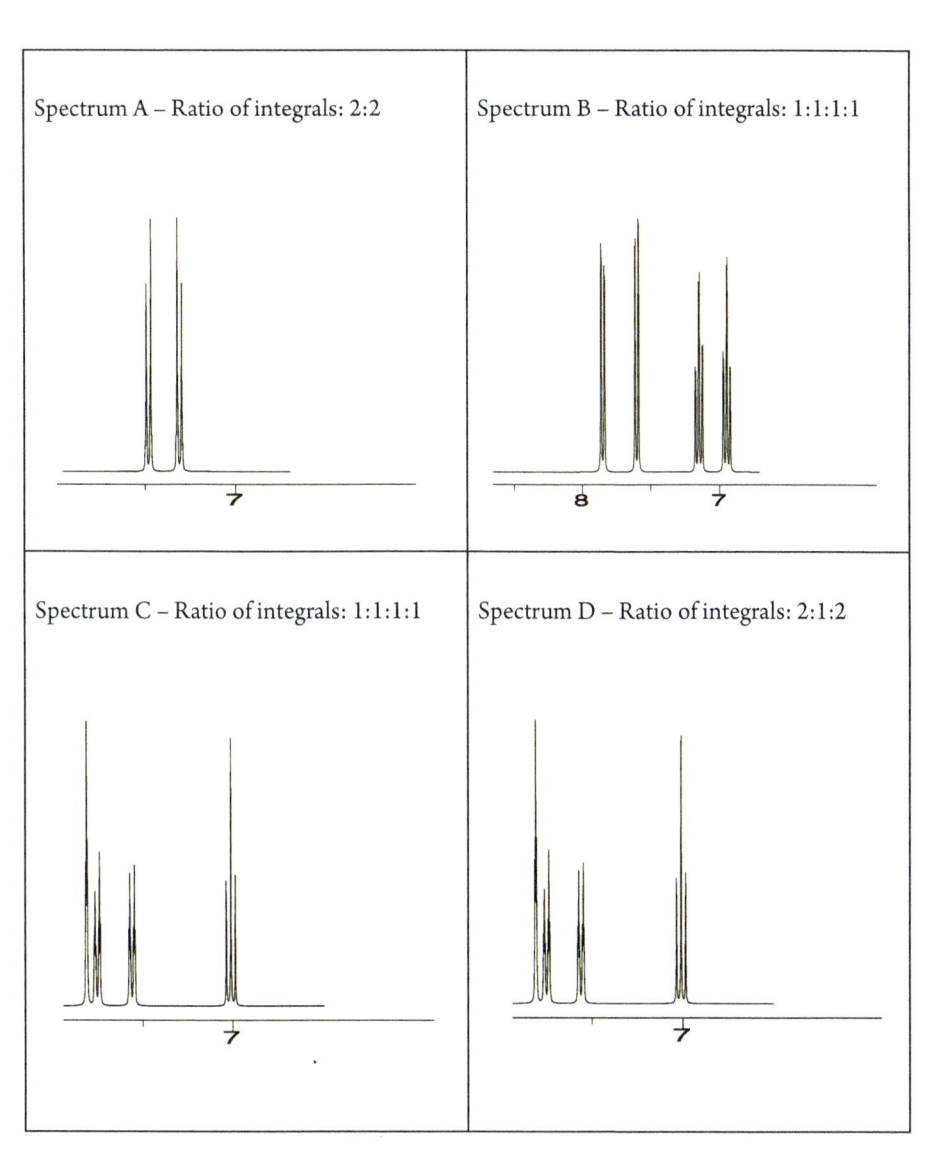

I) 3-bromo-1-iodobenzene   II) iodobenzene   III) 1-bromo-4-chlorobenzene   IV) 1-iodo-2-bromobenzene

Spectrum A – Ratio of integrals: 2:2

Spectrum B – Ratio of integrals: 1:1:1:1

Spectrum C – Ratio of integrals: 1:1:1:1

Spectrum D – Ratio of integrals: 2:1:2

D. Chemical shift

Circulation of electrons around a nucleus creates a local magnetic field that shields the nucleus from the external magnetic field. The extent of shielding depends on the local chemical environment. Thus, NMR signals show a chemical shift. Since the extent of shielding is proportional to the external magnetic field, we use independent units for the chemical shifts δ.

In NMR spectroscopy, the scale is generally displayed from highest to lowest frequency, which is proportional to the chemical shift. The higher electron density around a nucleus leads to a weaker magnetic field (i.e., **shielded**) resulting in a smaller chemical shift. On the other hand, a decrease in the electron density leads to an increased **effective magnetic field** (**deshielded**), with these protons resonating at a higher frequency with a higher chemical shift. The deshielding effect is usually caused by electron-withdrawing functional groups, such as carbonyl, hydroxyl, halogens, and amines. Therefore we can gather information on which protons are in proximity to these functional groups. The designations low field (downfield) and high field (upfield) are historical designations that are commonly used in the scientific literature and represent the high frequency and low frequency regions, respectively. These designations are shown in Figure 9.7.

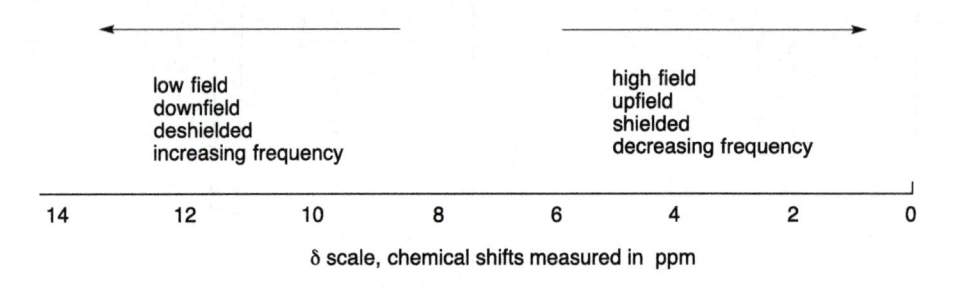

```
←——————————————              ——————————————→
    low field                    high field
    downfield                    upfield
    deshielded                   shielded
    increasing frequency         decreasing frequency
|———————————————————————————————————————————|
14      12      10      8      6      4      2      0
            δ scale, chemical shifts measured in ppm
```

FIGURE 9.7.   Important designations used in NMR.

The chemical shifts of protons on carbon in organic molecules fall in several distinct regions depending on the nature of adjacent carbon atoms and the substituents on those carbons. To a first approximation, protons attached to sp^3 and sp carbons appear at 0-5 ppm, whereas those on sp^2 carbons appear at 5-10 ppm.

TABLE 9.4.   Important chemical shifts.

^{1}H NMR	^{13}C NMR
0 – 2 ppm: Alkane protons R$_3$CH	0 – 70 ppm: Alkyl carbon atoms
2 – 4.5 ppm: Protons near heteroatoms	110 – 140 ppm: Alkenyl/aromatic carbon atoms
4.5 –7 ppm: Alkenyl protons	170 – 210 ppm: Carbonyl groups
6.5 – 8 ppm: Aromatic protons	
9 – 10 ppm: Aldehyde protons	

Chemical shift requires the following general considerations:

1. Hybridization: sp^2 hybridized carbons and protons that bond to them have a higher chemical shift than sp^3 hybridized carbons and protons that bond to these carbons.

Due to the ring current generated within an aromatic ring, aromatic carbons tend to have a higher chemical shift than alkenyl carbon atoms.

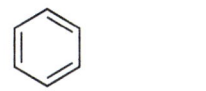

13C NMR: δ = 128 ppm
1H NMR: δ = 7.3 ppm

13C NMR: δ = 27 ppm
1H NMR: δ = 1.4 ppm

2. Substitution: the more substituted, the higher the chemical shift for both the carbon atom and any proton that is bonded to it:

Methyl carbon atom < 1° carbon atom < 2° carbon < 3° carbon atom < 4° carbon atom

increasing chemical shift

3. Electronegativity of the substituent: the higher the electronegativity of a substituent, the higher the chemical shift of the carbon that the substituent is bonded to and any hydrogen that may be bonded to this carbon.

	$(CH_3)_4C$	$(CH_3)_3N$	$(CH_3)_2O$	CH_3F
1H NMR:	δ = 0.9 ppm	δ = 2.1 ppm	δ = 3.2 ppm	δ = 4.1 ppm

Effects may be additive and may also affect the carbon atom and its protons, adjacent to the carbon bonded to the functional group.

Based on the above considerations, the ^{13}C NMR chemical shifts in 2-pentanone and 3-pentanone can be assigned as follows:

	2-pentanone	3-pentanone
13C NMR:	16.6, 45.4, 29.8, 13.6, 207.7	7.9, 35.6, 35.6, 7.9, 210.8

SPECTROSCOPY OF ORGANIC COMPOUNDS   183

Problems

9.5. For each compound, identify the signal of the indicated carbon atom in the ^{13}C NMR spectrum.

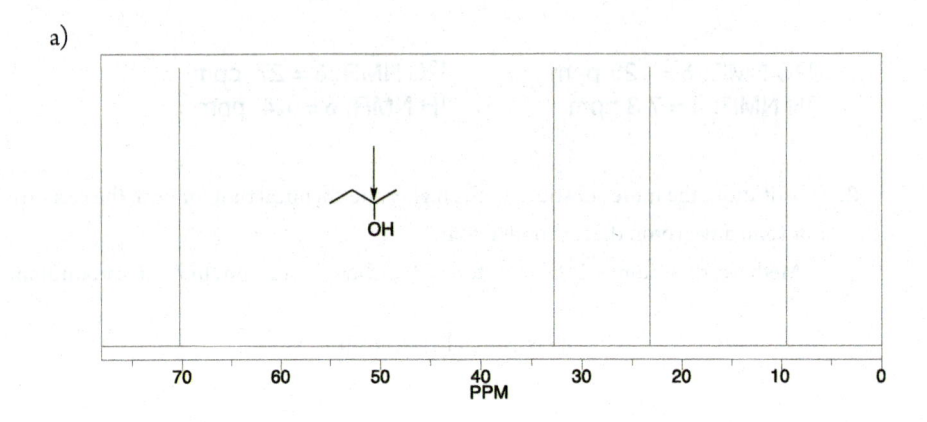

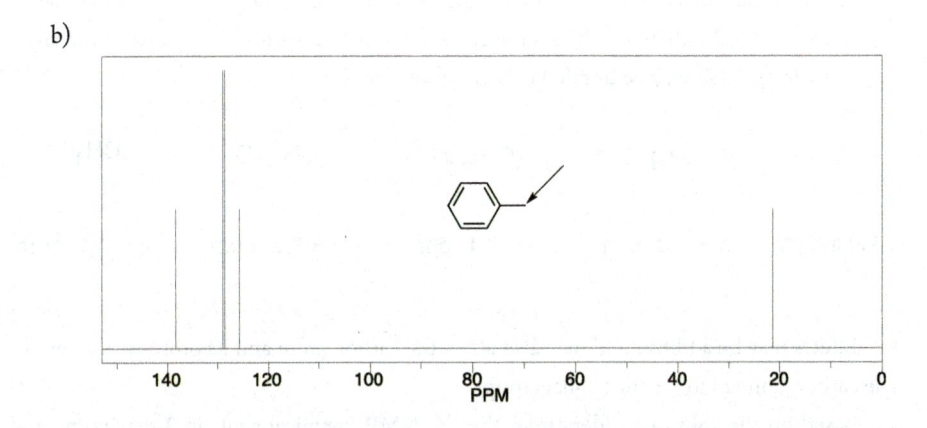

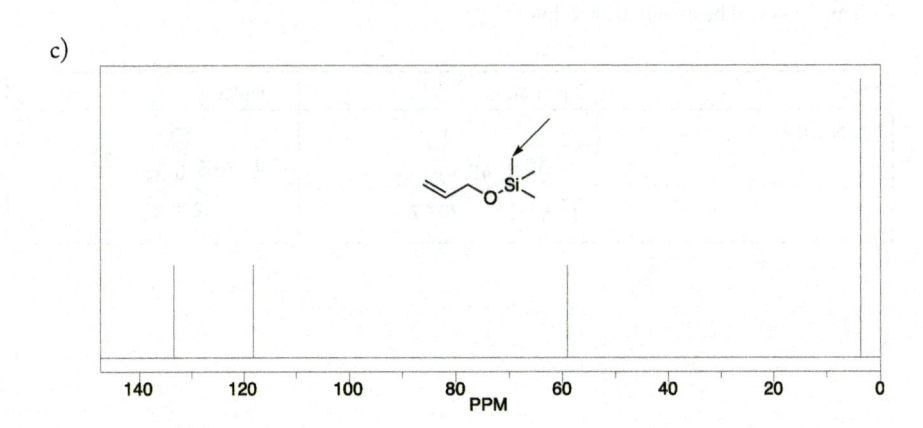

a)

b)

c)

9.6. Provide the structures, based on the molecular formula and the ^{1}H NMR spectrum.

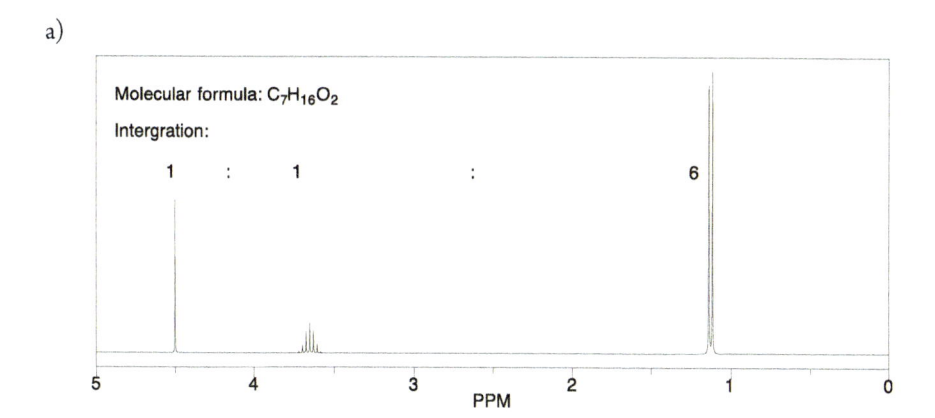

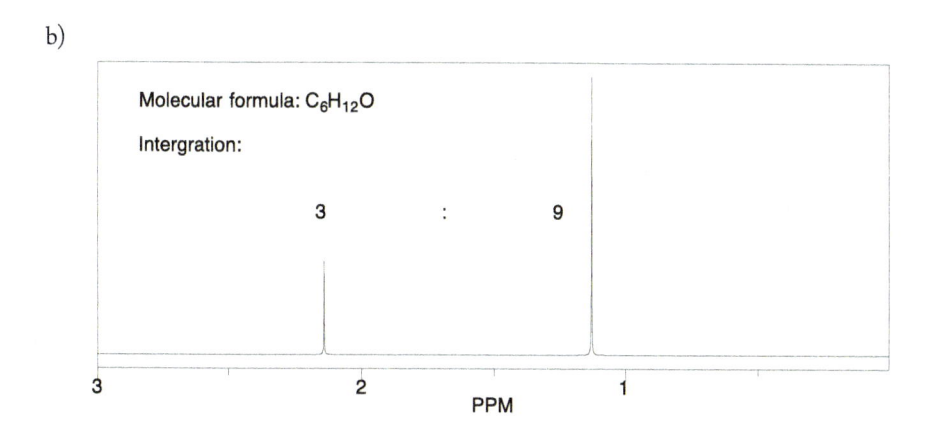

a)

Molecular formula: $C_7H_{16}O_2$

Intergration:

1 : 1 : 6

b)

Molecular formula: $C_6H_{12}O$

Intergration:

3 : 9

9.7. For each compound, identify the corresponding ^{1}H NMR spectrum.

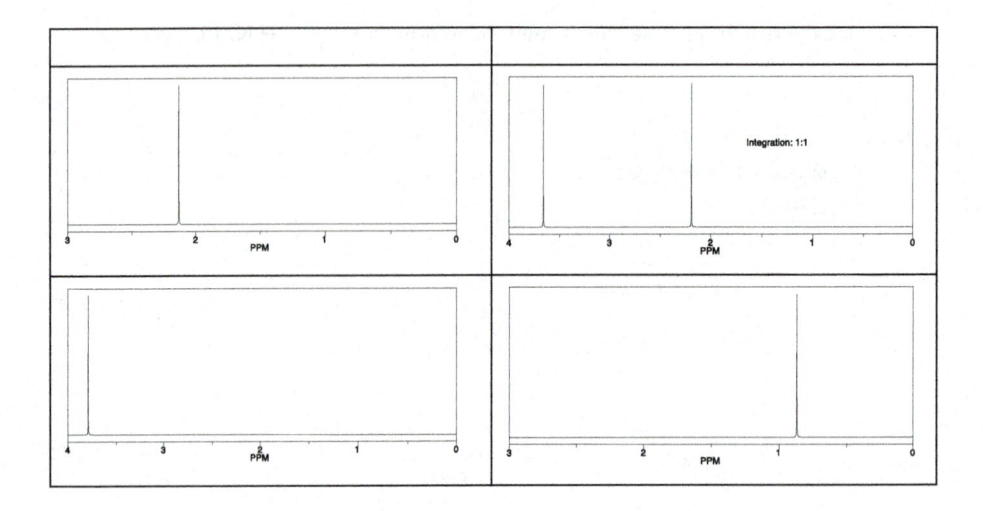

9.3. MASS SPECTROMETRY

9.3.1. The Mass Spectrometer

Mass spectrometry provides information concerning the mass of a molecule. It does not measure mass directly, but instead, it looks at the behavior of an ion derived from a sample of the molecule of interest.

A mass spectrometer converts molecules to ions so that they can be moved and manipulated by external magnetic and electrical fields. A mass spectrometer consists of three components:

1. An ion source: the ionized sample
2. A mass analyzer: sorts and separates ions according to their **mass over charge ratio** (m/z)
3. A detector: measures separated ions and displays the results on a chart

Because ions are very reactive and short-lived, their formation and manipulation must be conducted in a vacuum, which occurs at a very low pressure, within 10^{-5} to 10^{-8} torr. There are different ways of carrying out each of the component tasks listed above. Ionization is commonly achieved using a high energy beam of electrons. The ions are then accelerated and focused in an energy beam, which is bent under the influence of an external magnetic field. The ions are then detected and analyzed electronically.

9.3.2. Mass spectra

The mass spectra of 2-pentanone and 3-pentanone are shown in Figure 9.8. A mass spectrum is usually presented as a vertical bar graph in which each bar represents an ion with a specific mass-to-charge ratio (m/z); the length of the bar indicates the relative abundance of the ion. The mass spectra we are looking at in this course usually have a charge (z) of 1 per ion. Thus, most of the ions formed in a mass spectrometer will have an m/z value equivalent to the mass of the ion itself.

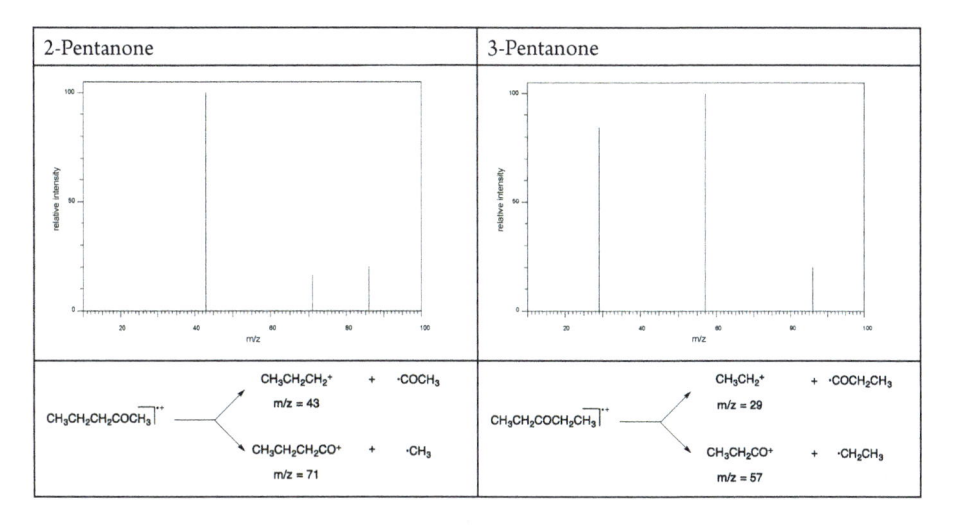

2-Pentanone	3-Pentanone
[mass spectrum]	[mass spectrum]
$CH_3CH_2CH_2COCH_3$ ⁺• → $CH_3CH_2CH_2^+$ + •$COCH_3$ m/z = 43 ↘ $CH_3CH_2CH_2CO^+$ + •CH_3 m/z = 71	$CH_3CH_2COCH_2CH_3$ ⁺• → $CH_3CH_2^+$ + •$COCH_2CH_3$ m/z = 29 ↘ $CH_3CH_2CO^+$ + •CH_2CH_3 m/z = 57

FIGURE 9.8. Mass spectra and fragmentation pattern of 2-pentanone and 3-pentanone.

The most intense ion is assigned an abundance of 100 and it is referred to as the **base peak**. The highest mass ion in a spectrum is normally considered to be the molecular ion and lower mass ions are fragments of the molecular ion. These fragments may provide additional structural information.

The ions present in the mass spectra of 2-pentanone and 3-pentanone can be converted into an atomic composition using the following relative isotopic masses: Hydrogen 1, Carbon 12 and Oxygen 16. Thus the molecular ion m/z = 86 corresponds to an ion with the formula $C_5H_{10}O^+$:

5 carbon atoms – relative isotopic mass 12	$5 \times 12 = 60$
10 hydrogen atoms – relative isotopic mass 1	$10 \times 1 = 10$
1 oxygen atom – relative isotopic mass 16	$1 \times 16 = 16$
Total:	86

We can calculate the formulas of all other ions in the mass spectra of 2-pentanone and 3-pentanone accordingly. Thus, in the mass spectrum of 2-pentanone the base peak (m/z = 43) corresponds to an ion with the formula $C_3H_7^+$, while for 3-pentanone the base peak (m/z = 57) corresponds to an ion with the formula $C_3H_5O^+$. Further, the ion at m/z = 71 in the spectrum of 2-pentanone can be calculated as $C_4H_7O^+$, while the ion at m/z = 29 in the spectrum of 3-pentanone corresponds to $C_2H_5^+$. For both ketones, the formation of these fragment ions can be easily explained by considering a cleavage of the carbon-carbon bond next to the carbonyl group in the molecular ion (Fig 9.7). The molecular ion of each ketone can fragment via two pathways resulting either in an alkyl carbocation or an acyl cation. Since the molecular ion is a radical cation, regardless of the pathway, the side product is always a radical that is not detected by the mass spectrometer. Therefore we can be fairly certain that in addition to an acyl fragment, 3-pentanone contains an ethyl group (m/z = 29), while 2-pentanone contains a propyl fragment (m/z = 43).

Nitrogen has a relative isotopic mass of 14. For example, the mass-to-charge ratio of a molecular ion of methyl amine (CH_3NH_2) is an odd number (m/z = 31). Thus, the presence of an odd number of nitrogen atoms in an organic molecule can be easily detected by mass spectrometry. Such a compound will always have an odd molecular ion.

The presence of chlorine or bromine in an organic molecule is detected by recognizing the intensity ratios of ions differing by 2. For chlorine the intensity ratio is 3:1, while for bromine this ratio is 1:1.

Problems

9.8. What is the structure of this compound?
IR: strong absorption at 3350 cm^{-1}
Mass spectrum: molecular ion m/z = 32.

9.9. For each of the following compounds identify the corresponding mass spectrum:

Mass spectrum A:

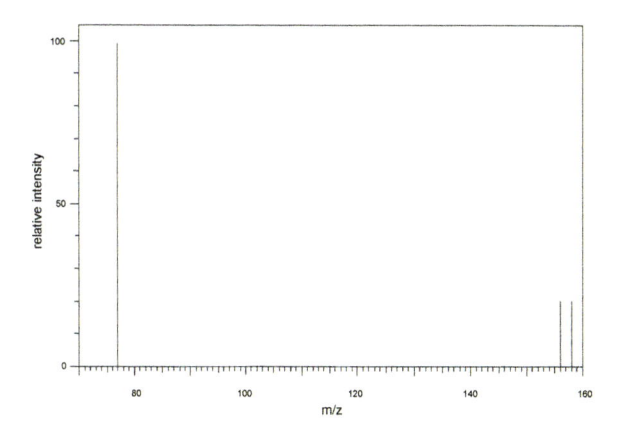

Mass spectrum B:

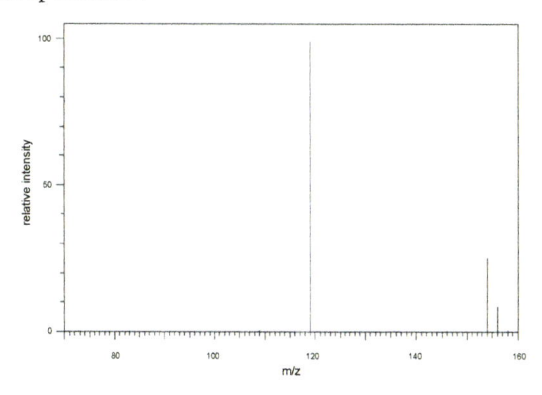

Mass spectrum C:

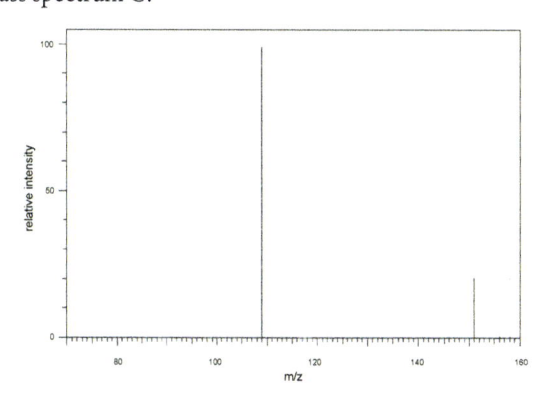

9.10. What are the structures of these two compounds?

The mass spectrum shows a molecular ion with m/z = 88, indicating that these compounds are isomers.

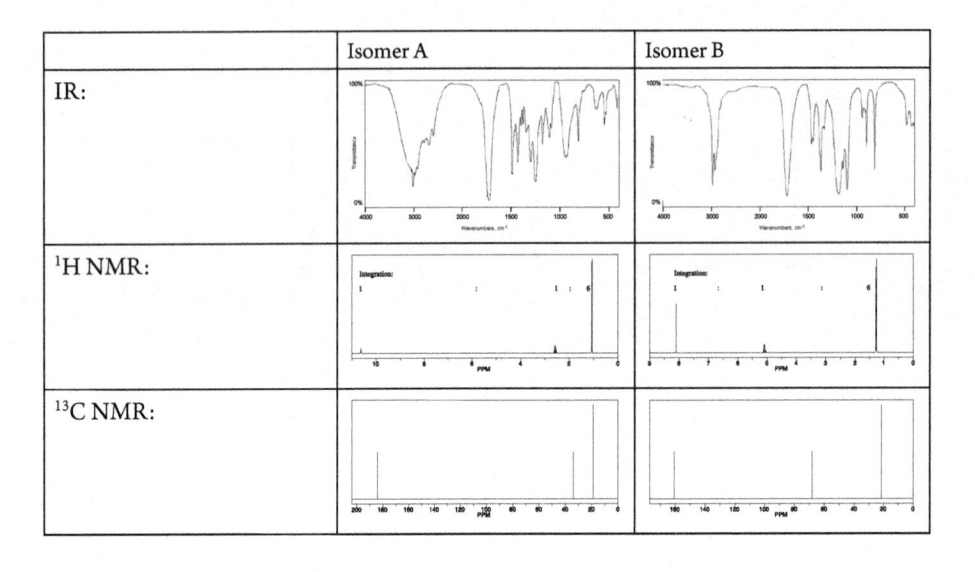

	Isomer A	Isomer B
IR:		
¹H NMR:		
¹³C NMR:		

9.11. Identify these compounds:

a) Mass spectrum - Molecular Ion m/z = 99; IR spectrum: Characteristic bands at 2260 cm^{-1} and 1754 cm^{-1}; ¹³C NMR – 4 signals; ¹H NMR: all signals are singlets.

A) $\diagup\!\!\diagdown\!\!\diagup$OH   B) H$_3$CO−C(=O)−CH$_2$−C≡CH   C) CH$_3$−C(=O)−CH$_2$−OH   D) H$_3$CO−C(=O)−CH$_2$−CN

b) The mass spectrum shows 4 major peaks: m/z = 159, m/z = 157, m/z = 78 and m/z = 51.

A) Cl−CN   B) CH$_3$Br   C) chlorobenzene   D) 4-bromopyridine

c) The ratio of integrals in the ¹H NMR spectrum is 2:2:3.

A) CH$_3$−C(=O)−O−CH$_2$CH$_2$CH$_3$ B) H−C(=O)−O−CH$_2$CH$_2$CH$_3$ C) (CH$_3$)$_2$CH−C(=O)−CH$_3$ D) Cl−C(=O)−O−CH$_2$CH$_2$CH$_3$

190   ORGANIC CHEMISTRY

d) The structure contains diastereotopic protons.

A) B) C) D)

e) The ¹H NMR spectrum shows a septet, and other signals.

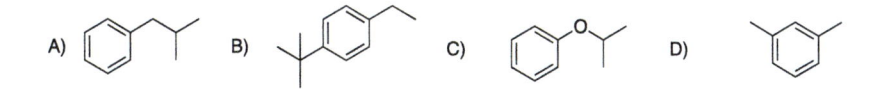

A) B) C) D)

CHAPTER 10

Organometallic Compounds and Transition Metal Catalysis

Organometallic compounds are defined as organic compound molecules with a carbon-metal bond. This is a really broad area of chemistry since metals (i.e., main group metals and transition metals) predominate in the periodic table. All metals are less electronegative than carbon, and therefore organometallic molecules are characterized by the presence of a partial or full negative charge on the carbon atom, which generally acts as a base or a nucleophile in organic reactions. The structural features of organometallic compounds are discussed in inorganic chemistry courses, while organic chemistry deals mainly with the applications of these compounds as reagents in organic reactions. Organometallic compounds are widely used as reagents and catalysts in research laboratories and in industry for making pharmaceuticals, polymers, and many other important products. Because of the importance and utility of organometallic compounds, research in the area has been awarded numerous Nobel Prizes, including the recent awards in 2001 (William S. Knowles, Ryoji Noyori, and K. Barry Sharpless), 2005 (Yves Chauvin, Robert H. Grubbs and Richard R. Schrock), and 2010 (Richard F. Heck, Ei-ichi Negishi, and Akira Suzuki). This chapter provides an overview of essential reactions involving organometallic compounds as reagents or catalysts.

10.1. ORGANOLITHIUM AND ORGANOMAGNESIUM COMPOUNDS

Organolithium (RLi, alkyllithium) and organomagnesium compounds (RMgX, alkylmagnesium halides, X = Cl, Br, or I) are prepared by reacting the corresponding active metal (Li or Mg) with alkyl halide (RX) in an appropriate organic solvent, such as ether or pentane (Figure 10.1). These reactions result in the transfer of electrons from the metal to the more electronegative carbon, and therefore can be classified as oxidation of metal and reduction of carbon (see Section 6.5). These reactions are commonly referred to as **oxidative addition** of alkyl halides to a metal.

RCl + 2Li $\xrightarrow{\text{pentane}}$ RLi + LiCl
alkyllithium

RBr + Mg $\xrightarrow{\text{diethyl ether}}$ RMgBr
alkylmagnesium bromides
(Grignard reagents)

Examples of organometallic compounds:

CH_3Li (methyllithium)

$CH_3CH_2CH_2CH_2Li$ (butyllithium)

PhLi (phenyllithium)

CH_3CH_2MgBr (ethylmagnesium bromide)

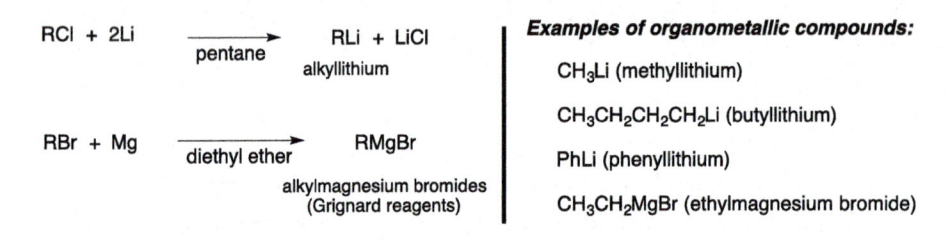

FIGURE 10.1. Preparation of organolithium and organomagnesium compounds.

Alkyllithium and alkylmagnesium reagents are extremely reactive and potentially dangerous compounds that react violently with water and can ignite spontaneously in air. These compounds are always prepared and handled as solutions in organic solvents under the atmosphere of inert gas (nitrogen or argon). Alkylmagnesium bromides are also known as **Grignard reagents**, named after the French chemist Victor Grignard, who was awarded Nobel Prize in 1912 for discovering these compounds.

The carbon-metal bonds in RLi and RMgBr are generally classified as polar covalent bonds (Section 1.1) with partial negative charge on the carbon and partial positive charge on the metal. However, because of the large difference in electronegativity between carbon (electronegativity value 2.5) and lithium (1.0) or magnesium (1.2), this bond is highly polarized and has a significant ionic character. In chemical reactions, alkyllithium and alkylmagnesium reagents act as the source of **carbanions**, the molecular ions bearing a formal negative charge on carbon atom (Figure 10.2). Carbanions are important reactive species that can function as strong bases and nucleophiles. Since carbanions are conjugate bases of hydrocarbons (pK_a = 50–60), they can accept protons from any compounds with pK_a values below 50 (see Table 2.1 and the related text in Chapter 2). Figure 10.2. shows several examples of the essentially irreversible proton transfer reactions involving alkyllithium and alkylmagnesium compounds.

Alkyllithium and alkylmagnesium compounds can also act as nucleophiles in reactions with some electrophilic organic substrates such as epoxides (Section 6.5.2), carbonyl

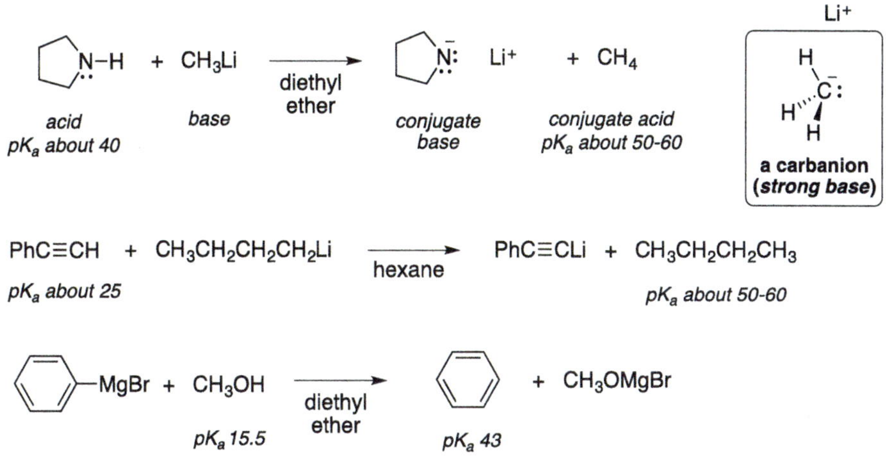

FIGURE 10.2. Examples of proton transfer reactions involving alkyllithium and alkylmagnesium compounds.

compounds (Chapter 11), and derivatives of carboxylic acids (Chapter 12). However, nucleophilic substitution reactions with alkyl halides (Chapter 5) are not common for RLi or RMgBr. The reactions of these reagents with alkyl halides usually produce alkenes by a β-elimination mechanism because of the extremely strong basicity of the carbanionic species.

Problems

10.1. Sort the following according to increasing basicity:

(isopropyllithium)   (tert-butyllithium)   CH₃Li   (ethyllithium)

10.2. COMPOUNDS OF TRANSITION METALS

Transition metals occupy groups three through twelve of the periodic table. The bonds in compounds and complexes of transition metals may involve electrons occupying d and f orbitals, which makes their chemistry much more complicated compared to carbon and other elements of the second period. The electronegativity values of transition metals are significantly higher those that of lithium and magnesium (for example, the electronegativity of copper is 1.9 and palladium is 2.2), and because of lower carbon-metal bond polarity, organic compounds of transition metals are not characterized by high basicity. In chemical reactions, organic compounds of transition metals act as effective carbon nucleophiles, especially useful in nucleophilic substitution reactions with alkyl halides, resulting in the formation of new carbon-carbon bonds (**coupling reactions**).

10.2.1. Organocopper compounds

Lithium diorganocuprates (R_2CuLi), which are also known as **Gilman reagents**, belong to a particularly important class of organocopper compounds. Lithium diorganocuprates are produced by the reaction of two molar equivalents of organolithium compound (RLi) with one equivalent of copper(I) salt (CuI or CuBr) in ether solution (Figure 10.3).

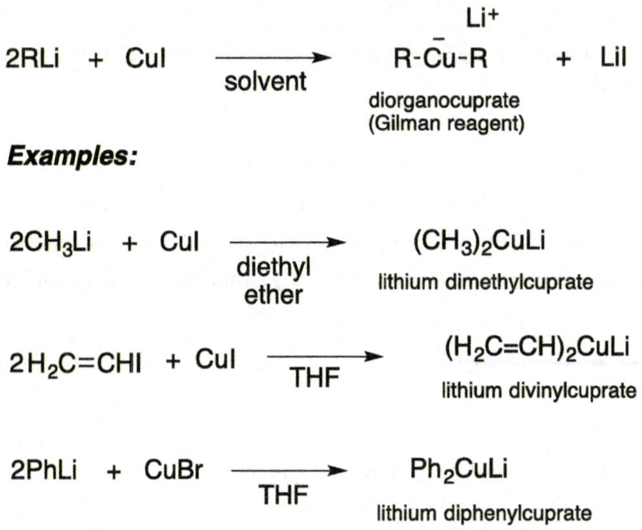

$$2RLi + CuI \xrightarrow{solvent} [R-Cu-R]^- Li^+ + LiI$$

diorganocuprate (Gilman reagent)

Examples:

$$2CH_3Li + CuI \xrightarrow{diethyl\ ether} (CH_3)_2CuLi$$
lithium dimethylcuprate

$$2H_2C=CHI + CuI \xrightarrow{THF} (H_2C=CH)_2CuLi$$
lithium divinylcuprate

$$2PhLi + CuBr \xrightarrow{THF} Ph_2CuLi$$
lithium diphenylcuprate

FIGURE 10.3. Preparation of lithium diorganocuprates (Gilman reagents).

Gilman reagents serve as effective sources of carbon nucleophiles ($R:^-$) in coupling reactions with organic halides. In contrast to alkyllithium and alkylmagnesium reagents, organocopper reagents are not strong bases and therefore, their nucleophilic substitution reactions with alkyl halides do not produce alkenes as by-products of β-elimination process. Gilman reagents also react as nucleophiles in the nucleophilic ring-opening reactions of epoxides (Section 6.5.2). Several examples of lithium diorganocuprate reactions with organic substrates are shown in Figure 10.4. Note that these reactions use only one of the two R groups in R_2CuLi as a nucleophile.

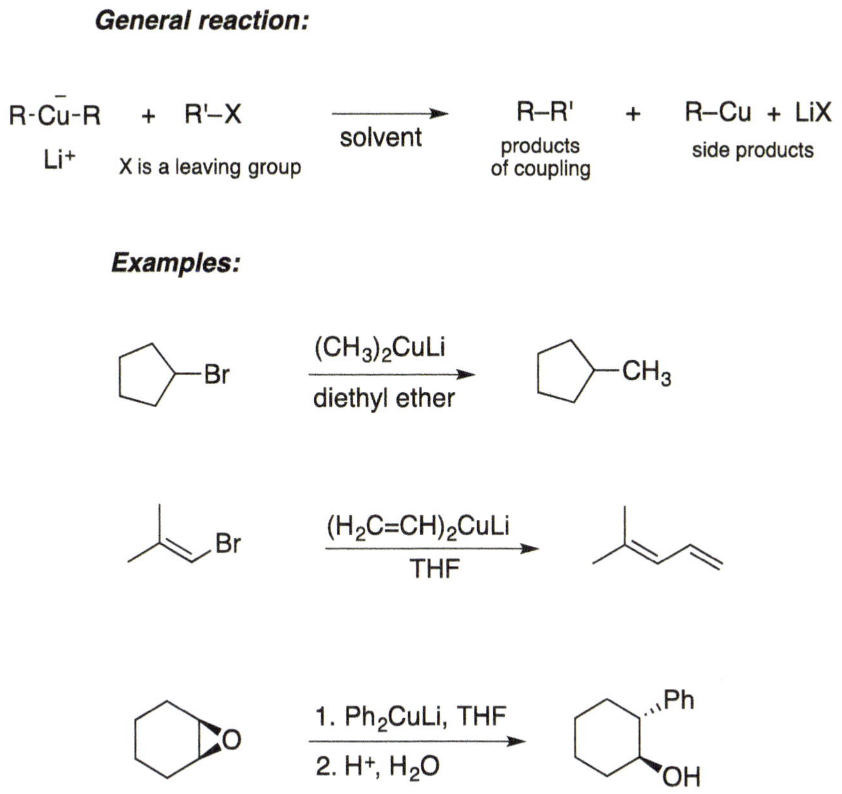

General reaction:

R-Cu-R + R'-X $\xrightarrow{\text{solvent}}$ R-R' + R-Cu + LiX
Li⁺   X is a leaving group   products   side products
                                            of coupling

Examples:

cyclopentyl-Br $\xrightarrow[\text{diethyl ether}]{(CH_3)_2CuLi}$ cyclopentyl-CH₃

(CH₃)₂C=CH-Br $\xrightarrow{(H_2C=CH)_2CuLi}_{THF}$ (CH₃)₂C=CH-CH=CH₂

cyclohexene oxide $\xrightarrow[\text{2. H⁺, H}_2\text{O}]{\text{1. Ph}_2\text{CuLi, THF}}$ trans-2-phenylcyclohexanol (Ph, OH)

FIGURE 10.4. Reactions of lithium diorganocuprates as nucleophiles.

Problems

10.2. Provide the Gilman reagent that gives the indicated products in high yield:

a) cyclohexyl-I → ? → cyclohexyl-CH(CH₃)₂ (isobutyl cyclohexane)

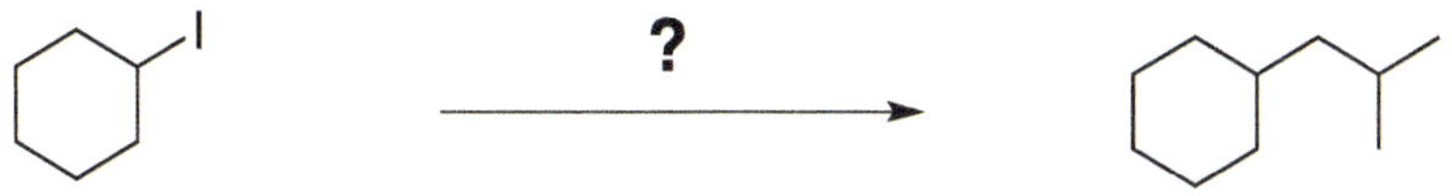

b) Ph-Br → ? → Ph-CH=CH₂ (styrene)

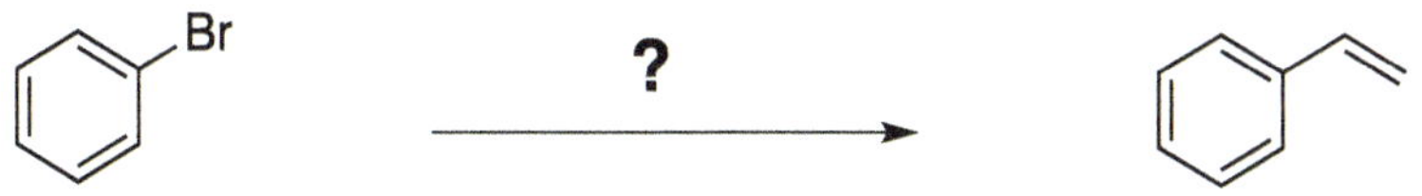

10.2.2. Palladium-catalyzed coupling reactions

In contrast to organocopper compounds, which are required in stoichiometric amounts, palladium compounds can promote coupling reactions when used in catalytic amounts. Mechanisms of these catalytic reactions are very complex and involve oxidative addition of an organic substrate to palladium, followed by exchange of ligands (ligands are the substituents on metal center), and finally, elimination (also called **reductive elimination**) of the product of ligand coupling. Palladium-catalyzed coupling reactions (also called **cross-coupling**) have attracted a significant amount of interest in recent years, and research on this topic was awarded the Nobel Prize in Chemistry in 2010.

Numerous variations of palladium-catalyzed couplings have been developed in the last 30-40 years. These reactions are usually named after the chemists who initially reported them in the scientific literature. All palladium-catalyzed coupling reactions involve a source of electrophilic carbon (an organic halide), a source of nucleophilic carbon (e.g., organoborane, alkene, alkyne, organotin, organosilicon, organomagnesium compounds, etc.), a palladium catalyst (e.g., palladium acetate, tetrakis (triphenylphosphine) palladium, and other palladium complexes), and some other additives, such as organic or inorganic bases and solvents. Figure 10.5 shows several examples of important coupling reactions.

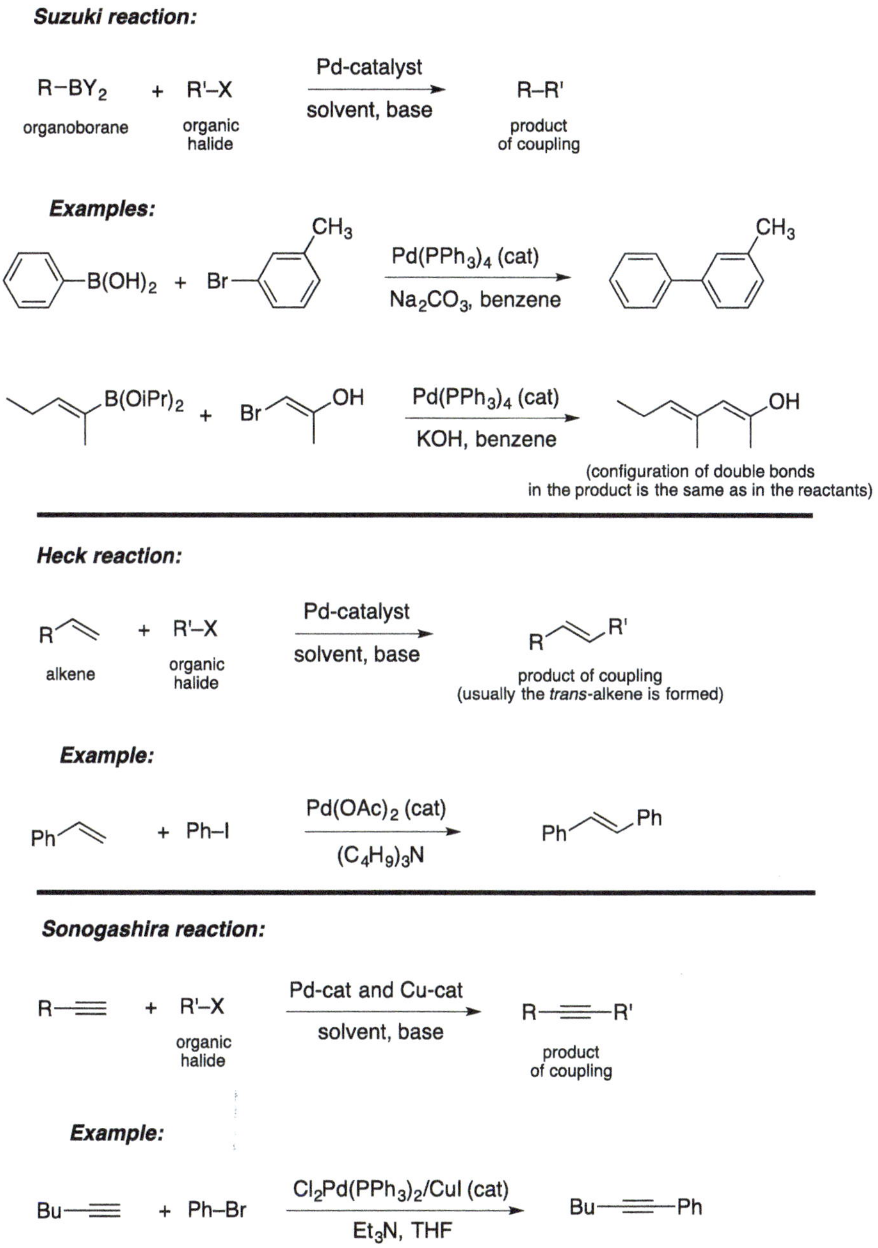

FIGURE 10.5. Palladium-catalyzed coupling reactions.

Problems

10.3. Provide the products of the following transformations:

a)

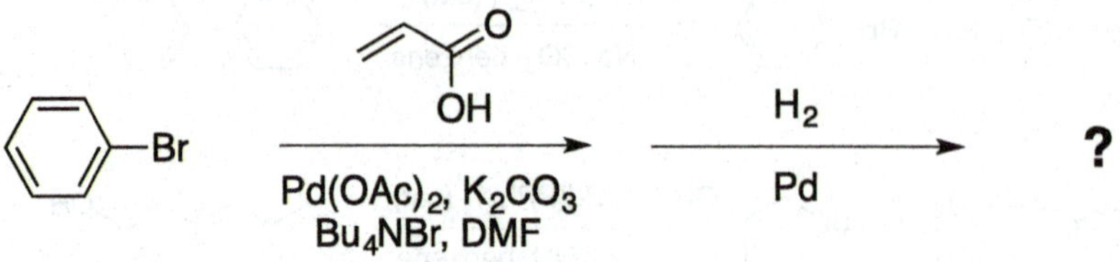

PhBr + CH₂=CH-COOH →[Pd(OAc)₂, K₂CO₃, Bu₄NBr, DMF] →[H₂, Pd] ?

b)

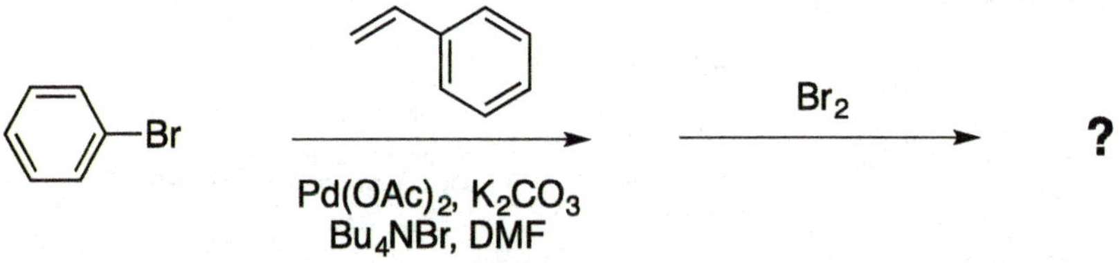

PhBr + CH₂=CH-Ph →[Pd(OAc)₂, K₂CO₃, Bu₄NBr, DMF] →[Br₂] ?

c)

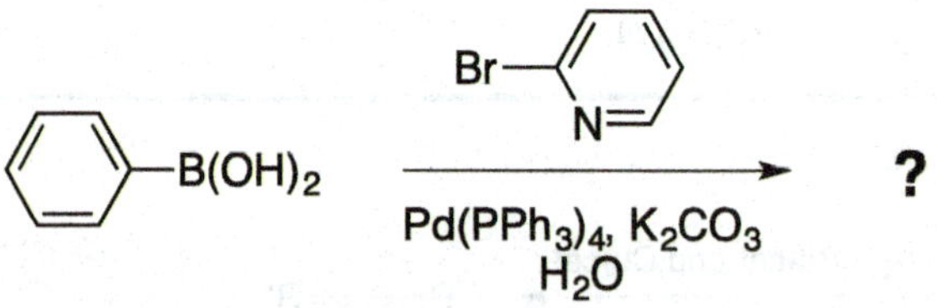

PhB(OH)₂ + 2-bromopyridine →[Pd(PPh₃)₄, K₂CO₃, H₂O] ?

10.2.3. Carbene intermediates and the Simmons-Smith reaction

Carbenes (R_2C:) are important intermediates in organic reactions, and consist of an uncharged carbon atom with two substituents and an unshared electron pair. The carbon in carbene has only six valence electrons, which makes it unstable and highly reactive. The electronic structure of carbene is best explained by sp^2 hybridization of the carbon atom, with a vacant, unhybridized p orbital and two bonding electron pairs, and one unshared electron pair occupying the sp^2 orbital (Figure 10.6).

The simplest carbene, methylene (H$_2$C:), can be generated by decomposition of diazomethane (CH$_2$N$_2$) in the presence of heat or light. In chemical reactions, carbene can act as an electrophile (i.e., accepting electrons to the vacant p orbital), and also as a nucleophile (i.e., donating the unshared electronic pair). The most typical reaction of carbenes is the addition to alkenes, yielding cyclopropanes (Figure 10.6).

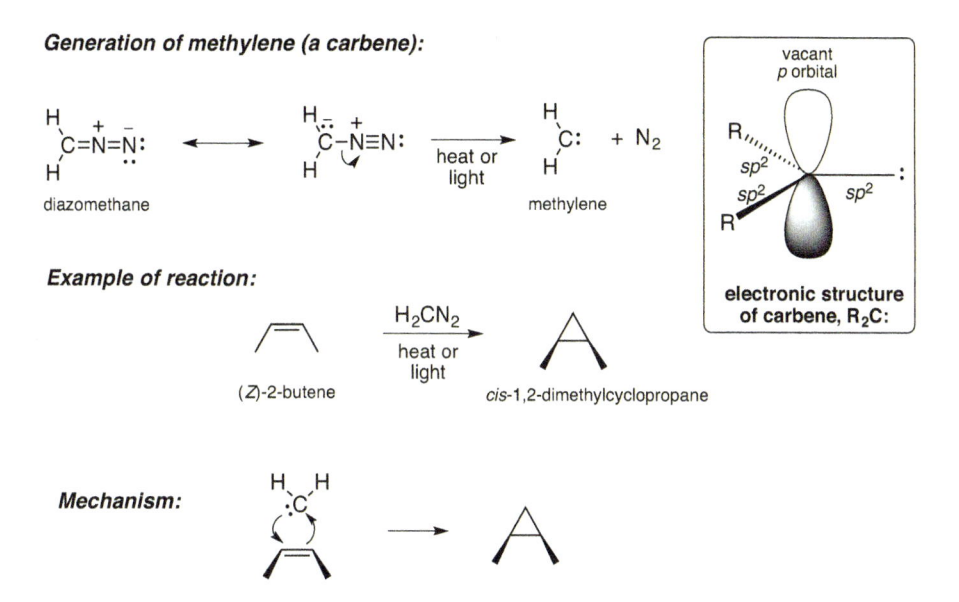

FIGURE 10.6. Generation of methylene and its reaction with alkenes.

The preparation of cyclopropanes using the toxic and explosive diazomethane gas is a dangerous and inconvenient procedure. A much more practical approach to the synthesis of cyclopropanes involves the reaction of alkenes with a diiodomethane and zinc-copper couple, Zn(Cu), which is known as the **Simmons-Smith reaction**. The Simmons-Smith reaction involves an intermediate organozinc compound (ICH$_2$ZnI), which acts as an immediate precursor to carbene (known as a **carbenoid**). This is a stereoselective *syn*-addition and the relative configuration of substituents in the alkene is retained in the cyclopropane product (Figure 10.7).

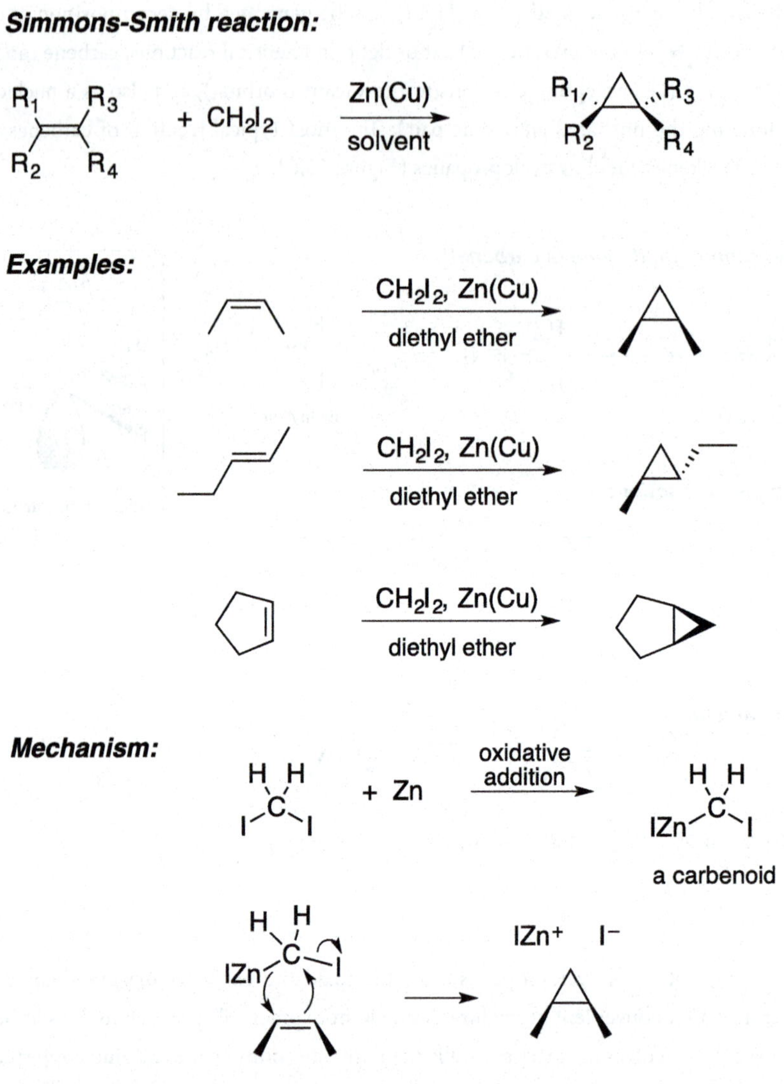

FIGURE 10.7. Simmons-Smith reaction.

Different representatives of carbene intermediates, such as dichlorocarbene (Cl$_2$C:) or dibromocarbene (Br$_2$C:), can be generated by the reaction of chloroform (CHCl$_3$) or bromoform (CHBr$_3$) with a strong base, such as *t*-BuOK. Analogous to methylene (H$_2$C:), dichlorocarbene or dibromocarbene can react stereoselectively with alkenes, yielding the corresponding 1,1-dichlorocyclopropanes or 1,1-dibromocyclopropanes.

202   ORGANIC CHEMISTRY

Problems

10.4. Propose a mechanism for the formation of dichlorocarbene (Cl$_2$C:) by the reaction of chloroform (CHCl$_3$) with *t*-BuOK.

10.5. How could you synthesize the following molecules, using an alkene and the Simmons-Smith reagent?

a)

b)

10.2.4. Alkene metathesis

Alkene metathesis is an organic reaction between two molecules of alkene, resulting in exchange of the carbons by breaking and rebuilding the carbon-carbon double bonds (Figure 10.8). This reaction is catalyzed by metal complexes named **Grubbs' catalysts** (complexes of ruthenium) and **Schrock catalysts** (complexes of molybdenum and tungsten). Three scientists, Yves Chauvin, Robert H. Grubbs, and Richard R. Schrock, were collectively awarded the 2005 Nobel Prize in Chemistry for the discovery and study of alkene metathesis. This reaction has a complex mechanism involving coordination of alkene molecules on the metal, followed by redistribution of bonds and elimination of final products. The catalysts and intermediates in the metathesis reaction are called **metallocarbenes**, which are the complexes of transition metals with a metal-carbon double bond such as Ru=CHR.

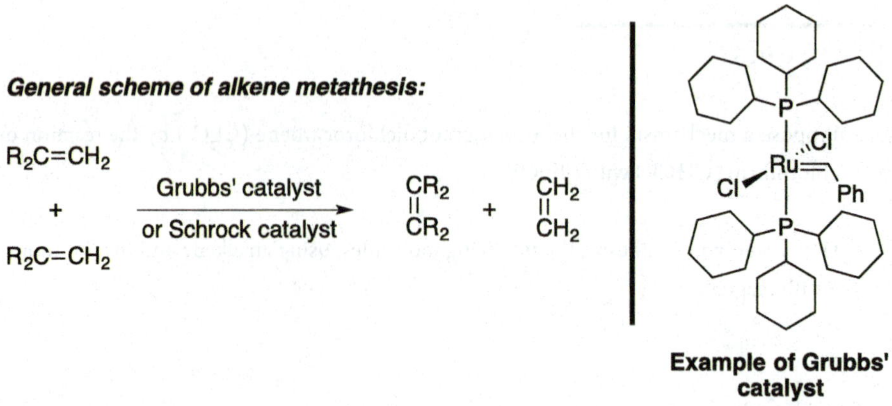

General scheme of alkene metathesis:

R₂C=CH₂
  +   →(Grubbs' catalyst or Schrock catalyst)→   CR₂=CR₂ + CH₂=CH₂
R₂C=CH₂

Example of Grubbs' catalyst

FIGURE 10.8. Alkene metathesis.

Several examples of metathesis reactions are shown in Figure 10.9. Many of these reactions occur in the synthesis of complex organic molecules in industry. A particularly important industrial application is the ring-opening metathesis polymerization (ROMP), during the course of which a cycloalkene is converted to a corresponding unsaturated polymeric chain (Figure 10.9).

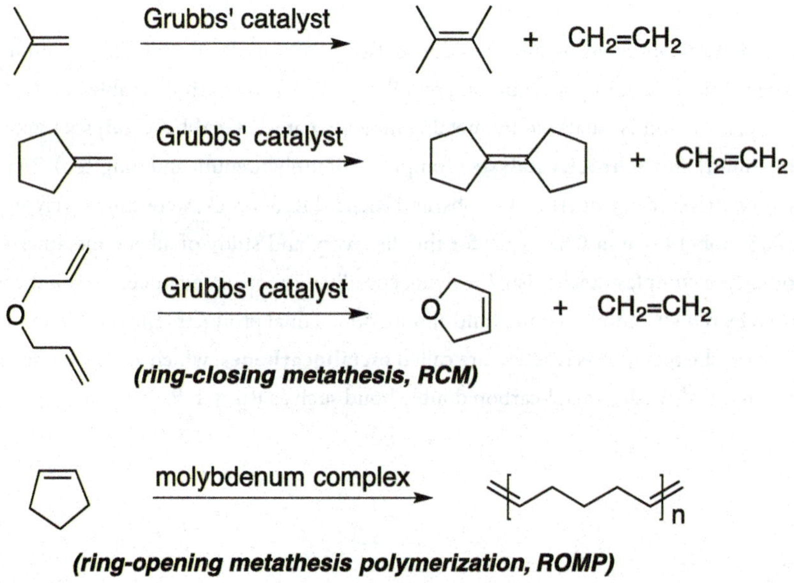

(ring-closing metathesis, RCM)

(ring-opening metathesis polymerization, ROMP)

FIGURE 10.9. Examples of metathesis reactions.

204   ORGANIC CHEMISTRY

Problems

10.6. What are the products of the following transformations?

a)

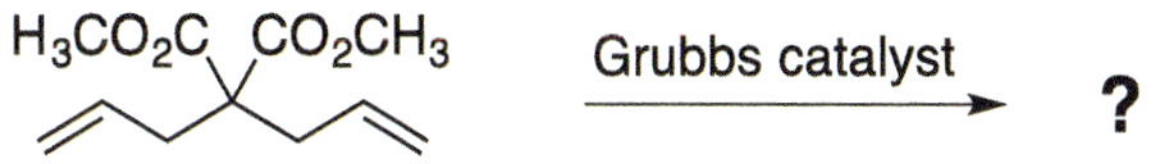

H_3CO_2C CO_2CH_3 on a di-allyl carbon → Grubbs catalyst → ?

b)

2 MeO_2C—(CH$_2$)$_8$—CH=CH$_2$ → Grubbs catalyst → ?

c)

methyl vinyl ketone + CH$_2$=CH–Bu → Grubbs catalyst → ?

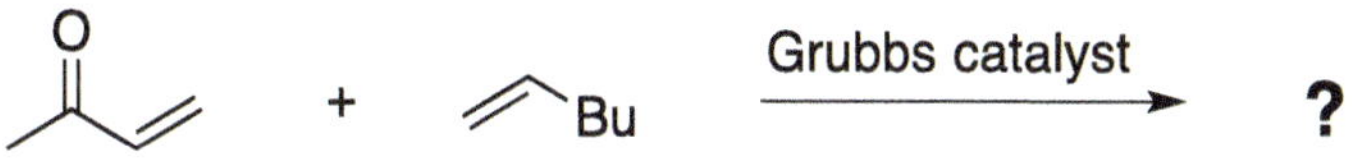

CHAPTER 11

Aldehydes and Ketones

Aldehydes and **ketones** are generally defined as compounds with a **carbonyl** functional group (C=O) bonded to hydrogen or carbon atoms (Section 1.3). Aldehydes (RHC=O) have at least one hydrogen bonded to the carbon of the carbonyl group, while ketones (R$_2$C=O) have two alkyl substituents. Aldehydes and ketones are very common in nature and play important role in our lives. In particular, the carbonyl group is present in many multifunctional biomolecules, such as carbohydrates. The simplest aldehyde, formaldehyde (H$_2$C=O), is an industrially important compound used as a starting material in the production of various resins and plastics. A saturated aqueous solution of formaldehyde (about 40% by volume) is known as formalin, which is used as a preservative for biological and medical specimens. The simplest ketone, acetone (Me$_2$C=O), is an industrial product and a common solvent with many practical uses. Aldehydes and ketones serve as precursors in the synthesis of various important organic molecules. This chapter discusses nomenclature, structural features, properties, and reactions of aldehydes and ketones.

11.1. NOMENCLATURE OF ALDEHYDES AND KETONES

According to the general rules of IUPAC nomenclature, the presence of the aldehyde or ketone functional groups in a molecule is indicated by the respective suffixes *-al* or *-one* for all compounds, when the carbonyl group is the functional group with the highest priority (see Table 1.4 and the related text in Chapter 1). In compounds where a higher priority group is present, the prefix *oxo-* is placed at the beginning of the name to indicate the position of the carbonyl group. The carbons of the parent chain should be numbered so that the carbonyl group has the lowest possible number, unless a higher priority group is present. For aldehydes, this is always carbon number 1 and there is no need to show this number in the name. In cyclic ketones, the carbon bearing carbonyl group is counted as number 1. The suffix *-carbaldehyde* is used to indicate the aldehyde group (–CHO) directly bonded as a substituent in cycloalkyl ring. Compounds with two aldehyde or two ketone groups are called *dials* or *diones* respectively. Several examples of aldehyde and ketone IUPAC names are shown in Figure 11.1.

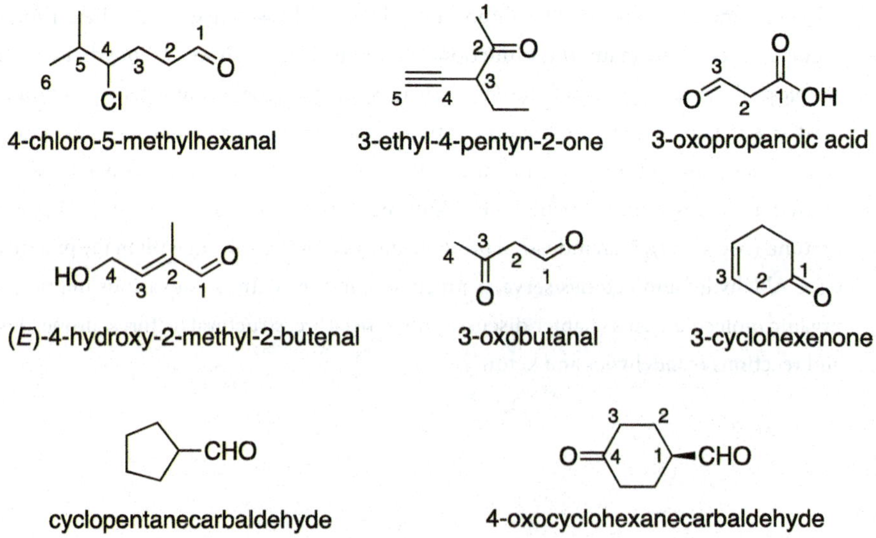

4-chloro-5-methylhexanal   3-ethyl-4-pentyn-2-one   3-oxopropanoic acid

(*E*)-4-hydroxy-2-methyl-2-butenal   3-oxobutanal   3-cyclohexenone

cyclopentanecarbaldehyde   4-oxocyclohexanecarbaldehyde

FIGURE 11.1. IUPAC nomenclature of aldehydes and ketones.

Important common names of aldehydes and ketones include the following: formaldehyde ($H_2C=O$), acetaldehyde (MeCHO), benzaldehyde (PhCHO), glyoxal (OHCCHO), acetone ($Me_2C=O$), benzophenone ($Ph_2C=O$), and acetophenone (PhMeC=O).

Problems

10.1 Provide the IUPAC names for the following compounds:

11.2. GENERAL CHARACTERISTICS OF ALDEHYDES AND KETONES

The carbonyl group is composed of sp² hybridized carbon and oxygen atoms connected by one σ bond and one π bond. The carbonyl carbon atom has a trigonal planar geometry (Section 1.4). Because of a significant difference in the electronegativity values of carbon (2.5) and oxygen (3.5), the carbonyl group is highly polar, which can be illustrated by the presence of a minor resonance contributor with positive charge on carbon and negative charge on oxygen (Figure 11.2). The presence of this minor resonance contributor results in high reactivity of carbonyl carbon toward the addition of nucleophilic reagents and basicity of carbonyl oxygen. The most typical reactions of aldehydes and ketones involve **nucleophilic addition** to the carbonyl carbon as the key step in the reaction mechanism. Strong nucleophiles such as organolithium and organomagnesium compounds can add directly to the carbonyl group, yielding alkoxide anions, which can be further converted to alcohols by treatment with water or an acid. Reactions of weak nucleophiles require the presence of a strong acid for the initial protonation of carbonyl oxygen, resulting in the formation of the highly electrophilic protonated carbonyl species (Figure 11.2).

Because of the polar character of carbonyl group, aldehydes and ketones in liquid state are characterized by dipole-dipole attraction between molecules (Section 1.5). Aldehydes and ketones have higher boiling points compared to nonpolar molecules of comparable mass. However, alcohols have higher boiling points than carbonyl compounds because of hydrogen bonding between molecules. For example, acetaldehyde (CH_3CHO) has a bp = +20° C, which is much higher compared to propane $CH_3CH_2CH_3$ (bp = −42 °C), but lower than ethanol CH_3CH_2OH (bp = +78 °C). Simple aldehydes and ketones are soluble in water because they can interact with molecules of H_2O as acceptors of hydrogen bonds.

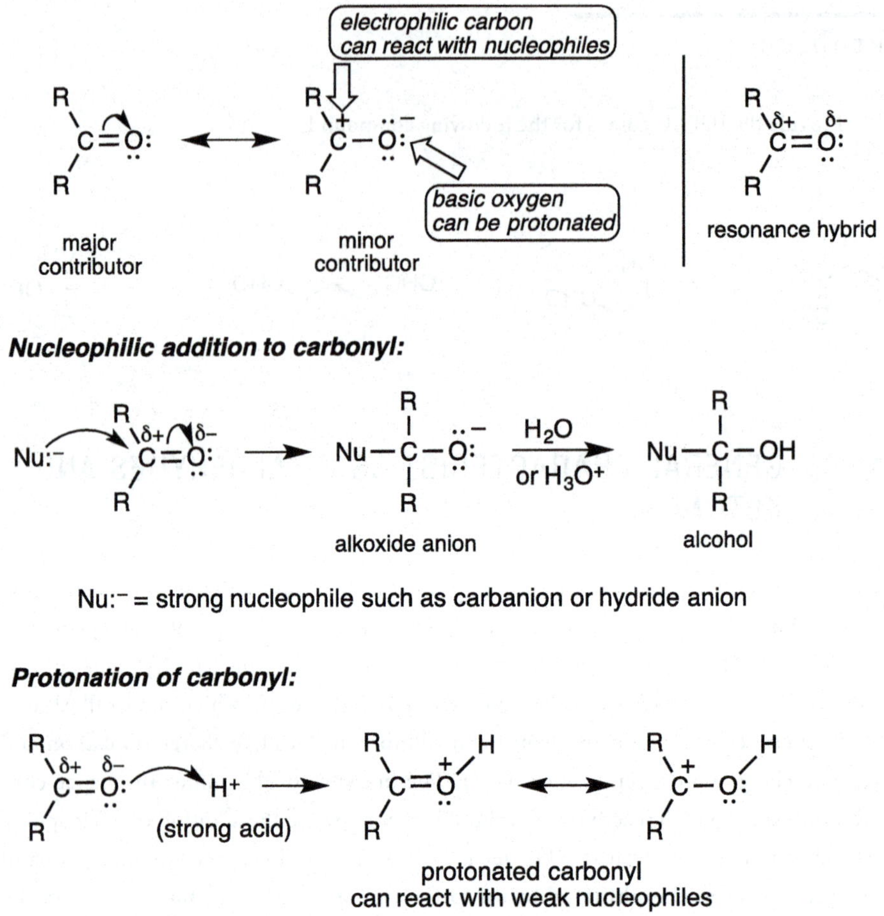

FIGURE 11.2. General reactivity of carbonyl groups.

The preparation of aldehydes and ketones by oxidation of alcohols (Section 8.7), acid-catalyzed hydration of alkynes (Section 7.4.3), and ozonolysis of alkenes (Section 6.5.3) has been discussed in previous chapters.

Problems

11.2. Why are ketones, in general, less reactive towards nucleophilic addition to the carbonyl group than aldehydes?

11.3. For each of the following provide the IUPAC name of the major product:

a) 1-methylcyclohexene + 1. O₃, 2. (CH₃)₂S → ?

b) pent-1-yne + H₂SO₄, H₂O, HgSO₄ (catalyst) → ?

c) cyclohexylacetylene + 1. BH₃, THF; 2. H₂O₂, NaOH, H₂O → ?

d) 3-hydroxy-3-methylbutan-1-ol (HO-C(CH₃)₂-CH₂-CH₂-OH) + IBX, DMSO (solvent) → ?

11.3. REACTIONS OF ALDEHYDES AND KETONES WITH CARBON NUCLEOPHILES

Typical carbon nucleophiles include carbanions (Section 10.1), alkynyl anions (Section 7.3), and cyanide anions (NC⁻, Section 5.1). These anions are strong nucleophiles (Section 5.1), which can add directly to the carbonyl group of aldehydes and ketones to yield alkoxide anions according to the mechanism shown in Figure 11.2. The actual reagents, sources of these anions, are the corresponding metal derivatives such as alkyllithium and alkylmagnesium compounds, sodium or lithium acetylides, and sodium or potassium cyanide. Reactions of the strongly basic RLi, RMgBr, and metal acetylides with aldehydes or ketones must be performed in an aprotic organic solvent, such as ether, followed by careful addition of an aqueous acid, usually a diluted HCl solution, in order to protonate the initially formed alkoxide intermediate (Figure 11.2). Cyanide anions are weaker bases that cannot be protonated by water (the pK_a of HCN is 9.2, while the pK_a of H₂O is 15.7), and therefore, the reactions of NaCN or KCN can be carried out in aqueous solutions. Figure 11.3 shows several examples of anionic carbon nucleophiles reactions with aldehydes and ketones.

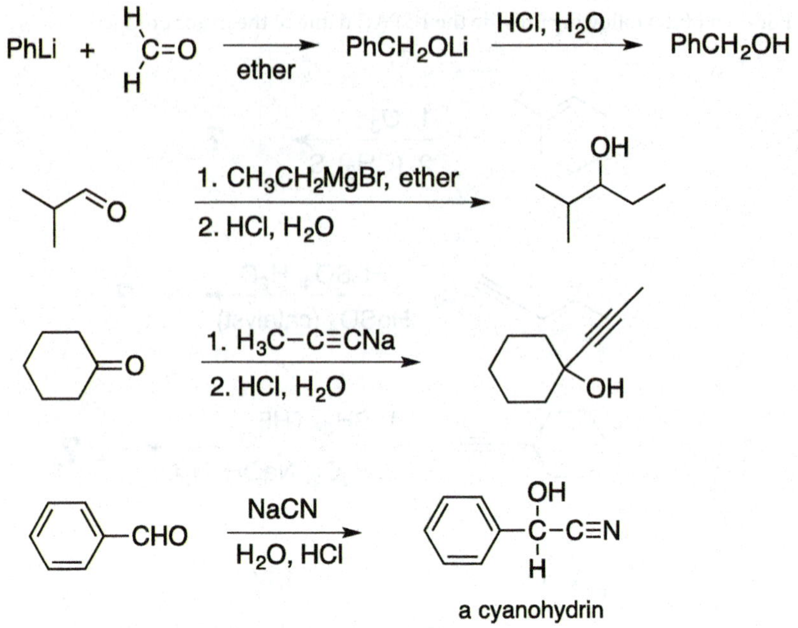

FIGURE 11.3. Reactions of anionic carbon nucleophiles with aldehydes and ketones.

Various alcohols are obtained as the final products in these reactions. Depending on the structure of the carbonyl compound, the reaction product can be a 1° alcohol (from $H_2C=O$), 2° alcohol (from an aldehyde), or 3° alcohol (from a ketone). The addition of cyanide anions to carbonyl compounds form the so-called cyanohydrins, which are important industrial compounds.

Problems

11.4. Propose a mechanism for each reaction shown in Figure 11.3.

11.5. Use organolithium or Grignard reagents to synthesize each of the compounds:

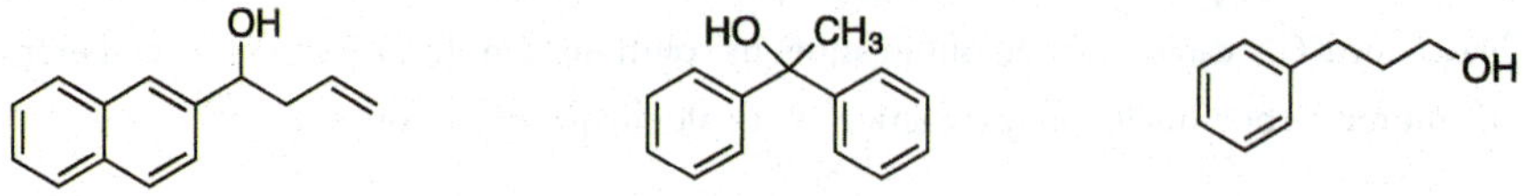

11.4. REACTION WITH HYDRIDE ANION

Metal hydrides, such as sodium hydride (Na⁺ :H⁻) or lithium hydride (Li⁺ :H⁻), are formed by the reaction of the corresponding metal with hydrogen gas. Since hydrogen is ignificantly more electronegative that sodium or lithium, the pair of bonding electrons in metal hydrides is completely shifted to hydrogen, resulting in an ionic compound composed of a metal cation and **hydride anion** (H:⁻). Hydride anions are strong bases (see Table 2.1 in Chapter 2) and strong nucleophiles, especially reactive in nucleophilic addition to carbonyl compounds (Figure 11.2).

From a practical standpoint, NaH or LiH are not very convenient sources of hydride anions because they are insoluble in organic solvents and react explosively with water. Chemists have developed a convenient reagent that functions as a source of hydride anions by combining NaH with boron hydride (BH_3). BH_3 is a Lewis acid and can accept H:⁻, resulting in the formation of borohydride anions (BH_4^-). The resulting ionic compound, sodium borohydride ($NaBH_4$), is a safe and convenient source of hydride anion that can even be used in aqueous solutions or alcohols. Likewise, LiH can be converted to lithium aluminum hydride ($LiAlH_4$), which in contrast to LiH is soluble in ether, and can serve as an excellent source of hydride anions in organic solvents. Because of the lower electronegativity value of Al (1.5) than B (2.0), hydrogen atoms in $LiAlH_4$ have a relatively higher negative charge than $NaBH_4$, and therefore, lithium aluminum hydride is a more efficient source of hydride anions. In contrast to $NaBH_4$, $LiAlH_4$ can explosively react with water. Therefore, its reactions with aldehydes or ketones must be performed in an aprotic organic solvent, such as ether, followed by careful addition of an aqueous acid, usually a diluted HCl solution, in order to protonate the initially formed alkoxide intermediate. Figure 11.4 shows several examples of sodium borohydride and lithium aluminum hydride reactions with aldehydes and ketones.

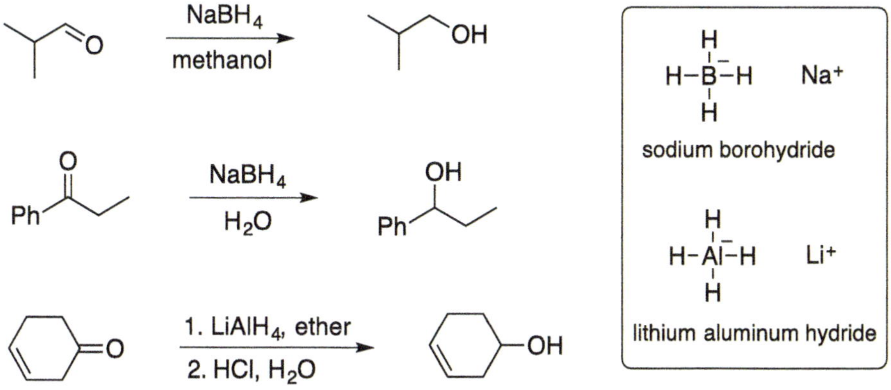

FIGURE 11.4. Reactions of sodium borohydride and lithium aluminum hydride with aldehydes and ketones.

All these reactions lead to the reduction of carbonyl compounds to the corresponding alcohols. The opposite reaction, the oxidation of alcohols to the corresponding carbonyl compounds, can be achieved by using appropriate oxidants (Section 8.7).

Problems

11.6. Suggest mechanisms for the reactions shown in Figure 11.4.

11.7. The reduction of 2-methylcyclohexanone leads to several stereoisomers. Name each stereoisomer and indicate which are diastereomers and which are enantiomers.

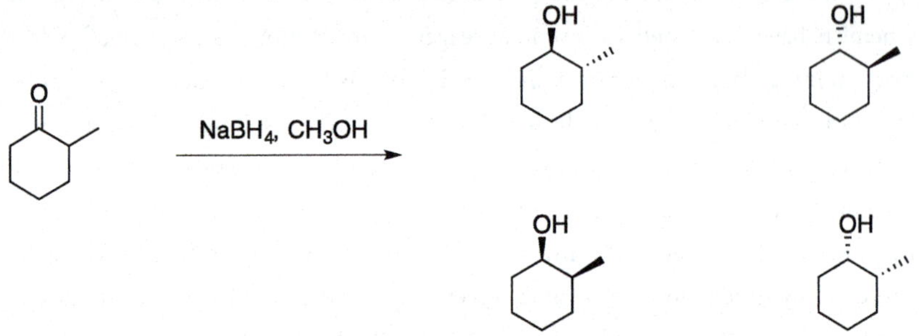

11.5. REACTION WITH PHOSPHONIUM YLIDES (THE WITTIG REACTION)

Ylides are generally defined as uncharged molecules containing a negatively charged carbon atom directly bonded to a positively charged atom of phosphorus or other element, such as sulfur. Phosphonium ylides can act as carbon nucleophiles in reactions with carbonyl compounds, which makes them useful reagents in organic synthesis. Phosphonium ylides are prepared by S_N2 reactions of triphenylphosphine with alkyl halides, followed by deprotonation

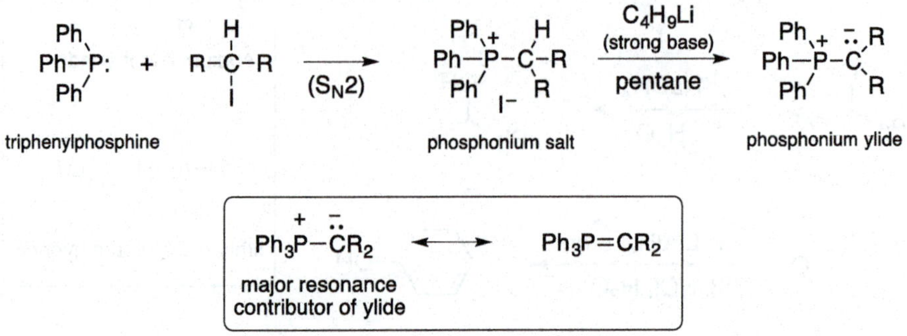

FIGURE 11.5. Preparation of phosphonium ylides.

of phosphonium salt, with a strong base such as alkyllithium (Figure 11.5). Phosphonium ylides are often shown in the form of a resonance contributor with a P=C double bond, although the dipolar contributor gives a better representation of the real molecule.

Phosphonium ylides react with the carbonyl group in aldehydes and ketones, yielding alkenes, a process called the **Wittig reaction**. This reaction was discovered in the 1950's by the German chemist Georg Wittig, who was awarded the Nobel Prize in Chemistry in 1979. Several examples of Wittig reactions and its mechanism are shown in Figure 11.6. The mechanism of this reaction includes initial nucleophilic addition of the anionic carbon ylide to a carbonyl group with formation of dipolar betaine, which then cyclizes to an oxaphosphetane intermediate. The highly unstable oxaphosphetane ring finally breaks, forming molecules of alkene and triphenylphosphine oxide. The driving force for this reaction is the formation of a very strong phosphorus oxygen double bond in Ph_3PO.

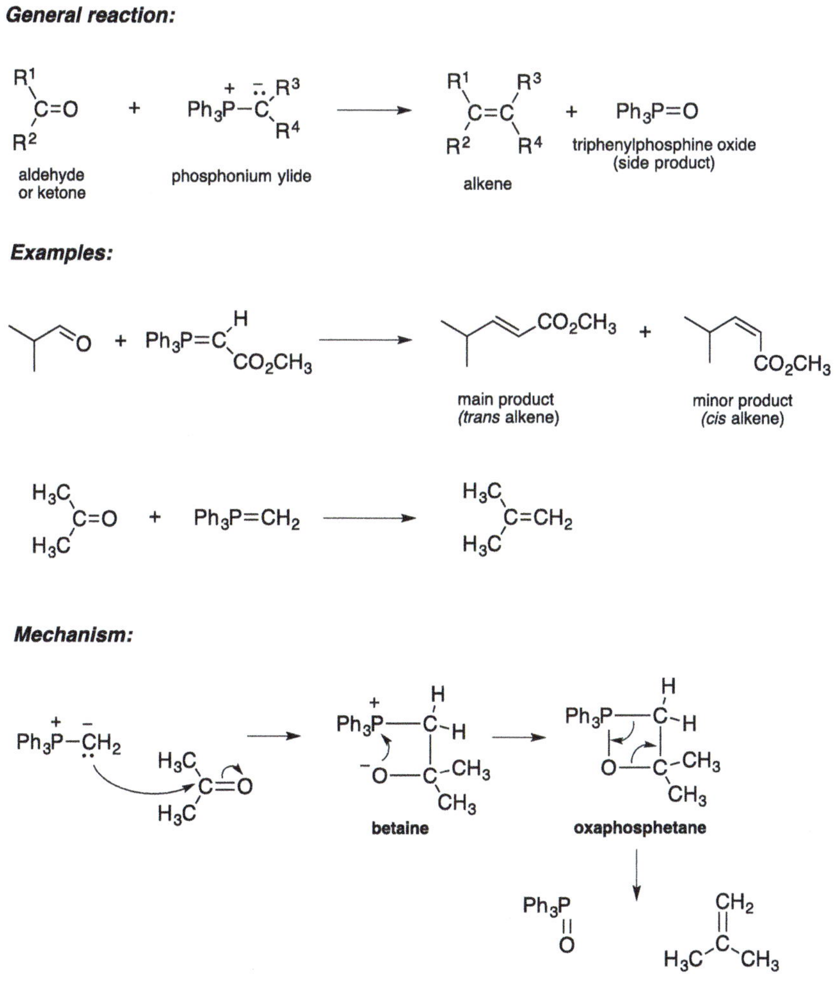

FIGURE 11.6. Reactions of phosphonium ylides with aldehydes or ketones (Wittig reactions).

The Wittig reaction is a valuable synthetic tool allowing a straightforward conversion of aldehydes or ketones to a diverse variety of substituted alkenes.

Problems

11.8. Each of the following can be prepared by Wittig reaction of a ketone with a phosphonium ylide. Provide the structure of the ketone and the ylide.

11.9. Draw the major products in the following reactions:

a)

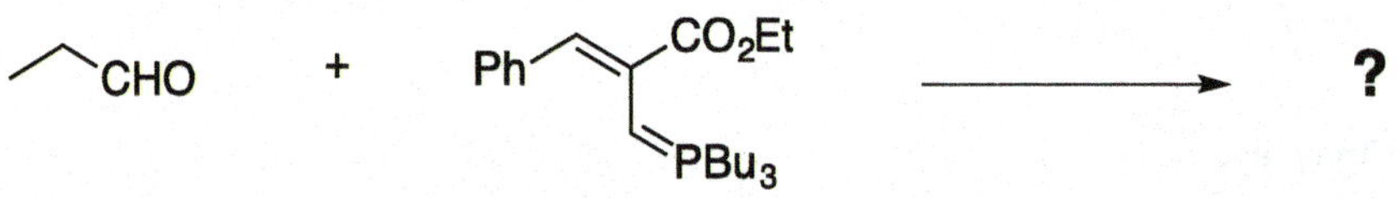

$\diagup\!\!\diagdown$CHO + Ph—C(CO$_2$Et)=PBu$_3$ $\longrightarrow$?

b)

H$_3$CO—C$_6$H$_4$—CHO $\xrightarrow{\text{Ph}_3\text{P}=\text{CHC(O)CH}_3}$?

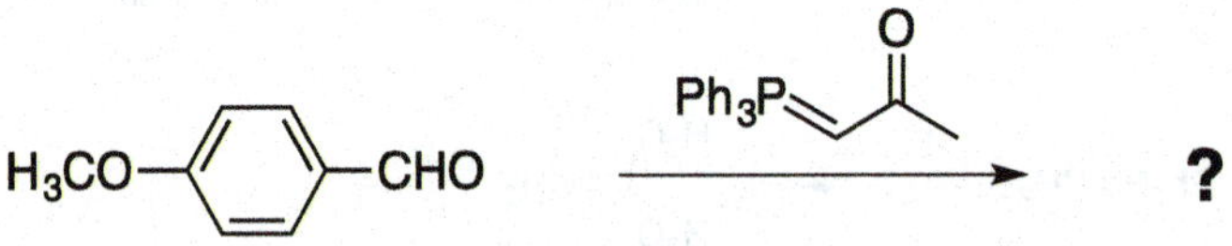

c)

camphor $\xrightarrow{\text{Ph}_3\text{P}=\text{CH}_2}$?

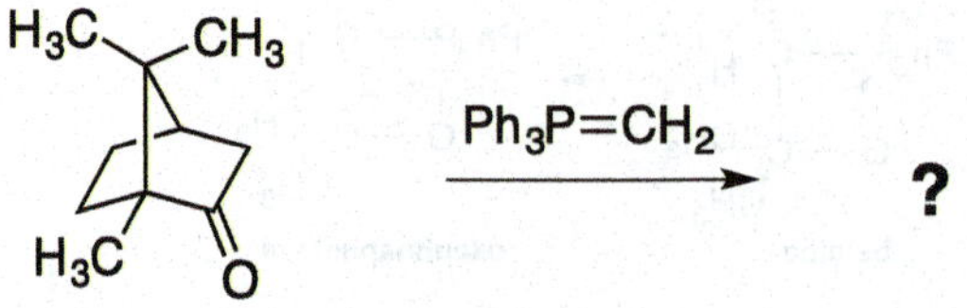

11.10. You have been assigned to synthesize methylenecyclohexane in high yield. You consider two approaches for making this compound, using cyclohexanone as a starting material:

Option I:

Ph₃PCH₃⁺ I⁻ →(BuLi) [cyclohexanone] → [methylenecyclohexane]

Option II:

[cyclohexanone] →(1. CH₃MgBr, ether; 2. HCl, H₂O) →(cat. TsOH, -H₂O) [methylenecyclohexane]

Why would you choose to pursue Option I in order to synthesize methylenecyclohexane from cyclohexanone in high yield?

11.6. REACTION WITH AMINES AND OTHER NITROGEN NUCLEOPHILES

Primary amines and related nitrogen compounds with a general formula RNH_2 are uncharged, moderate nucleophiles (Section 5.1). Reactions of nitrogen nucleophiles (RNH_2) with aldehydes or ketones ($R_2C=O$) produce the final products $R_2C=NR$ with an imine functional group (see Table 1.3 in Chapter 1). Dehydration of the initial products of nucleophilic addition of amines to carbonyl compounds form imines (Figure 11.7). Since amines are weaker nucleophiles than carbanions, their addition to a carbonyl group may require the presence of an acid catalyst in order to activate the carbonyl group by protonation (Scheme 11.2), and to facilitate elimination of water by an E1 mechanism (Chapter 5). Depending on the specific nitrogen nucleophiles, these reactions can produce a variety of products, as illustrated in Figure 11.7.

Hydrazones are key intermediates in the conversion of carbonyl functionality (C=O) into a methylene group (CH_2) by a reaction referred to as the **Wolff-Kishner** reduction (also known as Kishner-Wolff reduction), which was discovered independently by a Russian chemist, Nikolai Kishner, in 1911 and a German chemist, Ludwig Wolff, in 1912. Wolff-Kishner reductions are carried out by heating aldehydes or ketones with hydrazine (H_2NNH_2) and a base (KOH) in a high-boiling solvent, such as ethylene glycol ($HOCH_2CH_2OH$) or diethylene glycol ($HOCH_2CH_2OCH_2CH_2OH$). Mechanism of this reaction involves deprotonation of hydrazone, followed by the elimination of nitrogen and protonation of the carbanionic intermediate (Figure 11.8). Wolff-Kishner reductions have found practical

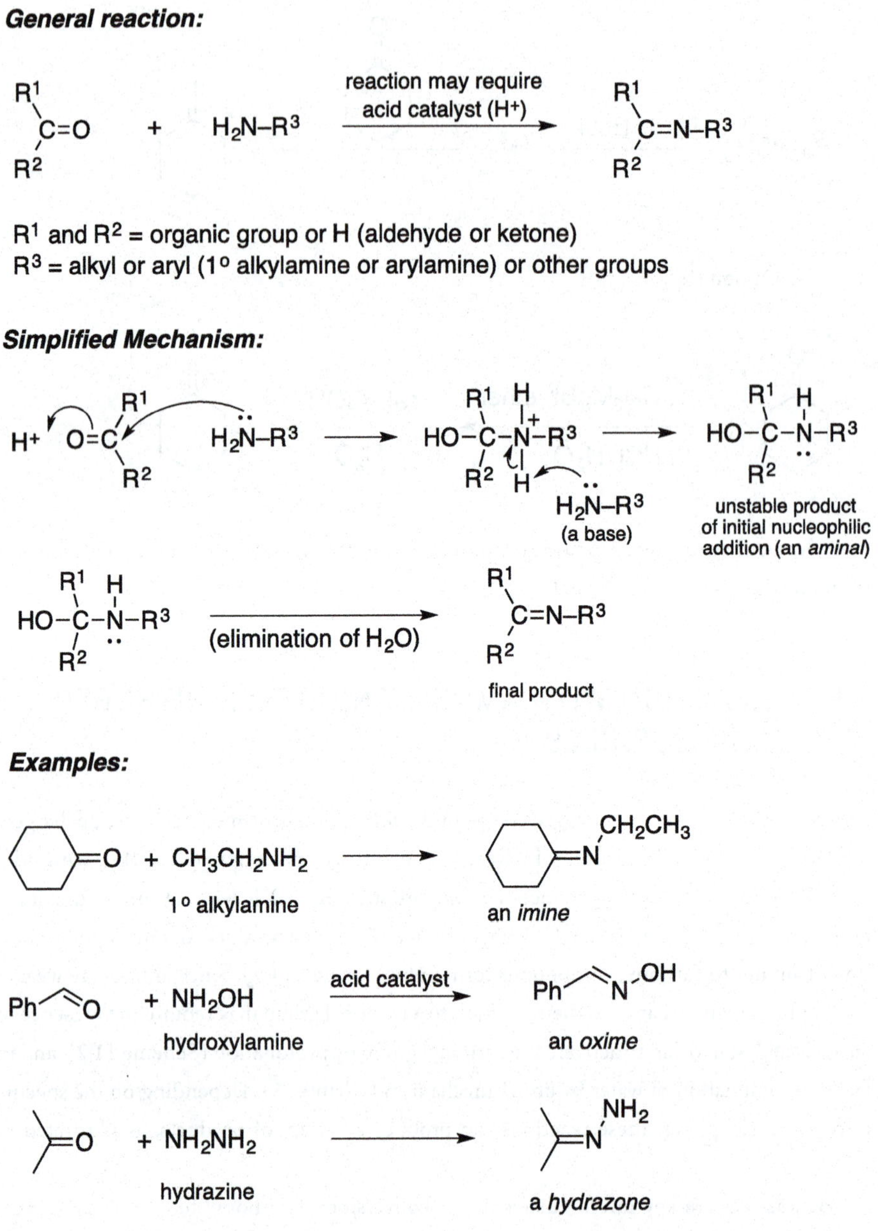

General reaction:

$$\begin{array}{c} R^1 \\ \diagdown \\ C=O \\ \diagup \\ R^2 \end{array} + H_2N-R^3 \xrightarrow{\text{reaction may require acid catalyst (H}^+\text{)}} \begin{array}{c} R^1 \\ \diagdown \\ C=N-R^3 \\ \diagup \\ R^2 \end{array}$$

R^1 and R^2 = organic group or H (aldehyde or ketone)
R^3 = alkyl or aryl (1° alkylamine or arylamine) or other groups

Simplified Mechanism:

[mechanism showing H⁺ protonation of C=O, attack by H₂N–R³, formation of HO–C(R¹)(R²)–N⁺H(R³)H intermediate, deprotonation by H₂N–R³ (a base) to give HO–C(R¹)(R²)–NH–R³, unstable product of initial nucleophilic addition (an *aminal*)]

$$\begin{array}{c} R^1\ H \\ \diagdown\ | \\ HO-C-N-R^3 \\ \diagup\ \ddot{} \\ R^2 \end{array} \xrightarrow{\text{(elimination of H}_2\text{O)}} \begin{array}{c} R^1 \\ \diagdown \\ C=N-R^3 \\ \diagup \\ R^2 \end{array}$$

final product

Examples:

cyclohexanone + $CH_3CH_2NH_2$ (1° alkylamine) ⟶ cyclohexylidene-N-CH₂CH₃ (an *imine*)

$Ph-CHO$ + NH_2OH (hydroxylamine) $\xrightarrow{\text{acid catalyst}}$ $Ph-CH=N-OH$ (an *oxime*)

(CH₃)₂C=O + NH_2NH_2 (hydrazine) ⟶ (CH₃)₂C=N-NH₂ (a *hydrazone*)

FIGURE 11.7. Reactions of primary amines and related nitrogen nucleophiles with aldehydes and ketones.

applications in the synthesis of complex organic molecules. Alternatively, aldehydes and ketones can be reduced to alkanes by reacting with Zn in aqueous hydrochloric acid (the **Clemmensen reduction**).

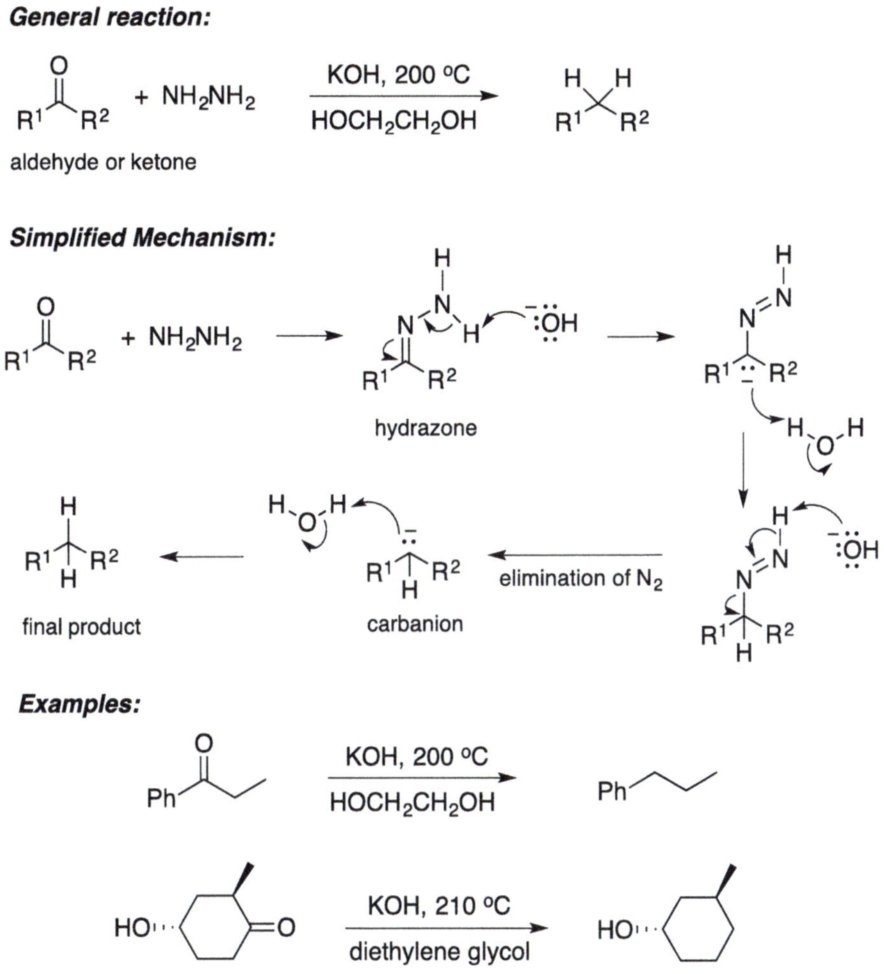

General reaction:

Simplified Mechanism:

Examples:

FIGURE 11.8. Wolff-Kishner reduction of carbonyl compounds.

Secondary alkylamines (R$_2$NH) react with aldehydes and ketones to produce enamines as final products. The carbon-carbon double bond in enamines is formed as a result of dehydration of the initial product of nucleophilic addition of the amine to the carbonyl compound (Figure 11.9). A practical use of enamines is their application as carbon nucleophiles in organic synthesis.

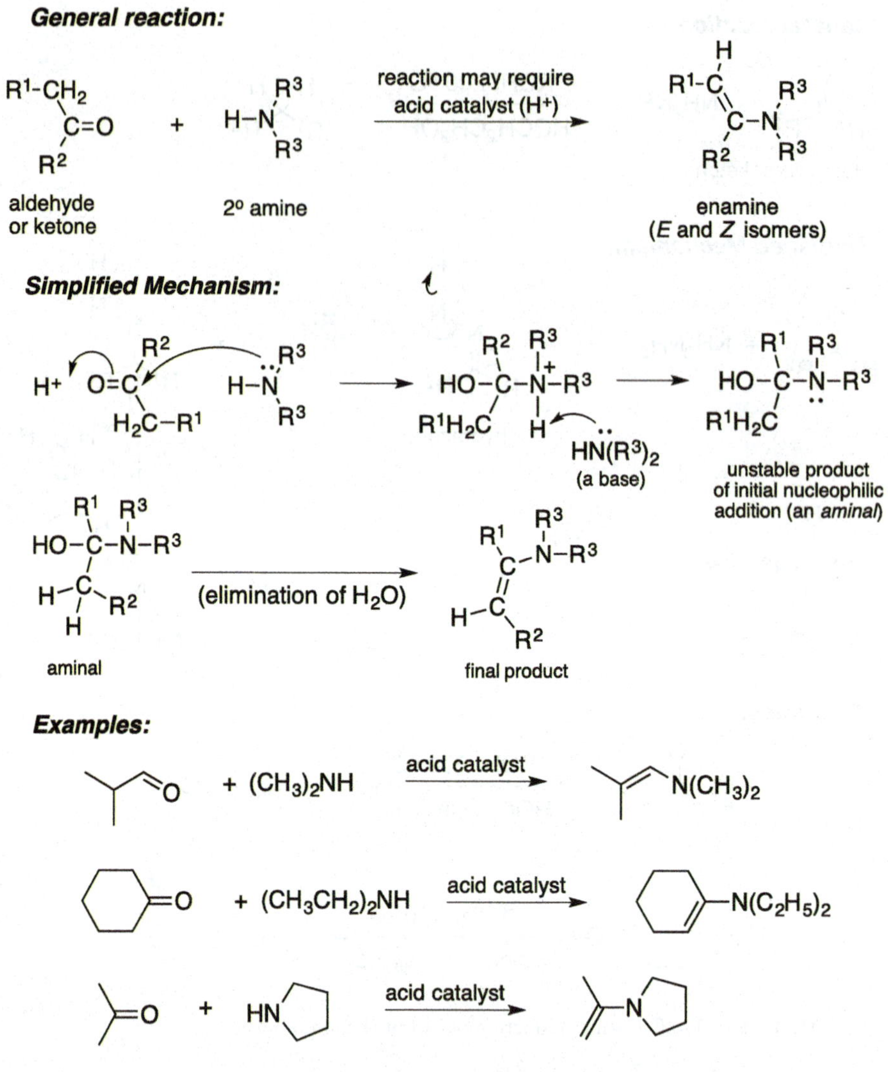

General reaction:

R¹-CH₂-C(R²)=O + H-N(R³)(R³) →(reaction may require acid catalyst (H⁺))→ enamine with R¹-CH=C(R²)-N(R³)(R³) and H on carbon

aldehyde or ketone   2° amine   enamine (*E* and *Z* isomers)

Simplified Mechanism:

H⁺ + O=C(R²)(H₂C-R¹) + H-N(R³)(R³) → HO-C(R²)(R¹H₂C)-N⁺(H)(R³)(R³) →[HN(R³)₂ (a base)]→ HO-C(R¹)(R¹H₂C)-N(R³)(R³)

unstable product of initial nucleophilic addition (an *aminal*)

HO-C(R¹)(R³)-N(R³)... H-C(R²)(H) →(elimination of H₂O)→ R¹C(=CH-R²)-N(R³)(R³)

aminal   final product

Examples:

(CH₃)₂CHCHO + (CH₃)₂NH →(acid catalyst)→ (CH₃)₂C=CH-N(CH₃)₂

cyclohexanone + (CH₃CH₂)₂NH →(acid catalyst)→ 1-(N,N-diethylamino)cyclohexene

(CH₃)₂C=O + HN(pyrrolidine) →(acid catalyst)→ CH₂=C(CH₃)-N(pyrrolidine)

FIGURE 11.9. Reactions of 2° alkylamines with aldehydes or ketones.

Problems

11.11. Provide the products of the following transformations:

a)

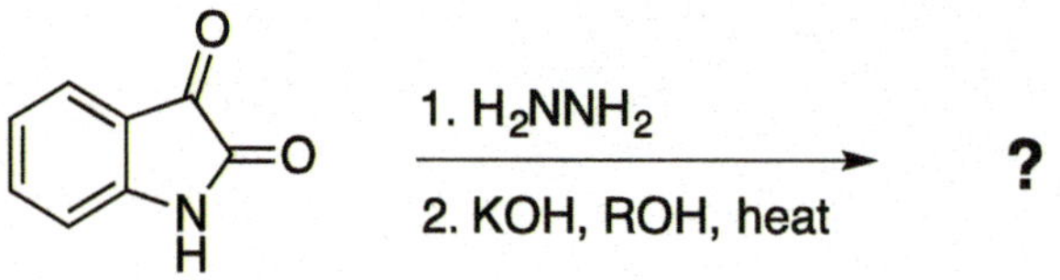

isatin + 1. H₂NNH₂; 2. KOH, ROH, heat → ?

b)

[structure: acetone + aniline (PhNH2), acid catalyst → ?]

c)

[structure: indan-1-one + (CH3)2NH, acid catalyst → ?]

d)

[structure: cyclohexanone + 2,4-dinitrophenylhydrazine (NO2, O2N, NHNH2), acid catalyst → ?]

11.7. REACTION WITH WATER AND ALCOHOLS

Water and alcohols are weak, uncharged nucleophiles requiring acid or base catalysis for their addition to the carbonyl group of aldehydes and ketones. Solutions of aldehydes and of ketones ($R_2C=O$) in water contain unstable hydrates [e.g., $R_2C(OH)_2$, also known as **geminal diols**] in equilibrium with the initial carbonyl compound. Similar reactions of aldehydes and ketones with alcohols in the presence of an acid or base initially produce **hemiacetals** as the products of nucleophilic addition of one molecule of alcohol to the carbonyl group (Figure 11.10). Hemiacetals are generally unstable compounds existing in equilibrium with the initial carbonyl compound. However, cyclic hemiacetals with a five- or six-membered ring have much higher stability and are very common in natural products, such as carbohydrates (see Chapter 16). The simplest five-membered cyclic hemiacetal is formed by cyclization of 4-hydroxybutanal (Figure 11.10).

In the presence of acid catalysts, hemiacetals can further react with a second molecule of alcohol, producing **acetal**. This is a reversible reaction in the presence of acid. However, acetals are generally more stable than hemiacetals and can be isolated as individual compounds. In

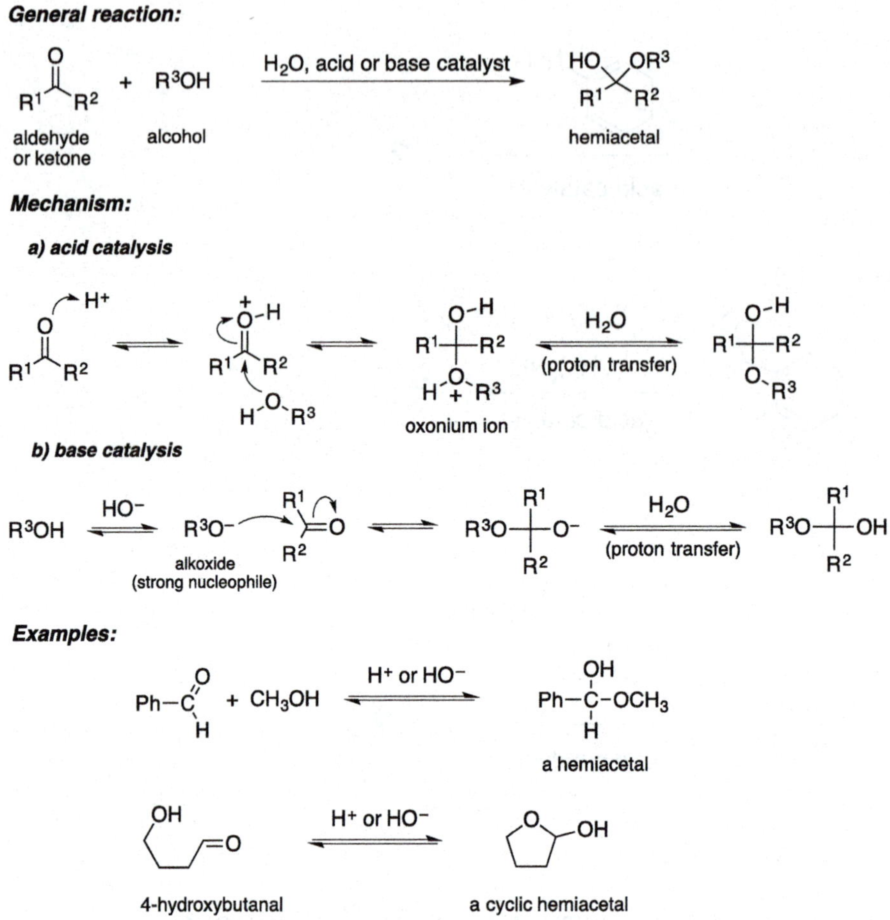

FIGURE 11.10. Formation of hemiacetals.

contrast to hemiacetals, acetals are stable under basic conditions. The principal step in acetal formation involves nucleophilic substitution of the protonated hydroxyl group by a molecule of alcohol (Figure 11.11). The acetals formed from cyclic hemiacetals are very common carbohydrate compounds (Chapter 16).

Reaction of diols (i.e., glycols) with aldehydes or ketones in the presence of acid gives cyclic acetals (Figure 11.12). Cyclic acetals are stable compounds under neutral or basic conditions. However, they can be converted back to the initial aldehydes or ketones by treatment with an excess of water in the presence of acid. In contrast to carbonyl compounds, cyclic acetals are unreactive to strong nucleophiles, such as hydride anions, organometallic compounds, and many other reagents. This property of cyclic acetals allows their use as a **protective group** that can be temporarily placed on carbonyl groups and removed after a sequence of chemical reactions involving other functional groups in a

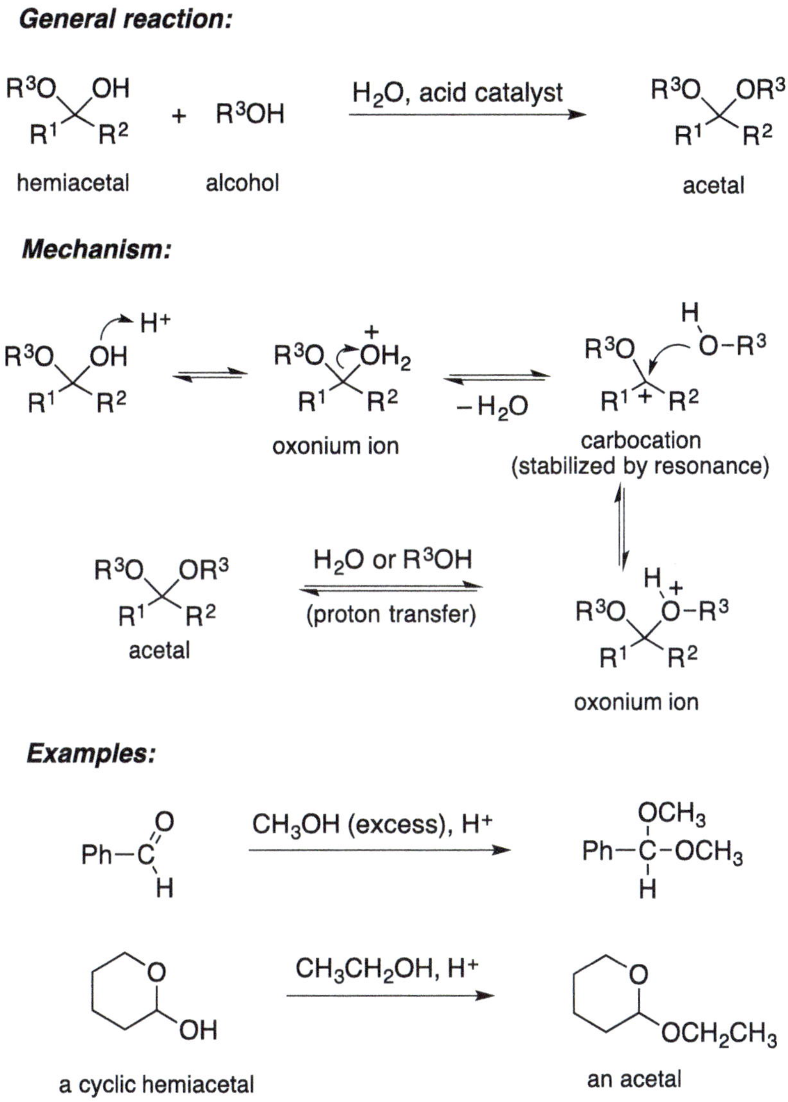

General reaction:

$$\underset{\text{hemiacetal}}{\overset{R^3OOH}{\underset{R^1R^2}{\times}}} + \underset{\text{alcohol}}{R^3OH} \xrightarrow{\text{H}_2\text{O, acid catalyst}} \underset{\text{acetal}}{\overset{R^3OOR^3}{\underset{R^1R^2}{\times}}}$$

Mechanism:

$$\overset{R^3OOH}{\underset{R^1R^2}{\times}} \xrightleftharpoons{H^+} \underset{\text{oxonium ion}}{\overset{R^3O\overset{+}{O}H_2}{\underset{R^1R^2}{\times}}} \xrightleftharpoons{-H_2O} \underset{\substack{\text{carbocation}\\\text{(stabilized by resonance)}}}{\overset{R^3OO-R^3H}{\underset{R^1\overset{+}{}R^2}{\times}}}$$

$$\underset{\text{acetal}}{\overset{R^3OOR^3}{\underset{R^1R^2}{\times}}} \xrightleftharpoons[\text{(proton transfer)}]{\text{H}_2\text{O or R}^3\text{OH}} \underset{\text{oxonium ion}}{\overset{R^3O\overset{+}{O}-R^3H}{\underset{R^1R^2}{\times}}}$$

Examples:

$$\text{Ph}-\underset{H}{\overset{O}{\overset{\|}{C}}} \xrightarrow{\text{CH}_3\text{OH (excess), H}^+} \text{Ph}-\underset{H}{\overset{OCH_3}{\underset{|}{\overset{|}{C}}}}-OCH_3$$

a cyclic hemiacetal → an acetal (with CH₃CH₂OH, H⁺)

FIGURE 11.11. Formation of acetals.

multifunctional molecule. An example of using the protective group for aldehyde groups in a sequence of reactions involving organolithium compound is shown in Figure 11.12. The direct preparation of an organolithium compound from 3-bromopropanal is impossible because the aldehyde group cannot exist in the same molecule with carbanion; therefore, it should be protected at the beginning of reaction sequence and unprotected at the end of the sequence.

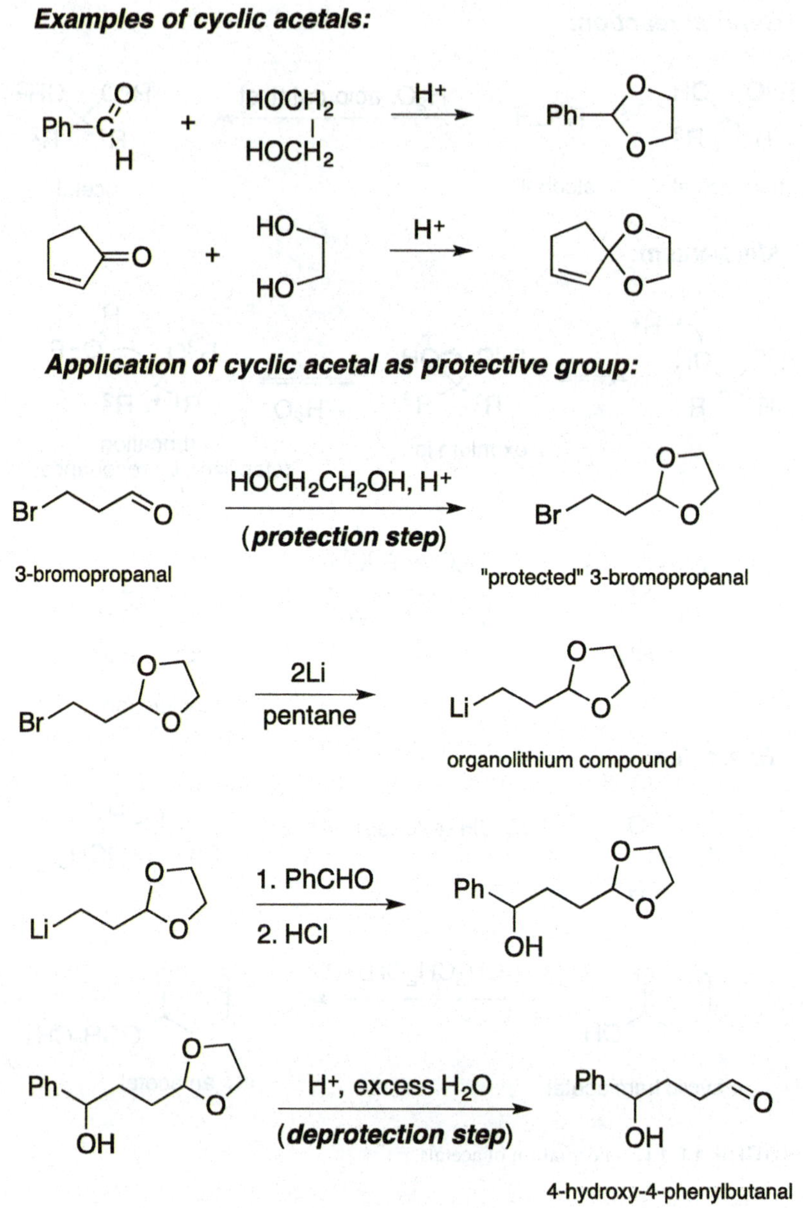

Examples of cyclic acetals:

Application of cyclic acetal as protective group:

FIGURE 11.12. Cyclic acetals as protective groups.

Problems

11.12. Suggest a mechanism for the formation of formaldehyde hydrate $H_2C(OH)_2$ in the reaction of formaldehyde with water.

11.13. Suggest mechanisms for the formation of five-membered cyclic hemiacetal from 4-hydroxybutanal (Figure 11.10), and six-membered cyclic hemiacetal from 5-hydroxypentanal, under conditions of acid catalysis and base catalysis.

11.14. Suggest mechanisms for the reactions shown in Figure 11.11.

11.15. Provide the structure of the missing reactants:

a)

b)

11.8. OXIDATION OF CARBONYL COMPOUNDS

Aldehydes are intermediate products during the course of the oxidation of 1° alcohols to the corresponding carboxylic acid using strong oxidants, such as chromium trioxide, chromic acid, and potassium permanganate (Section 8.7). The same strong oxidants can oxidize pure aldehydes to carboxylic acids by converting the aldehyde C–H bond to a carboxylic C–OH bond (Figure 11.13). Reactions of multifunctional molecules with strong oxidants are not selective and will affect other sensitive functional groups, such as alcohols. **Tollens' reagent**, consisting of an aqueous solution of silver nitrate and ammonia can selectively oxidate aldehyde groups in the presence of sensitive functional groups. This reagent, named after its discoverer Bernhard Tollens, is commonly used to determine the presence of aldehyde functional groups in complex molecules such as carbohydrates. A positive test with Tollens' reagent will show precipitation of elemental silver, producing a characteristic "silver mirror" on the inner surface of the reaction flask. Examples of selective oxidation with Tollens' reagent are shown in Figure 11.13.

In contrast to aldehydes, ketones are generally resistant to oxidation because of the absence of oxidizable C–H bonds at the carbonyl carbon. The C–C bond in ketones can be cleaved only at high temperature, using powerful oxidants, and produce a variety of different products.

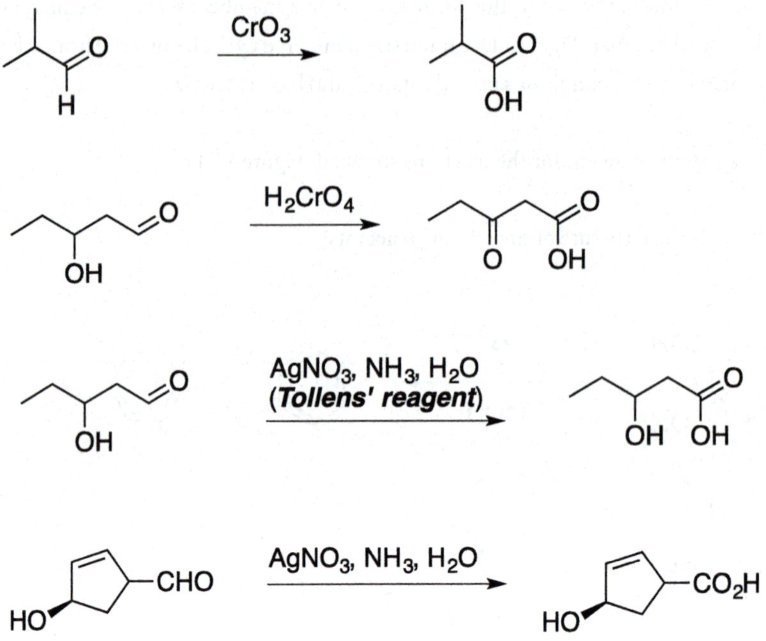

FIGURE 11.13. Oxidation of aldehydes.

Problems

11.16. Provide the major products in the following reactions:

a)

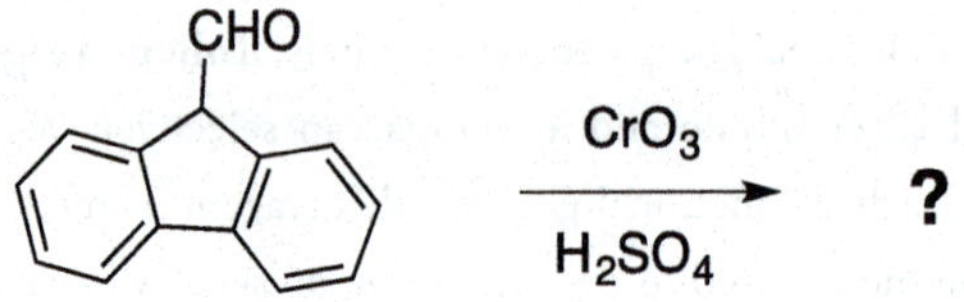

9H-fluorene-9-carbaldehyde $\xrightarrow[H_2SO_4]{CrO_3}$?

b)

thiophene-2-carbaldehyde $\xrightarrow[Na_2HPO_4]{KMnO_4}$?

226 ORGANIC CHEMISTRY

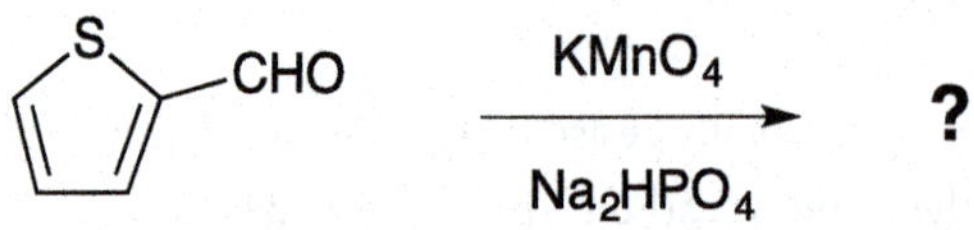

11.9. ENOLS AND ENOLATE ANIONS

A Greek letter (e.g., α, β, γ, δ) usually indicates the position of carbon atoms on alkyl chains attached to the carbonyl carbon in aldehydes or ketones. The **α position** is assigned to the carbon adjacent to the carbonyl carbon. Hydrogen atoms attached to the α-carbon (i.e., the α-hydrogens) are characterized by unusually high acidity (pK_a of about 20), which is much higher compared to the average acidity of C–H bonds in alkanes (pK_a = 50-60). The resonance stabilization of **enolate anions** formed by deprotonation of carbonyl compounds explains the high acidity of α-hydrogens (Figure 11.14). According to resonance rules, resonance contributors of enolate anions with negative charges on oxygen are the most important, major contributors (Section 1.6). Proton transfer from molecules of water to oxygen atoms of enolate anions form the **enol** tautomers of carbonyl compounds (see Section 7.3.3 for general definition of tautomers and tautomerization).

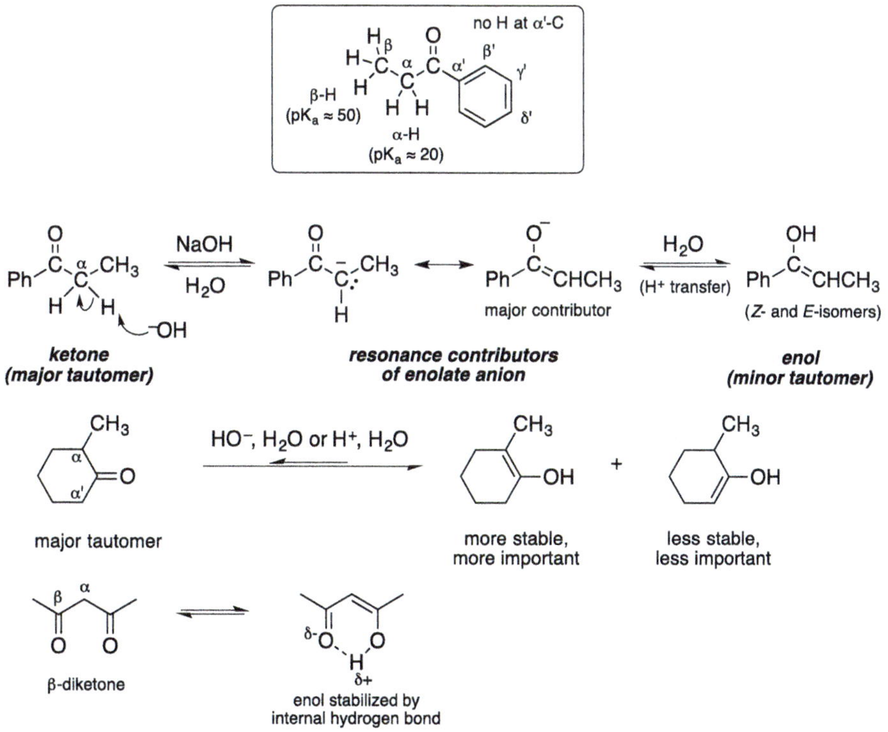

FIGURE 11.14. Keto-enol tautomeric equilibrium.

The equilibrium between tautomers can be catalyzed by a base, as shown in Figure 11.14, or by an acid (see Figure 7.9 in Chapter 7). Enol usually exists as a mixture of Z and E stereoisomers; enols with more substituents at the double bond are thermodynamically more stable (Figure 11.14). In general, the keto-form dominates in the tautomeric equilibrium (over 99.99999% of molecules in the equilibrium for simple ketones and aldehydes), while the enol form is a very minor contributor. However, in some carbonyl compounds, the enol form becomes more important in equilibrium due to additional stabilizing factors. For example, in the β-dicarbonyl compounds, about half of all molecules exist in the enol form because it is stabilized by intramolecular hydrogen bonding between the enolate O–H and the carbonyl group at the β-carbon (Figure 11.14).

The presence of enol form in aqueous solutions explains special reactivity of carbonyl compounds at the α position. Aldehydes and ketones react with halogens in water, forming products that result from the substitution of α-hydrogens with a halogens (Figure 11.15). The mechanism of α-halogenation reactions involves electrophilic addition of halogen to the enol form of carbonyl compound as shown in Figure 11.15.

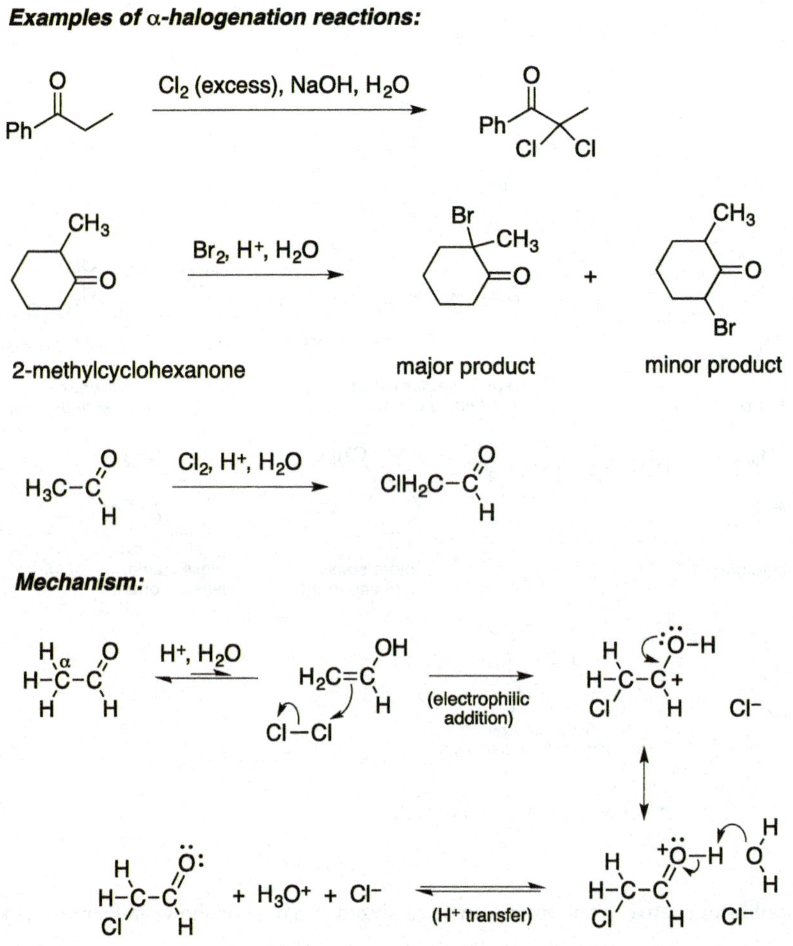

FIGURE 11.15. α-halogenation of aldehydes and ketones.

228   ORGANIC CHEMISTRY

Protons from a strong acid also can also react as an electrophile with enols. If the hydrogen isotope deuterium (D = ^{2}H) is used instead of proton (^{1}H), the reaction will result in the introduction of deuterium in the α-position of the carbonyl compound. A sequence of enolization steps and additions of D$^+$ will eventually replace all α-hydrogens in aldehyde or ketone with deuterium atoms (Figure 11.16). This is the so-called **deuterium exchange** reaction, used for the preparation of organic solvents in which hydrogen atoms are replaced with deuterium. Such deuterated solvents are used for the preparation of ^{1}H NMR samples because they do not show signals of protons in NMR spectrum that may overlap with signals of the dissolved organic compound.

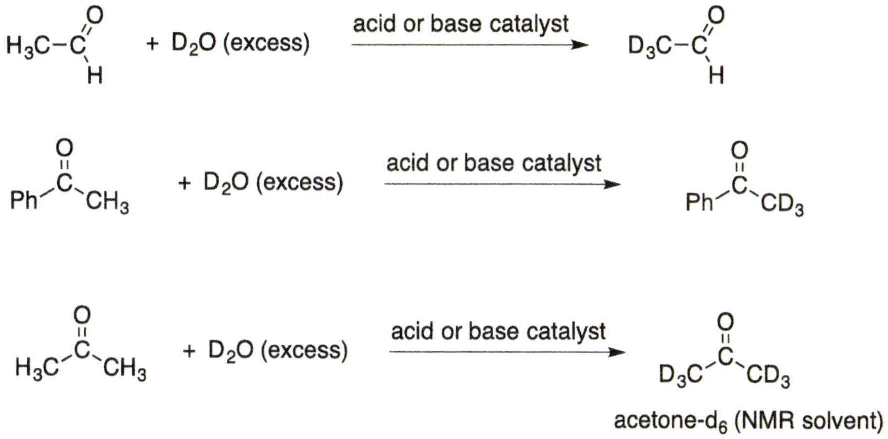

FIGURE 11.16. Deuterium exchange of α-hydrogens.

Problems

11.17. Suggest mechanisms of acid-catalyzed enolizations for 1-methylcyclopentanone.

11.18. Write detailed mechanism for deuterium exchange of α-hydrogens in acetaldehyde.

11.19. Explain the difference in the acidity of α-hydrogens in CH_3COCH_3 (pK_a = 30) and $CH_3COCH_2COCH_3$ (pK_a = 9).

11.10. REACTION OF ENOLATE ANION AS NUCLEOPHILE: ALDOL CONDENSATION

The enolate anion, formed by deprotonation of carbonyl compounds at the α-position (Figure 11.14), is a particularly important carbon nucleophile in organic synthesis. Common bases that are used for the α-deprotonation of carbonyl compounds include hydroxides

(NaOH or KOH), alkoxides (NaOEt in EtOH or *t*-BuOK in *t*-BuOH), amides such as lithium diisopropylamide [LiN(iPr)$_2$; LDA] and metal hydrides (NaH or KH). Enolate anion is a strong nucleophile, highly reactive in nucleophilic substitution or nucleophilic addition reactions. The major resonance contributor of enolate anion has negative charge on the oxygen atom, but enolate usually reacts as a carbon nucleophile because the oxygen atom is blocked by the closely associated counterion such as Na$^+$, K$^+$ or Li$^+$. Examples of nucleophilic substitution reactions (S$_N$2) of enolates with simple alkyl halides are shown in Figure 11.17.

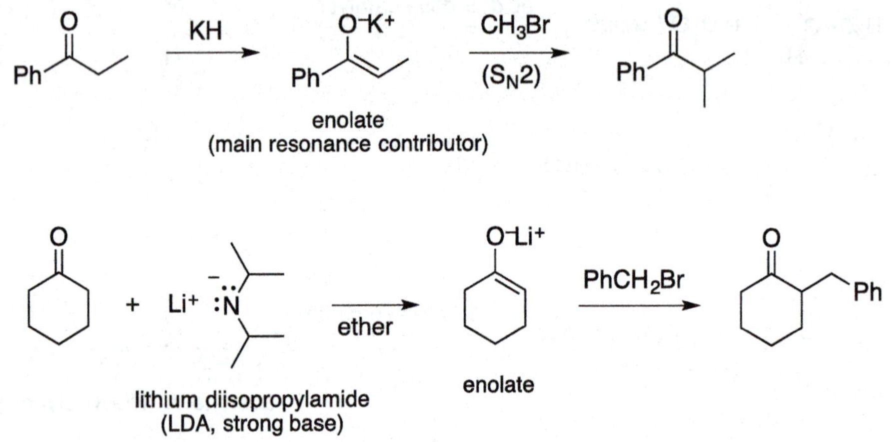

FIGURE 11.17. Nucleophilic substitution reactions of enolates with alkyl halides.

Nucleophilic addition of enolate anions to carbonyl groups of aldehydes or ketones is known as **aldol condensation**. The aldol condensation can involve two molecules of the same carbonyl compound, or two different carbonyl compounds. Condensation reactions are usually performed in the presence of a base that is required to generate the enolate nucleophile. The initial products of nucleophilic addition of enolate anions to carbonyl groups are the β-hydroxycarbonyl compounds or aldols, which are usually dehydrated by gentle heating to give α,β-unsaturated carbonyl compound as the final products. Several examples of aldol condensation reactions are shown in Figure 11.18.

Problems

11.20. Write detailed mechanisms for the reactions shown in Figure 11.17.

11.21. Write detailed mechanisms for reactions shown in Figure 11.18.

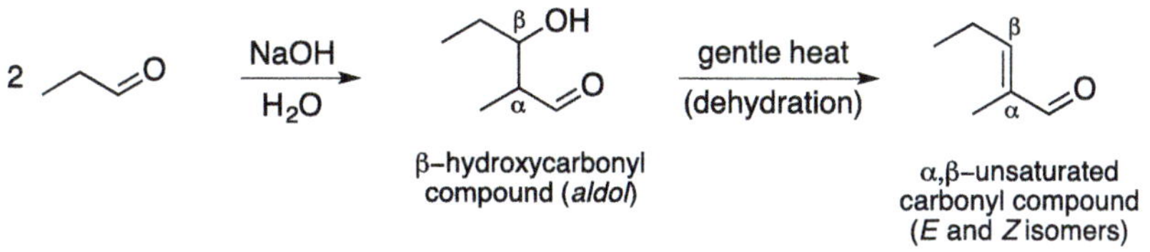

Simplified mechanism of aldol formation:

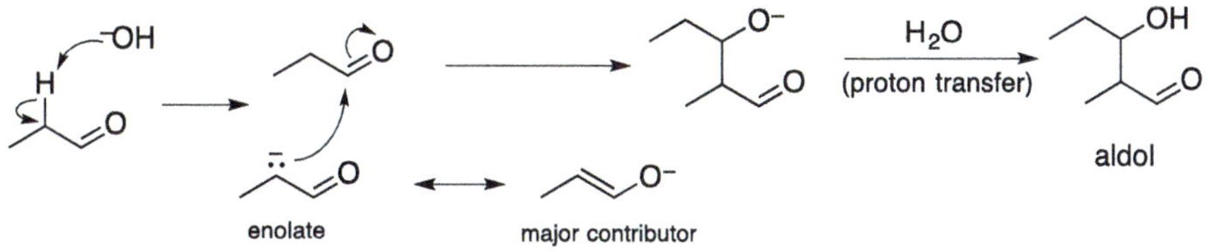

Additional examples of aldol condensation reactions:

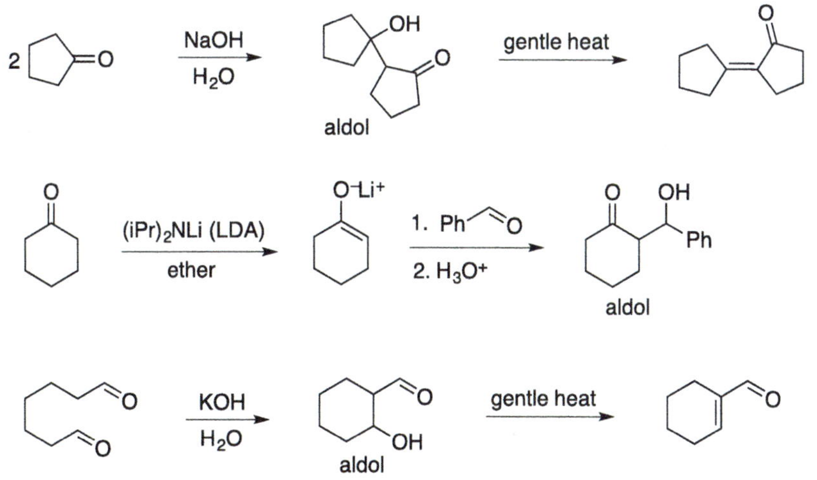

(***intramolecular aldol condensation***)

FIGURE 11.18. Aldol condensation reactions.

11.22. Which of the following compounds are able to undergo self-condensation by aldol reaction? Draw the products of these condensation reactions.

a) formaldehyde b) acetaldehyde c) benzaldehyde d) acetone

11.23. The synthesis of the product involves a mixed aldol reaction followed by an acetal formation. What is the structure of the reactant?

propanal (CH₃CH₂CHO) + ? (2 mol) →[buffer, pH 7] product (shown: 1,3-dioxane ring with methyl and OH substituents)

11.24. Why α-hydrogens in aldehydes are more acidic than the hydrogen atom directly attached to the carbon of the carbonyl group?

CHAPTER 12

Carboxylic Acids and their Derivatives

Carboxylic acids are generally defined as compounds with a carboxyl functional group (CO_2H) in their structure (Section 1.3). Carboxylic acids, including the amino acids that make up proteins, are very common in nature and play an important role in our lives. The aqueous solution acetic acid (CH_3CO_2H), known as vinegar, is a product of natural oxidation of ethanol in wine. Acetic acid and other simple carboxylic acids are large-scale industrial products used as starting materials in the production of polymers, pharmaceuticals, solvents, and food additives. Important carboxylic acid derivatives include acid chlorides, anhydrides, esters, and amides. All acid derivatives can be prepared from acids and can be converted back to acids via hydrolysis. Carboxylic esters and amides are common natural and synthetic compounds that are extremely important in our lives. This chapter presents nomenclature, structural features, properties, and reactions of carboxylic acids and their derivatives.

12.1. NOMENCLATURE OF CARBOXYLIC ACIDS AND THEIR DERIVATIVES

According to the general rules of IUPAC nomenclature, the carboxyl functional group (CO_2H) has the highest priority and its presence in a molecule is always indicated by the suffix *-oic acid* (see Table 1.4 and the related text in Chapter 1). The carbons of the parent chain are numbered so that the carboxyl group has the carbon number 1; there is no need to include this number in the name of compound. The suffix *-carboxylic acid* is used to indicate that the carboxyl group (CO$_2$H) is directly bonded as a substituent in a cycloalkyl ring. The suffix *-dioic acid* indicates the presence of two carboxyl groups in a molecule. Several examples of IUPAC aldehyde and ketone names are shown in Figure 12.1.

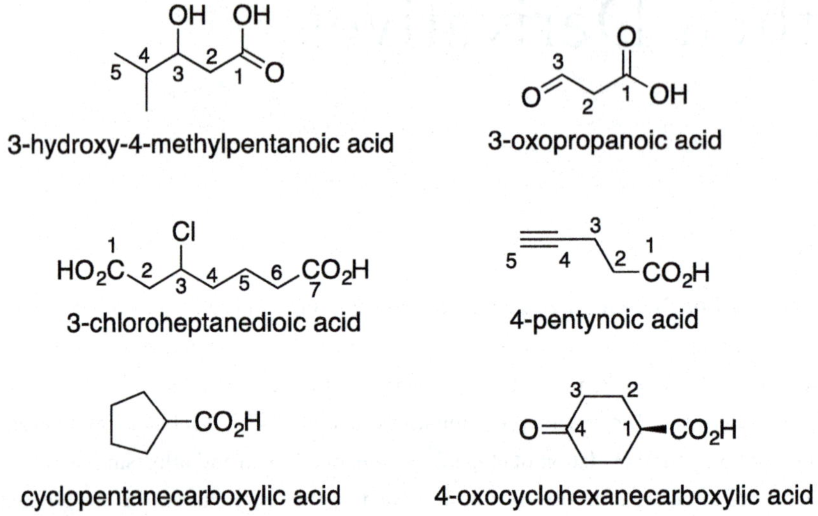

FIGURE 12.1. IUPAC nomenclature of carboxylic acids.

Common names of important carboxylic acids include formic acid (HCO$_2$H), acetic acid (CH$_3$CO$_2$H), benzoic acid (PhCO$_2$H), and oxalic acid (HO$_2$CCO$_2$H).

The most important derivatives of carboxylic acids are represented by acid chlorides (RCOCl), anhydrides (RCO$_2$COR), esters (RCO$_2$R'), and amides (RCONR'$_2$). All these derivatives have a common structural fragment, RCO, referred to as an **acyl group**, connected to halogen, oxygen, or nitrogen atoms (Figure 12.2). The names of acyl groups (RCO) are derived from the names of the corresponding alkanoic acids (RCOOH) by changing the ending *-ic acid* to the suffix *-yl*; for example, butanoyl for the acyl group C$_3$H$_7$CO present in butanoic acid, benzoyl (PhCO, standard abbreviation Bz) for benzoic acid, acetyl (CH$_3$CO, standard abbreviation Ac) for acetic acid, and formyl (HCO) for *formic acid*. The names of acyl groups are used in the nomenclature of acid halides (i.e., acyl halides). Names of anhydrides are created by changing the ending *acid* in alkanoic acid to *anhydride* (*alkanoic anhydrides*). In the names of esters, the group name *alkanoate* is used to indicate the acid fragment RCOO, connected to the alkyl group R' (*alkyl alkanoate*). The name alkanoate is also used for the anions of alkanoic acids (RCO$_2^-$); for example, the sodium salt of butanoic acid (C$_3$H$_7$CO$_2$Na) is called sodium butanoate. Amides are named by changing the ending

-oic acid, as in alkanoic acid, to the suffix -amide in the related amide (*alkanamide*). The presence of an alkyl group connected to the nitrogen atom of amide is indicated by the prefix *N-alkyl*. For two alkyl groups connected to the nitrogen of amide, the prefix *N,N-dialkyl* is used. Figure 12.2 provides general names of acid derivatives and specific examples.

Acid halides

R–C(=O)–X

acyl halide

Examples:

H₃C–C(=O)–Cl
acetyl chloride (AcCl)

Ph–C(=O)–Br
benzoyl bromide

C₃H₇–C(=O)–Cl
butanoyl chloride

Acid anhydrides

R–C(=O)–O–C(=O)–R

alkanoic anhydride

Examples:

H₃C–C(=O)–O–C(=O)–CH₃
acetic anhydride (Ac₂O)

C₃H₇–C(=O)–O–C(=O)–C₃H₇
butanoic anhydride

H₃C–C(=O)–O–C(=O)–Ph
acetic benzoic anhydride

Esters

R–C(=O)–O–R'

alkyl alkanoate

Examples:

H₃C–C(=O)–OC₂H₅
ethyl acetate

(isopropyl)–C(=O)–OMe
methyl 2-methylpropanoate

Ph–C(=O)–OiPr
isopropyl benzoate

Amides

R–C(=O)–N(H)(H)

alkanamide

Examples:

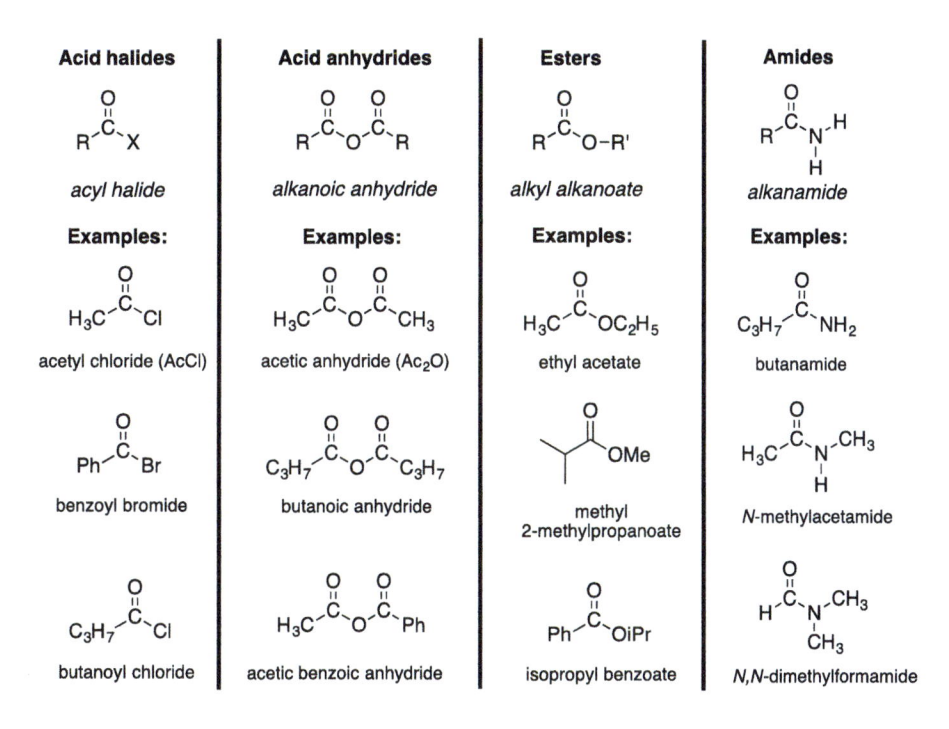

C₃H₇–C(=O)–NH₂
butanamide

H₃C–C(=O)–N(H)–CH₃
N-methylacetamide

H–C(=O)–N(CH₃)(CH₃)
N,N-dimethylformamide

FIGURE 12.2. Classification and naming of carboxylic acid derivatives.

Several important acid derivatives have cyclic structure. Examples and names of cyclic anhydrides, esters, amides, and imides are shown in Figure 12.3. Note that cyclic esters are generally named **lactones**, cyclic amides are named **lactams**, and compounds with two acyl groups connected to nitrogen are named **imides**.

succinic anhydride
(*cyclic anhydride*)

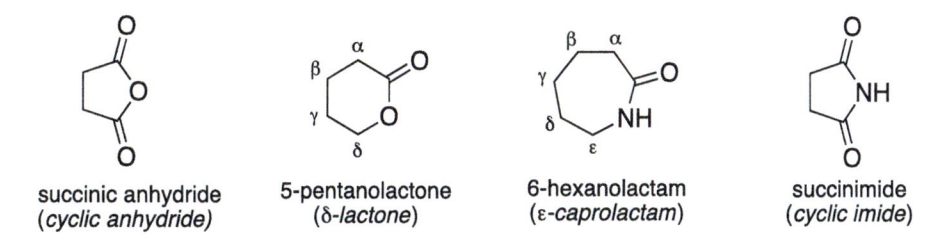

5-pentanolactone
(*δ-lactone*)

6-hexanolactam
(*ε-caprolactam*)

succinimide
(*cyclic imide*)

FIGURE 12.3. Examples of cyclic anhydrides, esters, amides, and imides.

The general principles of naming acid derivatives can be extended to sulfonic acids (RSO$_2$OH; see Section 8.6) and the organic derivatives of phosphoric acid (H$_3$PO$_4$). For example, alkyl esters of alkylsulfonic acids (RSO$_2$OR') are named alkyl sufonates and amides of alkylsulfonic acids (RSO$_2$NR'$_2$) are named alkyl sulfonamides.

The functional group of a nitrile (see Table 1.3 in Chapter 1), also known as the cyano group (CN), is present in organic nitriles (RC≡N). Organic nitriles are usually classified as acid derivatives despite the absence of the acyl group in their structure. Nitriles can be prepared from carboxylic acids and can be converted back to the corresponding acids via hydrolysis. Names of nitriles are also derived from the names of carboxylic acids; for example, acetonitrile (CH$_3$CN) is related to acetic acid and benzonitrile (PhCN) is a derivative of benzoic acid.

Problems

12.1. Provide the IUPAC names:

a) [structure with OH and COOH] b) [structure with N(CH$_3$)$_2$] c) [structure with H and NH$_2$] d) [Fischer projection with COOH, OH, HO, H, COOH]

e) [structure] f) [cyclic anhydride structure] g) [structure with Cl] h) [structure with CN]

12.2. The following compounds are listed as ingredients on the labels of cosmetics and household chemicals. Use web-based resources to determine the structures of these compounds.

a) Glyceryl stearate   b) Ethyl acetate   c) Citric acid   d) Cocamide MEA

12.2. ACIDITY AND PHYSICAL PROPERTIES OF CARBOXYLIC ACIDS AND THEIR DERIVATIVES

Carboxylic acids have much higher acidity (typical pK$_a$ values range from 3 to 5) compared to alcohols (typical pK$_a$ = 16–18) because of the resonance stabilization of carboxylate anions (see Section 2.3 for a detailed explanation). The presence of a substituent in the alkyl chain of carboxylic acid (RCO$_2$H) has a relatively weak effect on its acidity, depending on the electron-withdrawing or electron-donating inductive effects and the distance of substituent from the oxygen atoms of carboxylic group (Section 2.3). For example, fluoroacetic acid (pK$_a$ = 2.59) is more acidic than acetic acid (pK$_a$ = 4.76) because the negative charge on the oxygen atoms of the fluoroacetate anion is slightly more delocalized, owing to the shift of electronic density to the more electronegative fluorine atom (Figure 12.4). Chloroacetic acid (pK$_a$ = 2.85) is less acidic than fluoroacetic acid because Cl is a less electronegative element than F. Dichloroacetic acid (pK$_a$ = 1.48) is stronger than chloroacetic acid because it has two electron-withdrawing substituents. 3-Chloropropanoic acid (pK$_a$ = 3.98) is weaker than chloroacetic acid because the chlorine atom is farther away from the oxygen atoms of the carboxylic group. Propanoic acid (pK$_a$ = 4.86) is a slightly weaker acid than acetic acid (pK$_a$ = 4.76) because it has an additional electron-donating methyl group in the carbon chain (Figure 12.4). In general, all carboxylic acids are much weaker acids than the strong inorganic acids, such as HCl, HBr, HI, and H$_2$SO$_4$ (see Table 2.1). However, they form stable salts in reactions with strong bases.

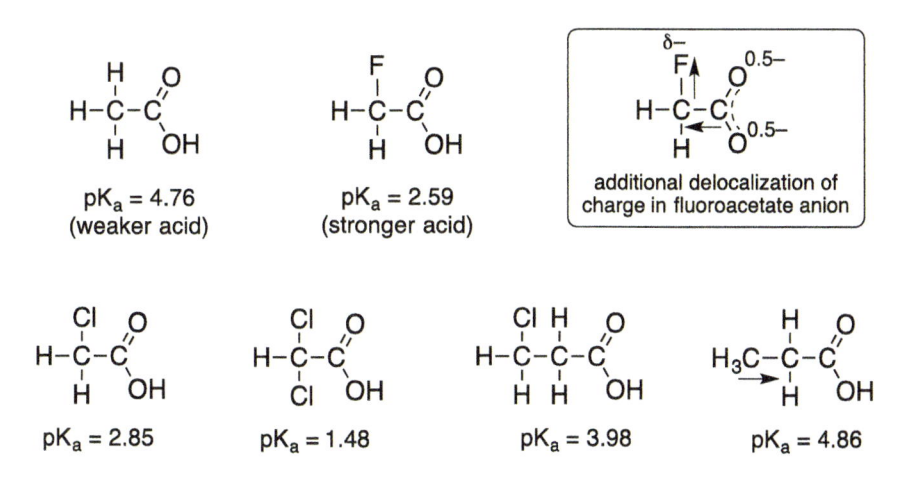

FIGURE 12.4. Effect of substituents on the acidity of carboxylic acids.

The N–H bond of amides and imides is also characterized by higher acidity than the N–H bonds of amines (the pK$_a$ value of RNH$_2$ is about 38). Analogous to the acidity of carboxylic acids, the higher acidity of amides and imides is explained by the resonance stabilization of the corresponding anions (Figure 12.5).

Carboxylic acids have higher boiling points than alcohols, aldehydes, and ketones of comparable size. For example, acetic acid has bp = +118 °C, which is much higher, compared to acetaldehyde (bp = +20 °C) and ethanol (bp = +78 °C). The high boiling points of carboxylic acids are explained by their polarity and the presence of strong hydrogen bonding between molecules. Amides have even higher boiling points (for example, the bp of acetamide is +221

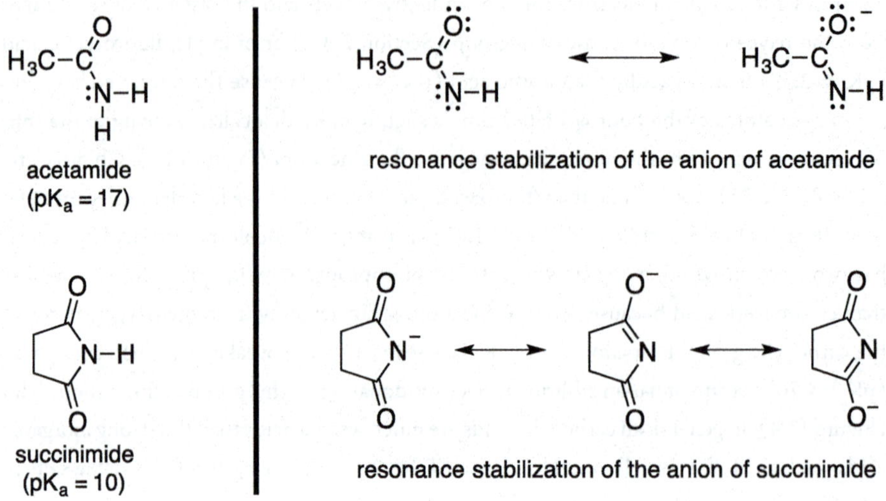

FIGURE 12.5. Acidity of amides and imides.

°C) because of the greater polarity of the amide functional group. Acid chlorides, anhydrides, esters, and nitriles have boiling points that are similar to those of aldehydes or ketones of comparable size. For example, acetyl chloride has a bp = +52 °C and ethyl acetate has a bp = +77 °C.

Problems

12.3. Sort the following compounds according to increasing boiling points:

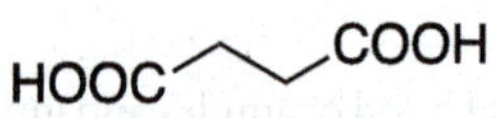

12.4. Rank these anions according to increasing basicity:

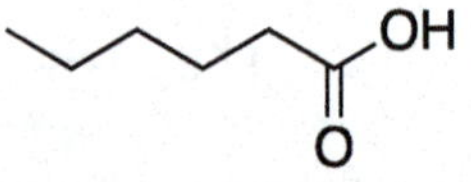

$ClCH_2COO^-$ FCH_2COO^- $CH_3CH_2COO^-$ CH_3COO^-

12.5. Sort the following carboxylic acids according to increasing acidity:

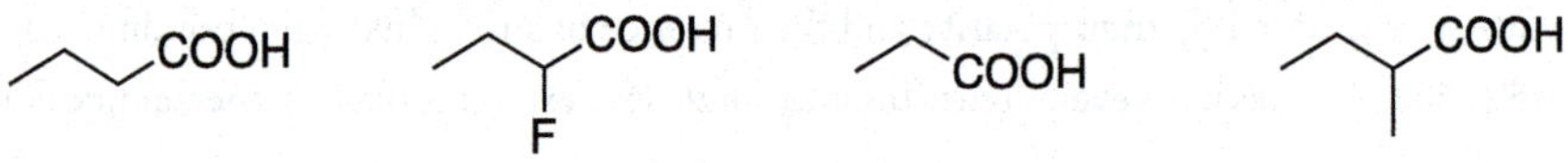

238   ORGANIC CHEMISTRY

12.3. GENERAL PRINCIPLES OF REACTIVITY OF ACYL DERIVATIVES: NUCLEOPHILIC ACYL SUBSTITUTION REACTIONS

Similar to the carbonyl group in aldehydes and ketones, the electrophilic carbon atom of the acyl group in acyl compounds can react with nucleophiles and the oxygen atom can be protonated (see Figure 11.2 and the related text in Section 11.2). However, there is a principal difference in the reactivity of ketones or aldehydes and acyl derivatives. Acyl derivatives (RCOY) have a potential leaving group (Y) that can be replaced by an appropriate nucleophile, while the carbon or hydrogen substituent in ketones or aldehydes in general is not a leaving group and cannot be replaced by another group. The reactions of carboxylic acids or acid derivatives resulting in the replacement of a substituent (the leaving group) at the carbon atom of an acyl group with a nucleophile are classified as **nucleophilic acyl substitution** reactions. All these reactions proceed according to a mechanism involving two general steps: (1) addition of a nucleophile (Nu) to the carbon of the acyl group with the formation of a tetrahedral intermediate, and (2) elimination of leaving group (Y) from the tetrahedral intermediate, forming the final substitution product (Figure 12.6). In addition to these two principal steps, several additional proton transfer steps are usually involved in the reactions of specific acyl compounds.

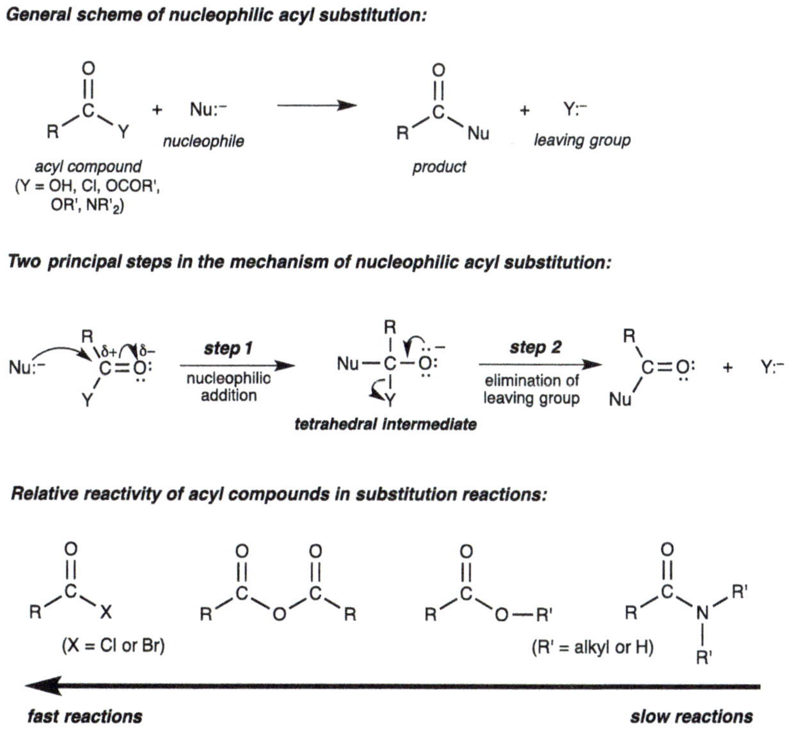

FIGURE 12.6. Nucleophilic acyl substitution reactions.

Reactions of acyl halides with nucleophiles are much faster than the reactions of other acyl derivatives, and amides are the least reactive acyl compounds. The reactions of carboxylic acids, esters, and amides with weak nucleophiles, such as alcohols or water, require the presence of a strong acid as a catalyst. The highest reactivity of acyl chlorides is explained by two factors: 1) the relatively high positive charge on the carbon atom of the acyl group, due to the high electronegativity of chlorine (this feature is essential in step 1 of the mechanism), and 2) the high stability of the chloride anion, which promotes a high leaving group ability for Cl, an important feature of the second step of the mechanism (Figure 12.6). Amides have the lowest reactivity due to the lower electronegativity of the nitrogen atom and poor leaving group ability of the amino group. Charge distribution in an amide molecule can be described by three resonance contributors, as shown in Figure 12.7. Because of the lower electronegativity of nitrogen (3.0) compared to oxygen (3.5), the contributor with a negative charge on O and a positive charge on N is particularly important in the overall structure of amides. The contributor with a positive charge on carbon has smaller input to the real structure (the resonance hybrid) of amides, which explains low reactivity of these compounds towards nucleophilic reagents. It is also important to note that the C–N bond in the resonance hybrid has a partial double bond character (Figure 12.7), which is an important structural feature of amides. Because of the structural rigidity and resistance to hydrolysis, amides are widespread in nature and find use in technology as structural materials. Amide linkage is the principal structural component of proteins, known as the peptide bond. Polyamides (known as nylons) are very robust synthetic polymeric materials (see Section 12.8).

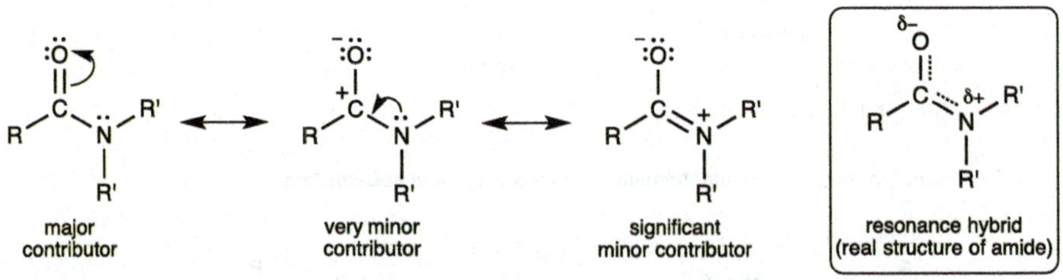

FIGURE 12.7. Resonance contributors of amides.

Problems

12.6. Provide the resonance contributor that explains why N,N-dimethylformamide (DMF) shows 3 signals in the ^{13}C NMR.

12.7. Sort these compounds according to their relative reactivity towards the nucleophilic acyl substitution:

12.4. PREPARATION AND INTERCONVERSION OF ACID DERIVATIVES

All acid derivatives can be prepared starting from carboxylic acids in one or two synthetic steps, by nucleophilic acyl substitution reactions. The hydroxyl fragment (OH) in a carboxylic group (COOH) is a poor leaving group and usually requires activation by protonation or by conversion to a better leaving group during the course of reactions. Acid derivatives can be converted back to acids by reaction with water (hydrolysis). Hydrolysis of the most reactive acyl compounds (acid chlorides and anhydrides) proceeds quickly and does not require any catalyst, while hydrolysis of esters and amides requires acid or base catalysis.

12.4.1. Acid chlorides

Acid chlorides are prepared by the reaction of carboxylic acids with thionyl chloride (SOCl$_2$). Analogous to the reaction of alcohols (see Figure 8.5 and the related text in Section 8.4), the mechanism involves the initial conversion of the hydroxyl group to a better leaving group, which is then replaced by a chloride anion via nucleophilic acyl substitution (Figure 12.8).

Acid chlorides are the most reactive acyl derivatives and readily participate in nucleophilic acyl substitution reactions with various neutral or anionic nucleophiles. Several examples of such reactions are shown in Figure 12.9.

Problem

12.8. Propose mechanisms for the reactions shown in Figure 12.9.

12.4.2. Anhydrides

Acid anhydrides can be prepared by nucleophilic acyl substitution of acid chlorides with anions of carboxylic acids (Figure 12.10). A more practical method is based on the

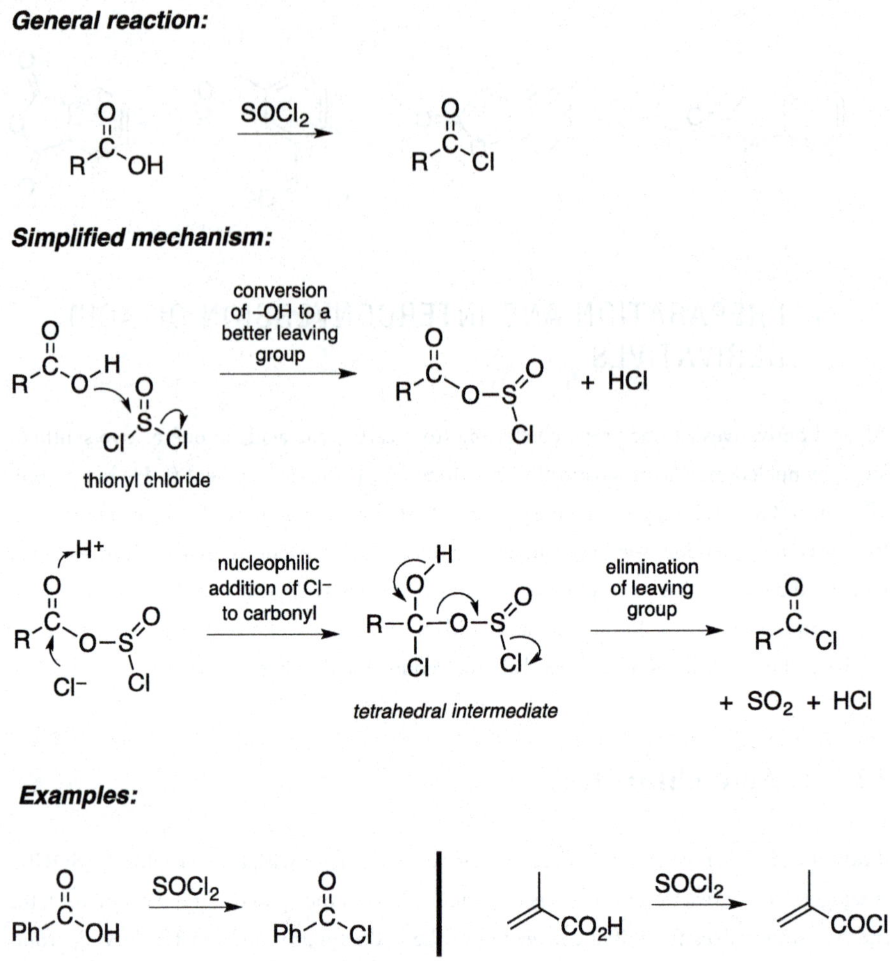

General reaction:

Simplified mechanism:

Examples:

FIGURE 12.8. Preparation of acid chlorides.

dehydration of carboxylic acids, using phosphorus pentoxide (P_2O_5) as a dehydrating reagent. Cyclic anhydrides are formed upon heating of the corresponding dicarboxylic acids.

Acid anhydrides readily participate in nucleophilic substitution reactions with various nucleophiles in the absence of acid catalysis. In particular, hydrolysis of an anhydride yields carboxylic acid and a similar reaction with alcohol yields an ester. Acetic anhydride is an industrial chemical widely used for preparing important esters of acetic acid such as cellulose acetate.

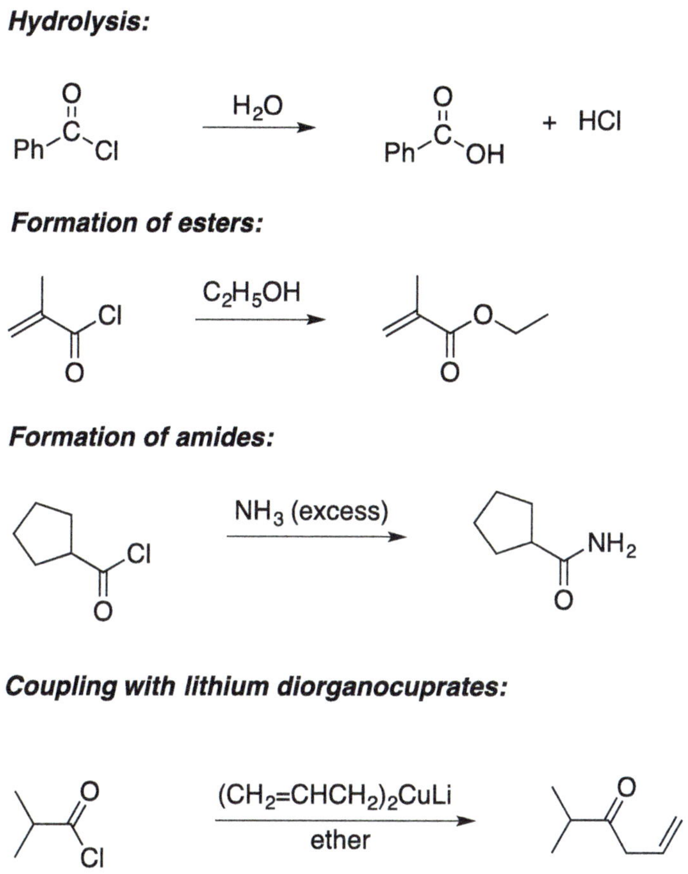

Hydrolysis:

Ph-C(=O)-Cl →[H₂O] Ph-C(=O)-OH + HCl

Formation of esters:

CH₂=C(CH₃)-C(=O)-Cl →[C₂H₅OH] CH₂=C(CH₃)-C(=O)-O-C₂H₅

Formation of amides:

cyclopentyl-C(=O)-Cl →[NH₃ (excess)] cyclopentyl-C(=O)-NH₂

Coupling with lithium diorganocuprates:

(CH₃)₂CH-C(=O)-Cl →[(CH₂=CHCH₂)₂CuLi, ether] (CH₃)₂CH-C(=O)-CH₂-CH=CH₂

FIGURE 12.9. Reactions of acid chlorides.

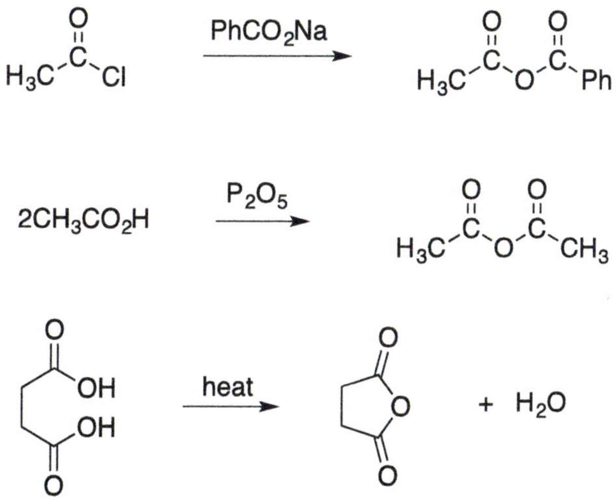

H₃C-C(=O)-Cl →[PhCO₂Na] H₃C-C(=O)-O-C(=O)-Ph

2 CH₃CO₂H →[P₂O₅] H₃C-C(=O)-O-C(=O)-CH₃

HOOC-CH₂-CH₂-COOH →[heat] succinic anhydride + H₂O

FIGURE 12.10. Preparation of acid anhydrides.

12.4.3. Esters

Esters can be prepared by the reaction of acid chlorides or anhydrides with alcohols (see previous sections). Carboxylic acids can react directly with alcohols in the presence of catalytic amounts of a strong acid (such as H_2SO_4), forming esters. This is a particularly important reaction known as **Fischer esterification,** named after the German chemist Emil Fischer, who was a recipient of the Nobel Prize in Chemistry in 1902. The acid catalyst in this reaction is needed to increase the electrophilic character of carboxylic carbon by the initial protonation of the carboxylic oxygen atom. The general scheme and mechanism of Fischer esterification are shown in Figure 12.11. In addition to the two principal steps of nucleophilic acyl substitution (Section 12.3), this mechanism includes several proton transfer steps to activate the carboxylic group, and also to improve the leaving group ability of a hydroxyl group in the elimination step.

Fischer esterification is a reversible reaction with about equal amounts of carboxylic acid, alcohol, ester, and water in the mixture under equilibrium conditions. The equilibrium can be shifted towards the formation of ester by using excess alcohol or by removing water from the reaction mixture. The reverse reaction in this equilibrium, acid-catalyzed hydrolysis, can be performed by reacting esters with an excess of water in the presence of a strong acid. All steps and intermediates in the mechanism of acid-catalyzed ester hydrolysis follow mechanistic steps of the esterification reaction (Figure 12.11) in reverse order.

Esters can also be hydrolyzed using an aqueous base (Figure 12.12). Basic hydrolysis is an irreversible reaction because the carboxylate anion produced is completely unreactive in nucleophilic acyl substitution.

Basic hydrolysis of esters is commonly known as a **saponification reaction.** Saponification is an industrial process that produces soap (a salt of fatty acid) and glycerol from fats (triglycerides of fatty acids) by treatment with a strong base (Figure 12.12).

Problems

12.9. ^{18}O is a heavy isotope of oxygen used to determine the mechanism of a reaction. Based on the mechanism shown in Figure 12.11, what is the fate of this oxygen atom?

PhCOOH + $H_3C-^{18}O-H$ →(cat. H+) ?

General reaction:

$$R-C(=O)-OH + R'OH \underset{\text{hydrolysis}}{\overset{\text{esterification}, H^+ (cat)}{\rightleftharpoons}} R-C(=O)-OR' + H_2O$$

carboxylic acid + alcohol ⇌ ester

Mechanism:

Step 1 — activation of carboxylic group by initial protonation: the carboxylic acid is protonated on the carbonyl oxygen to give a resonance-stabilized cation; *protonated carboxyl is more electrophilic.*

Step 2 — nucleophilic addition of alcohol to activated carboxyl, followed by proton transfer (loss of H$_2$O-bound proton to give H$_3$O$^+$), yielding the *tetrahedral intermediate*.

Step 3 — proton transfer (H$_3$O$^+$) onto one of the hydroxyl groups, then elimination of leaving group (H$_2$O) to give the resonance-stabilized *protonated ester*.

Step 4 — proton transfer to H$_2$O gives the neutral ester $R-C(=O)-O-R'$ + H$_3$O$^+$.

Examples:

$$Ph-C(=O)-OH \xrightarrow[H_2SO_4 \text{ (cat)}]{CH_3OH \text{ (excess)}} Ph-C(=O)-OCH_3$$

4-hydroxybutanoic acid $\xrightarrow{H^+ \text{ (cat)}}$ γ-butyrolactone

FIGURE 12.11. Fischer esterification.

12.10. Use Fischer esterification to prepare the following esters:

- PhCH$_2$C(=O)OCH$_3$ (methyl phenylacetate)
- H$_3$C–C(=O)–O–CH$_2$Ph (benzyl acetate)
- ortho-C$_6$H$_4$(C(=O)OCH(CH$_3$)$_2$)(CHO) (isopropyl 2-formylbenzoate)

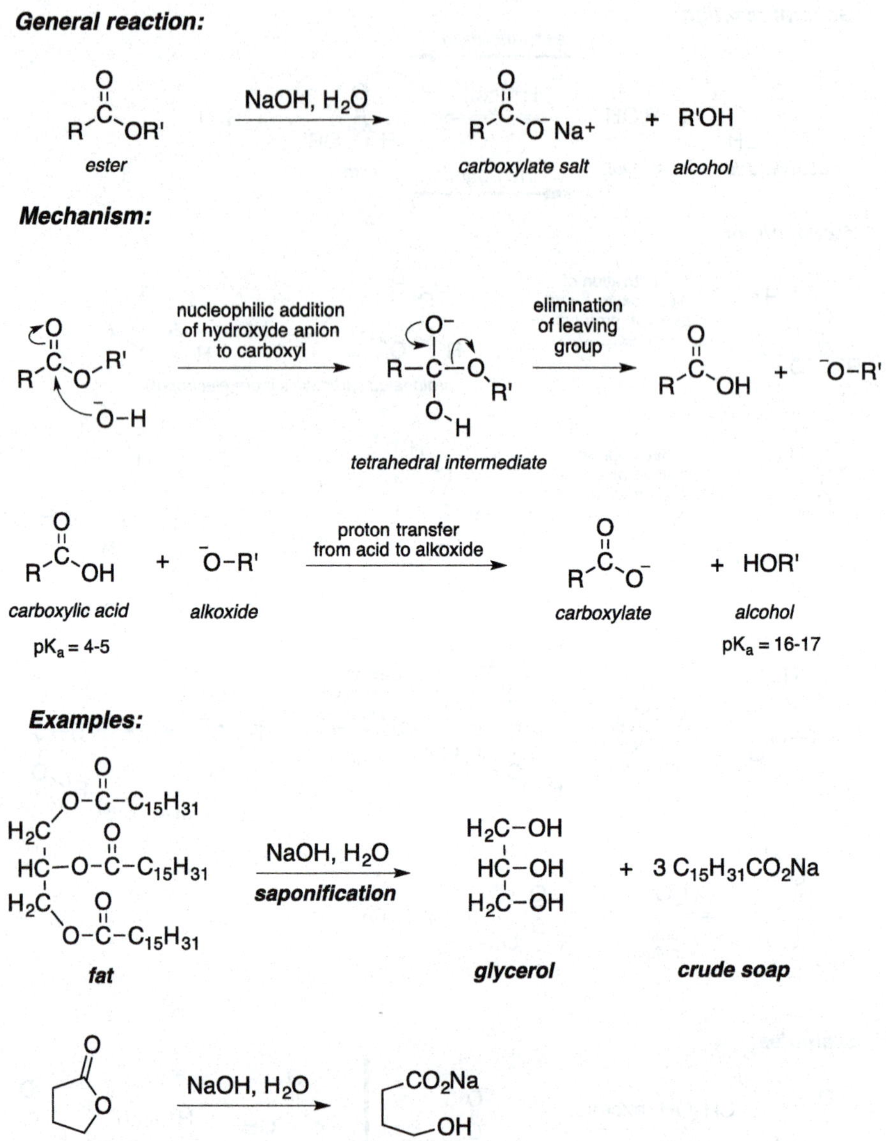

FIGURE 12.12. Hydrolysis of esters using aqueous base.

12.11. Esters can be hydrolyzed under acidic as well as under basic conditions. Why is the hydrolysis under basic conditions irreversible, while under acidic conditions it is reversible?

12.4.4. Amides

Amides can be prepared by reactions of any other, more reactive acyl derivatives (see Figure 12.6) with ammonia or 1° or 2° amines. A practically important approach involves the reaction of acid chlorides with appropriate amines (Figure 12.13). Two equivalents of the amine are required in order to remove HCl formed as the reaction by-product. The mechanism of this reaction involves replacement of a chloride anion with a neutral nitrogen nucleophile, followed by proton transfer and elimination steps. This is an irreversible reaction because of the low reactivity of the amide product in acyl nucleophilic substitution.

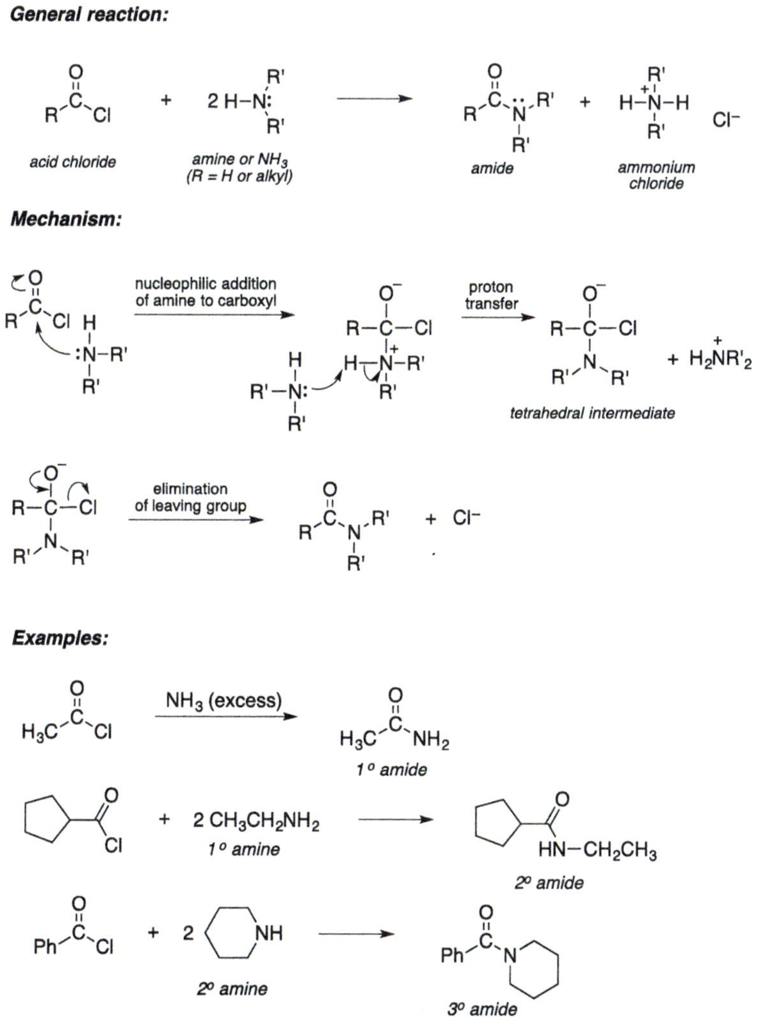

FIGURE 12.13. Preparation of amides.

Heating with an aqueous acid or base solution can hydrolize amides. Hydrolysis of an amide with one equivalent of acid produces ammonium salt as a by-product. This is an irreversible reaction because an ammonium ion cannot act as a nucleophile. Basic hydrolysis is also an irreversible reaction because the produced carboxylate anion is completely unreactive with amines in the nucleophilic acyl substitution. The mechanisms of amide hydrolysis is similar to that of ester hydrolysis, with a molecule of amine acting as a leaving group. Figure 12.14 shows several examples of amide hydrolysis reactions.

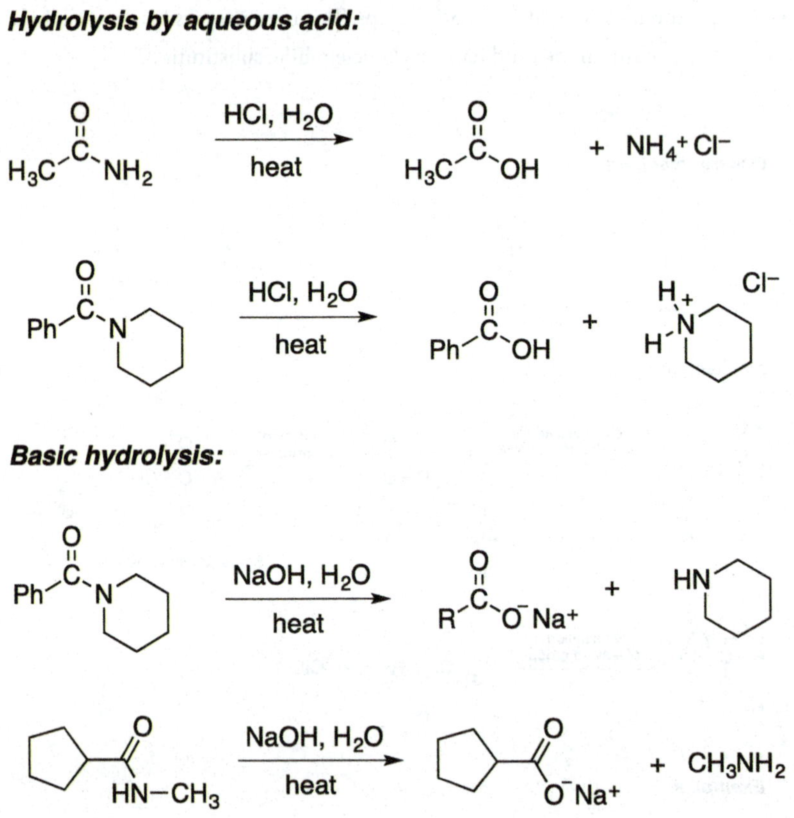

Hydrolysis by aqueous acid:

$$H_3C-CONH_2 \xrightarrow[\text{heat}]{HCl, H_2O} H_3C-COOH + NH_4^+ Cl^-$$

$$Ph-CO-N(\text{piperidine}) \xrightarrow[\text{heat}]{HCl, H_2O} Ph-COOH + \text{H}_2N^+(\text{piperidine}) \, Cl^-$$

Basic hydrolysis:

$$Ph-CO-N(\text{piperidine}) \xrightarrow[\text{heat}]{NaOH, H_2O} R-COO^-Na^+ + HN(\text{piperidine})$$

$$\text{cyclopentyl-CO-NH-CH}_3 \xrightarrow[\text{heat}]{NaOH, H_2O} \text{cyclopentyl-COO}^-Na^+ + CH_3NH_2$$

FIGURE 12.14. Hydrolysis of amides.

Primary amides (RCONH$_2$) can be dehydrated by heating in the presence of strong dehydrating reagents, such as P$_2$O$_5$, producing nitriles (RC≡N). A more practically useful method of nitrile formation is based on the S$_N$2 reaction of alkyl halides with a cyanide anion (see Figure 5.3 in Chapter 5). Nitriles can be hydrolyzed back into amides, and eventually to the corresponding acids, by heating in the presence of aqueous acid or base.

Problems

12.12. Propose mechanisms for the reactions shown in Figure 12.14.

12.13. Under acidic conditions, amides are protonated. For example:

Using resonance structures, explain the protonation of an amide occurs on the carbonyl group and not on the nitrogen atom?

12.14. Both esters and amides can be hydrolyzed to form carboxylic acid under acidic conditions. While the mechanisms of these reactions are very similar, the hydrolysis of amides under these conditions require at least an equimolar amount of acid, while for the hydrolysis of esters, a catalytical amount is fully sufficient. Explain.

12.5. REACTION WITH HYDRIDE ANION

Carboxylic acids react with lithium aluminum hydride (LiAlH$_4$; see Section 11.4) to produce the corresponding 1° alcohols as final products (Figure 12.15). The reaction is carried out under anhydrous conditions in ether or THF solution, and starts from deprotonation of the acid by a hydride anion, yielding an aluminum carboxylate complex. In the next step, the second hydride anion replaces the aluminum oxyanion via nucleophilic acyl substitution. This process produces aldehyde as the initial product of reaction. In most cases, it is impossible to isolate aldehyde as the final product due to the high reactivity of the aldehyde carbonyl group toward sources of the hydride anion in the reaction mixture. In the final reaction step, the third hydride anion adds to the aldehyde carbonyl group, producing alkoxide anions (Section 11.4). The alkoxide is converted to the final alcohol by careful addition of aqueous acid solution. Sodium borohydride (NaBH$_4$) is a less reactive source of hydride anions (Section 11.4) and does not react with carboxylic acids. However, borane (BH$_3$ as a complex with THF; see Section 6.4) can reduce acids to alcohols, analogous to LiAlH$_4$.

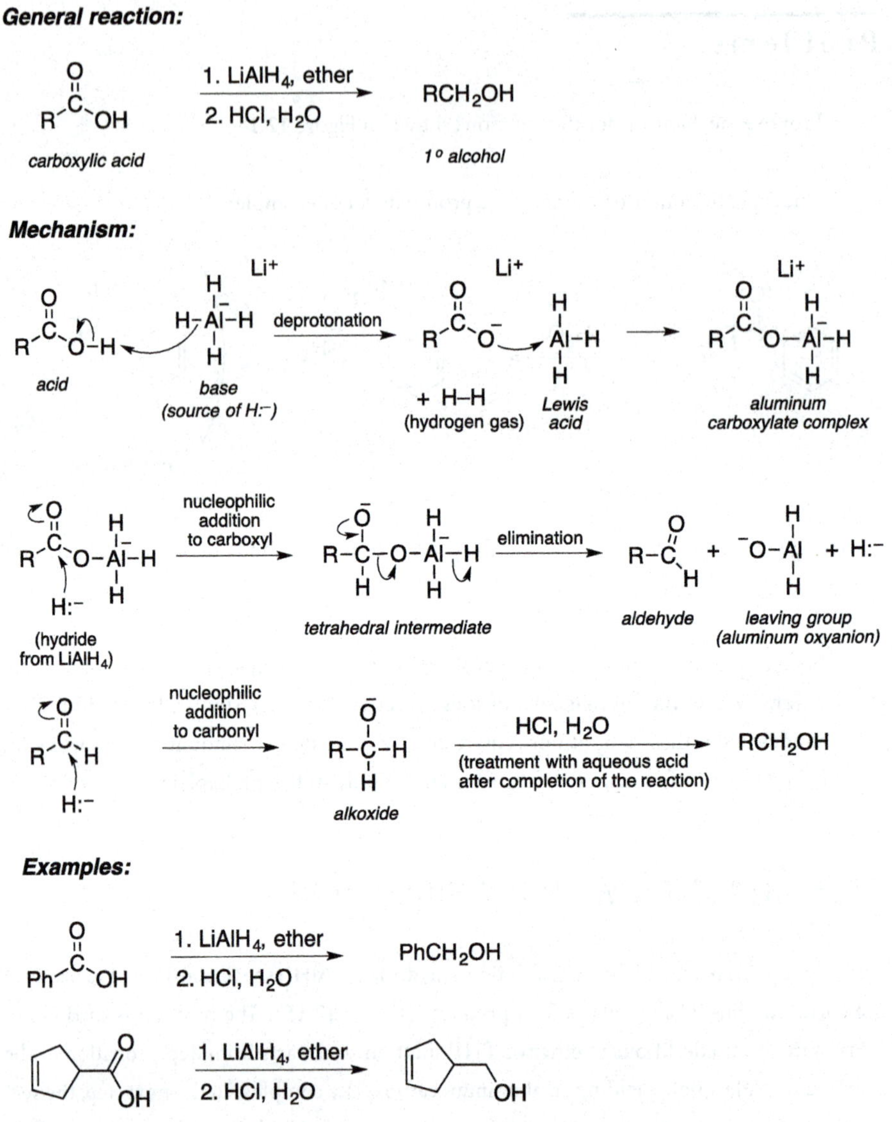

General reaction:

$$\underset{\text{carboxylic acid}}{R-\overset{O}{\underset{\|}{C}}-OH} \quad \xrightarrow[\text{2. HCl, H}_2\text{O}]{\text{1. LiAlH}_4, \text{ ether}} \quad \underset{\text{1° alcohol}}{RCH_2OH}$$

Mechanism:

[Mechanism diagram: carboxylic acid + LiAlH₄ → deprotonation gives H–H (hydrogen gas) and Lewis acid intermediate → aluminum carboxylate complex]

[Nucleophilic addition to carboxyl → tetrahedral intermediate → elimination → aldehyde + ⁻O–Al leaving group (aluminum oxyanion) + H:⁻]

[Nucleophilic addition to carbonyl → alkoxide → HCl, H₂O (treatment with aqueous acid after completion of the reaction) → RCH₂OH]

Examples:

$$\underset{}{Ph-\overset{O}{\underset{\|}{C}}-OH} \quad \xrightarrow[\text{2. HCl, H}_2\text{O}]{\text{1. LiAlH}_4, \text{ ether}} \quad PhCH_2OH$$

[cyclopentenyl-COOH] $\xrightarrow[\text{2. HCl, H}_2\text{O}]{\text{1. LiAlH}_4, \text{ ether}}$ [cyclopentenyl-CH₂OH]

FIGURE 12.15. Reduction of carboxylic acid to a 1o alcohol by LiAlH$_4$.

Esters react with LiAlH$_4$ under similar conditions to produce two alcohols: a molecule of a 1° alcohol, resulting from the addition of two hydride anions to the carbon of the acyl group; and a molecule of alcohol formed from the alkoxy group (⁻OR'), acting as a leaving group in the nucleophilic acyl substitution (Figure 12.16).

The analogous reaction of amides with LiAlH$_4$ forms amines as final products. The mechanism of this reaction is similar to the reduction of carboxylic acids, and involves nucleophilic substitution of the aluminum oxyanion by hydride anion (Figure 12.17). This reaction provides an important method of synthesizing 1°, 2°, and 3° amines from the

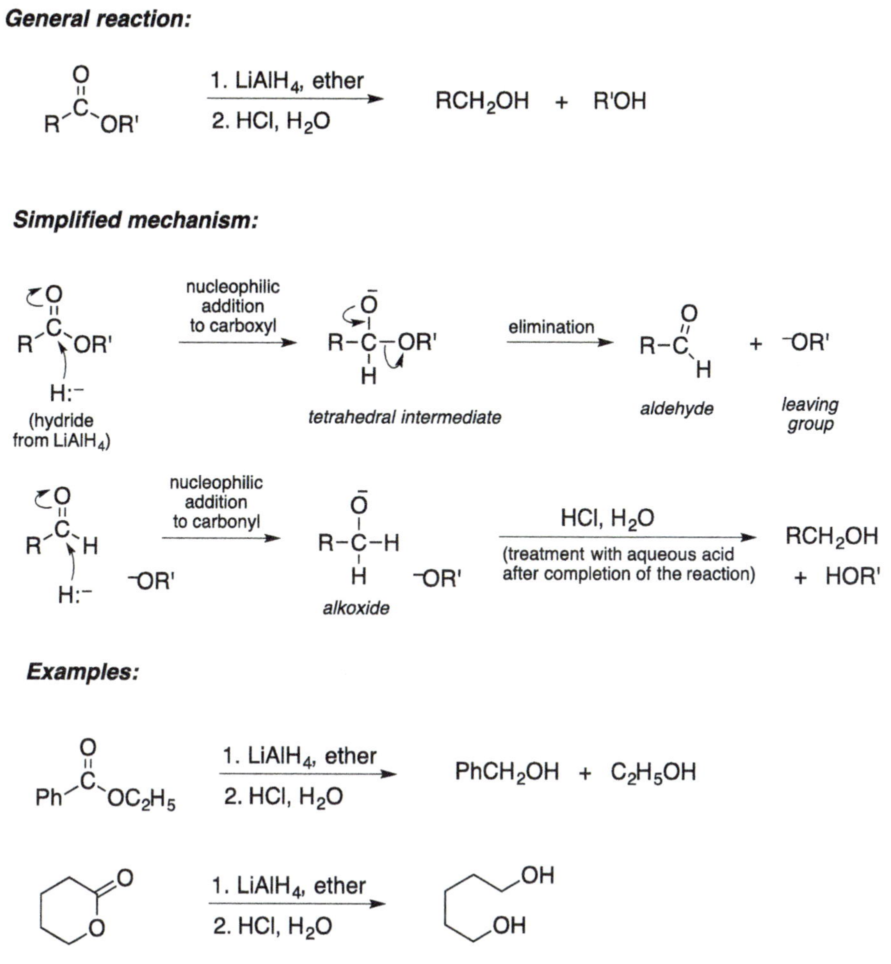

General reaction:

$$\underset{R}{\overset{O}{\parallel}}C-OR' \xrightarrow[\text{2. HCl, H}_2\text{O}]{\text{1. LiAlH}_4, \text{ ether}} RCH_2OH + R'OH$$

Simplified mechanism:

Step 1: nucleophilic addition to carboxyl — hydride (H:⁻ from LiAlH₄) adds to R−C(=O)−OR' giving tetrahedral intermediate R−C(H)(O⁻)−OR'; elimination gives aldehyde R−C(=O)−H + ⁻OR' (leaving group).

Step 2: nucleophilic addition to carbonyl — H:⁻ adds to R−C(=O)−H giving alkoxide R−C(H)(H)−O⁻ (with ⁻OR'); then HCl, H₂O (treatment with aqueous acid after completion of the reaction) gives RCH₂OH + HOR'.

Examples:

$$\underset{Ph}{\overset{O}{\parallel}}C-OC_2H_5 \xrightarrow[\text{2. HCl, H}_2\text{O}]{\text{1. LiAlH}_4, \text{ ether}} PhCH_2OH + C_2H_5OH$$

$$\text{(δ-valerolactone)} \xrightarrow[\text{2. HCl, H}_2\text{O}]{\text{1. LiAlH}_4, \text{ ether}} \text{HO−(CH}_2\text{)}_5\text{−OH}$$

FIGURE 12.16. Reactions of esters with LiAlH$_4$.

corresponding amides. Nitriles (RC≡N) are reduced by LiAlH$_4$ under similar conditions, producing 1° amines (RCH$_2$NH$_2$) as the final products.

It should be emphasized that only LiAlH$_4$ can serve as an effective source of hydride anions in reactions with carboxylic acids, esters, and amides. Sodium borohydride (NaBH$_4$) is a less reactive source of hydride anions (Section 11.4) and does not react with the less electrophilic acyl compounds. For example, the reaction of methyl 3-oxopropanoate with NaBH$_4$ yields methyl 3-hydroxypropanoate (Figure 12.18). Reaction of the same compound with lithium aluminum hydride results in the reduction of both aldehyde and ester functions to hydroxyl groups. A selective reduction of only ester function requires protection of the aldehyde group via cyclic acetal (see Section 11.7).

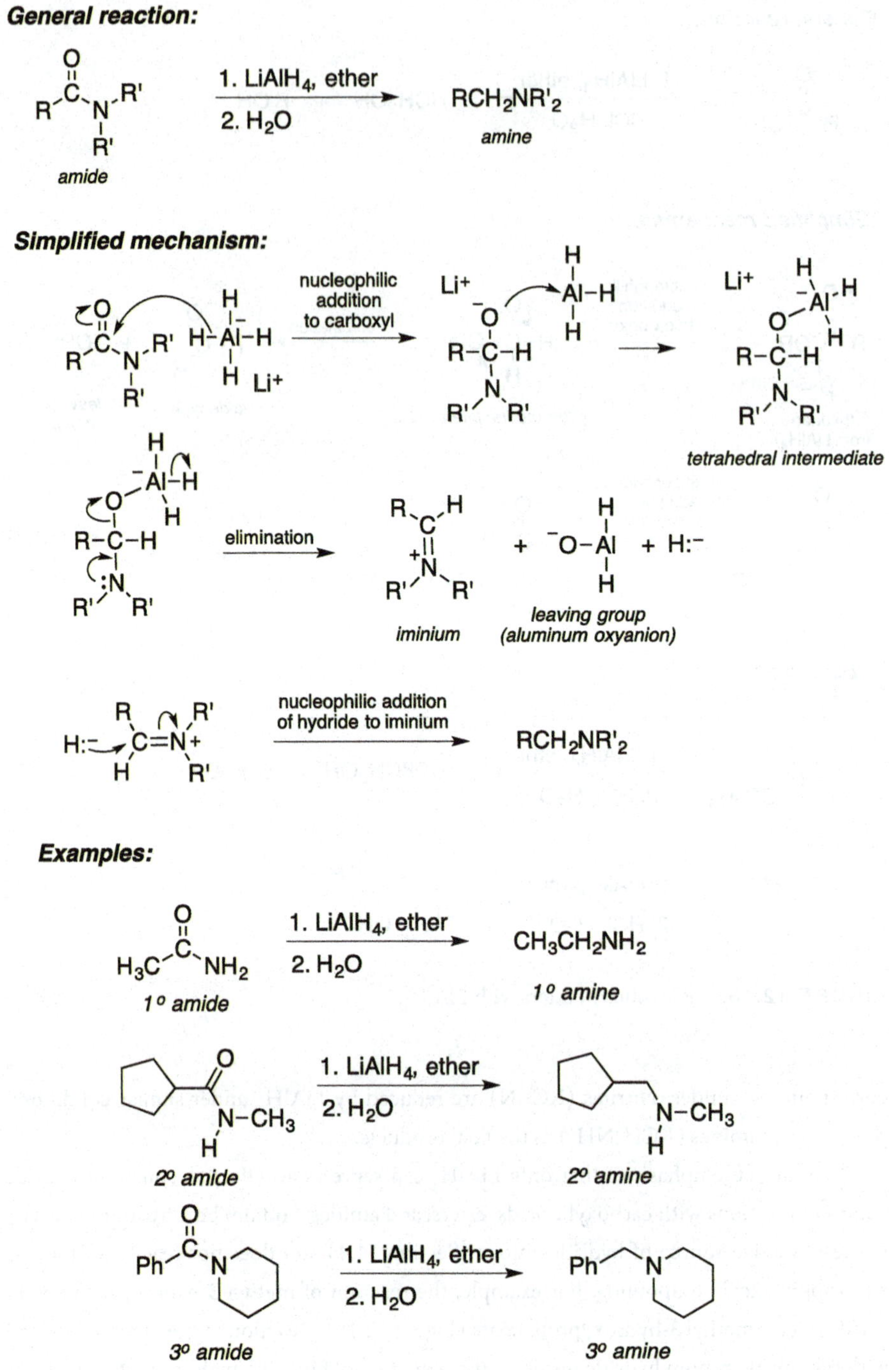

FIGURE 12.17. Reactions of amides with LiAlH$_4$.

252   ORGANIC CHEMISTRY

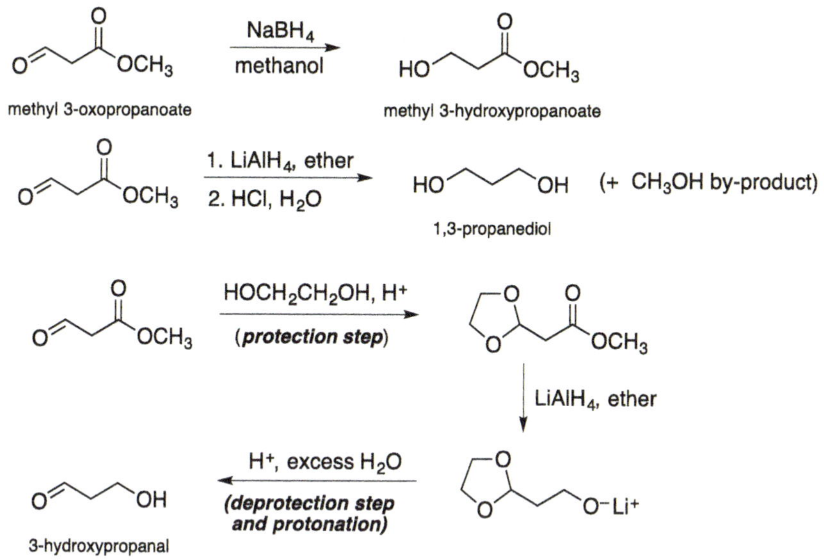

FIGURE 12.18. Selective reduction with NaBH4 and LiAlH$_4$.

Problems

12.15. Propose mechanisms for the reactions shown in Figure 12.18.

12.16. When isatin is reacted with LiAlH$_4$, both carbonyl groups are reduced. Treatment of the reaction mixture with diluted HCl results in the elimination of water and the formation of indole as major product. What is the structure of the product of the first step in this reaction sequence? What would be the expected outcome if you were to protect the ketone functional group?

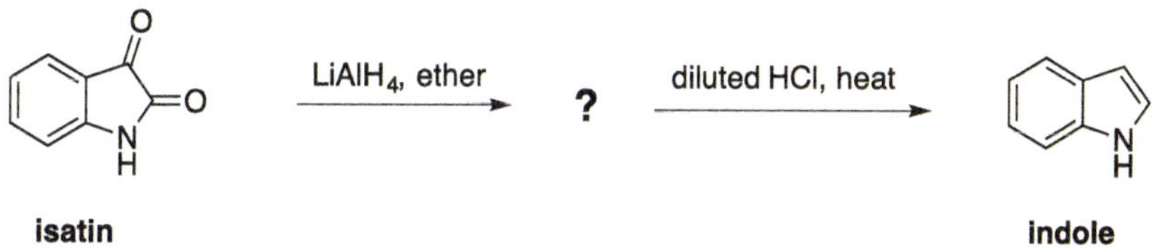

12.17. Provide either the starting material or the major products of the following sequence of reactions, as indicated:

a)

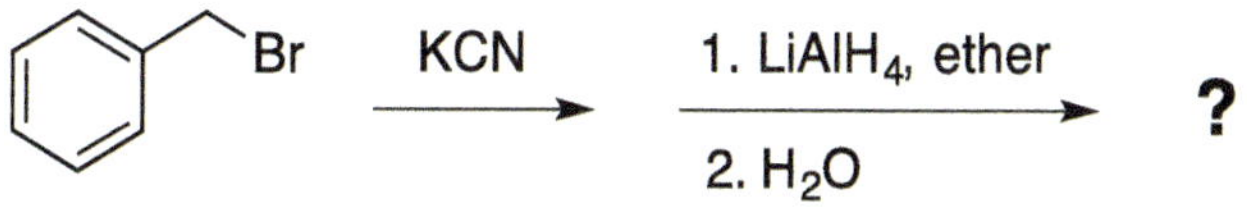

b)

? $\xrightarrow{\text{HN(CH}_3)_2 \text{ (excess)}}$ $\xrightarrow{\begin{array}{c}\text{1. LiAlH}_4\text{, ether}\\\text{2. H}_2\text{O}\end{array}}$ Ph-CH$_2$-N(CH$_3$)$_2$

c)

(cyclic lactam with NH, 6-membered ring) $\xrightarrow{\begin{array}{c}\text{1. LiAlH}_4\text{, ether}\\\text{2. H}_2\text{O}\end{array}}$ $\xrightarrow{\begin{array}{c}\text{CH}_3\text{I}\\\text{NaHCO}_3\text{ (base)}\end{array}}$?

12.6. REACTION WITH CARBANIONS

Carbanions are strong nucleophiles that can readily react with acyl compounds. From a practical viewpoint, reactions of esters with organolithium and organomagnesium reagents are the most useful reactions of this type. The reactions of esters with carbanions proceed analogously to the reactions with hydride anions (Section 12.5), producing two alcohols as the final products. One alcohol results from the addition of two carbanions to the carbon of the acyl group, and the second alcohol is formed by the alkoxy group ($^-$OR') acting as a leaving group in the nucleophilic acyl substitution (Figure 12.19). Reactions of the esters of formic acid (HCO$_2$R') with organometallic reagents produce 2° alcohols (R"$_2$HCOH), while 3° alcohols (RR"$_2$COH) are formed from the esters of all other carboxylic acids (RCO$_2$R').

Problems

12.18. Which of the following molecules cannot be prepared by treating ethyl acetate with a Grignard or organolithium reagent?

(four structures: t-butanol-like with OH; Ph$_2$C(OH)CH$_3$; CH$_2$=CH-C(OH)(CH$_3$)$_2$; (CH$_3$)$_2$C(OH)CH$_2$CH$_3$)

12.19. A common procedure for the preparation of carboxylic acids (RCOOH) consists of the treatment of Grignard reagents (RMgBr) with carbon dioxide (CO$_2$), followed by

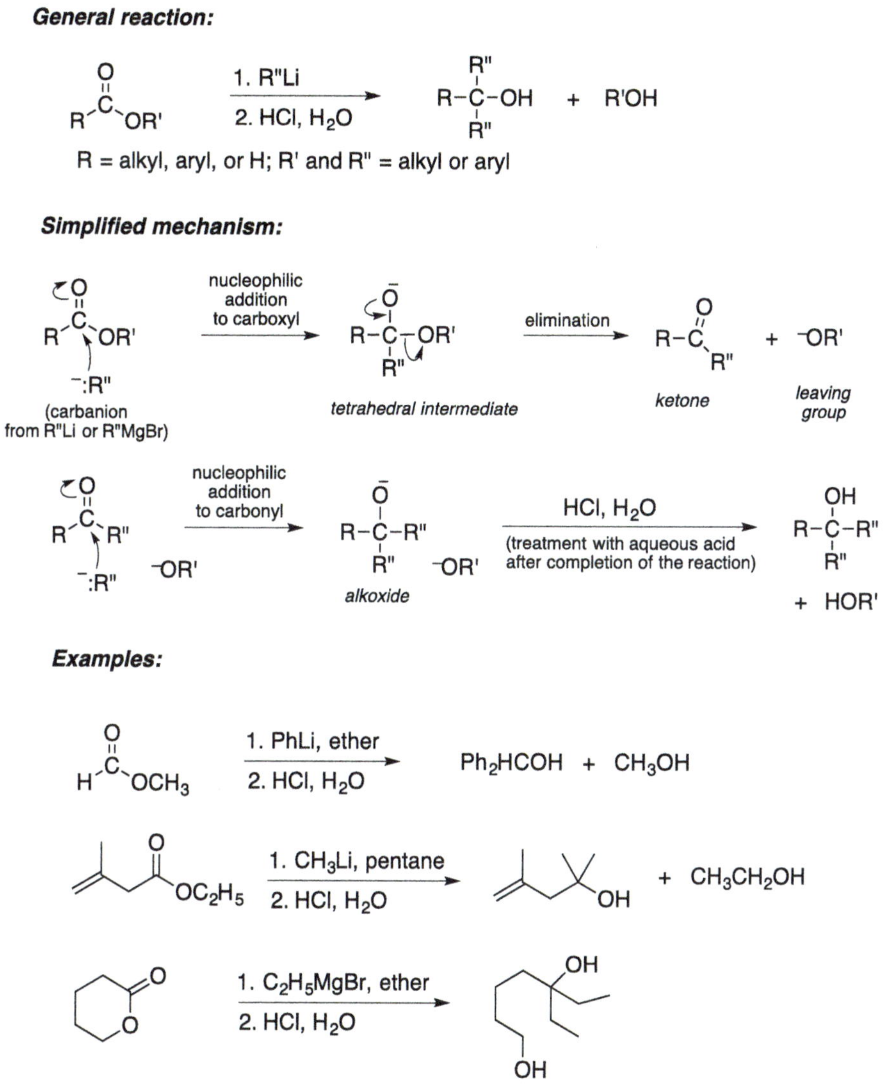

General reaction:

$$R-\underset{OR'}{\overset{O}{\overset{\|}{C}}} \xrightarrow[\text{2. HCl, H}_2\text{O}]{\text{1. R"Li}} R-\underset{R"}{\overset{R"}{\underset{|}{C}}}-OH + R'OH$$

R = alkyl, aryl, or H; R' and R" = alkyl or aryl

Simplified mechanism:

nucleophilic addition to carboxyl → tetrahedral intermediate → elimination → ketone + ⁻OR' (leaving group)

(carbanion from R"Li or R"MgBr)

nucleophilic addition to carbonyl → alkoxide → HCl, H₂O (treatment with aqueous acid after completion of the reaction) → R-C(OH)(R")R" + HOR'

Examples:

$$\underset{OCH_3}{\overset{O}{\overset{\|}{H-C}}} \xrightarrow[\text{2. HCl, H}_2\text{O}]{\text{1. PhLi, ether}} Ph_2HCOH + CH_3OH$$

$$\xrightarrow[\text{2. HCl, H}_2\text{O}]{\text{1. CH}_3\text{Li, pentane}} + CH_3CH_2OH$$

$$\xrightarrow[\text{2. HCl, H}_2\text{O}]{\text{1. C}_2\text{H}_5\text{MgBr, ether}}$$

FIGURE 12.19. Reaction of esters with organolithium or organomagnesium reagents.

the addition of a strong acid (aqueous HCl or H_2SO_4). Propose a mechanism for this reaction.

12.20. Provide the product of the following reaction:

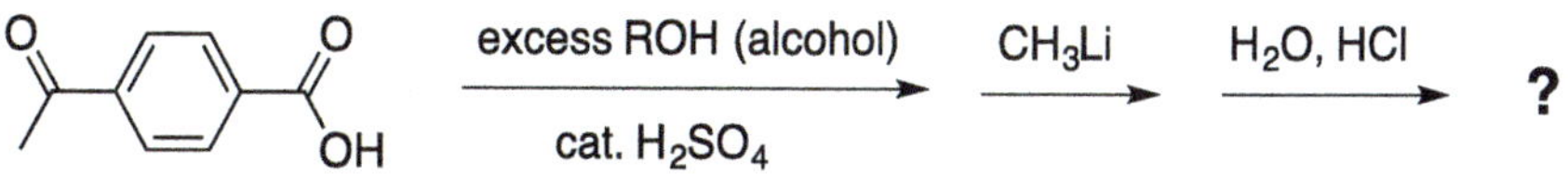

$$\xrightarrow[\text{cat. H}_2\text{SO}_4]{\text{excess ROH (alcohol)}} \xrightarrow{\text{CH}_3\text{Li}} \xrightarrow{\text{H}_2\text{O, HCl}} ?$$

12.21. Gilman reagents are less reactive than Grignard or organolithium reagents towards nucleophilic acyl substitution. In general, only acid chlorides react with these reagents and ketones are formed as products. For example:

For each of the following compounds propose a synthesis from an acid chloride and a Gilman reagent:

12.7. REACTION OF ENOLATE ANIONS DERIVED FROM ESTERS: CLAISEN CONDENSATION AND RELATED REACTIONS

Analogous to ketones and aldehydes (Section 11.9), acyl compounds have relatively acidic hydrogens at the carbon atom attached to the acyl group (the α-position). Similar to ketones, esters can be deprotonated at the α-position using common bases (alkoxides, lithium diisopropylamide, and metal hydrides), producing the corresponding enolate anions. Enolate anions of esters are strong nucleophiles, highly reactive in nucleophilic substitution, nucleophilic addition, and nucleophilic acyl substitution reactions.

Especially important is the nucleophilic acyl substitution reaction of enolate anions generated from ester producing a β-ketoester (Figure 12.20). This reaction is known as **Claisen condensation**, named after the German chemist Rainer Ludwig Claisen, who first published his work on condensation of esters in 1887. Claisen condensation can involve two molecules of the same ester or two different esters (crossed Claisen condensation). Molecules with two ester groups (diesters) can produce cyclic products of intramolecular condensation (**Dieckmann condensation**).

The products of Claisen condensation, β-ketoesters, can be hydrolyzed under mild conditions yielding β-ketoacids. β-ketoacids and β-dicarboxylic acids have low thermal stability and can easily eliminate a molecule of CO_2 upon gentle heating (Figure 12.21). The process of elimination of a molecule of carbon dioxide from carboxylic acid is called **decarboxylation**.

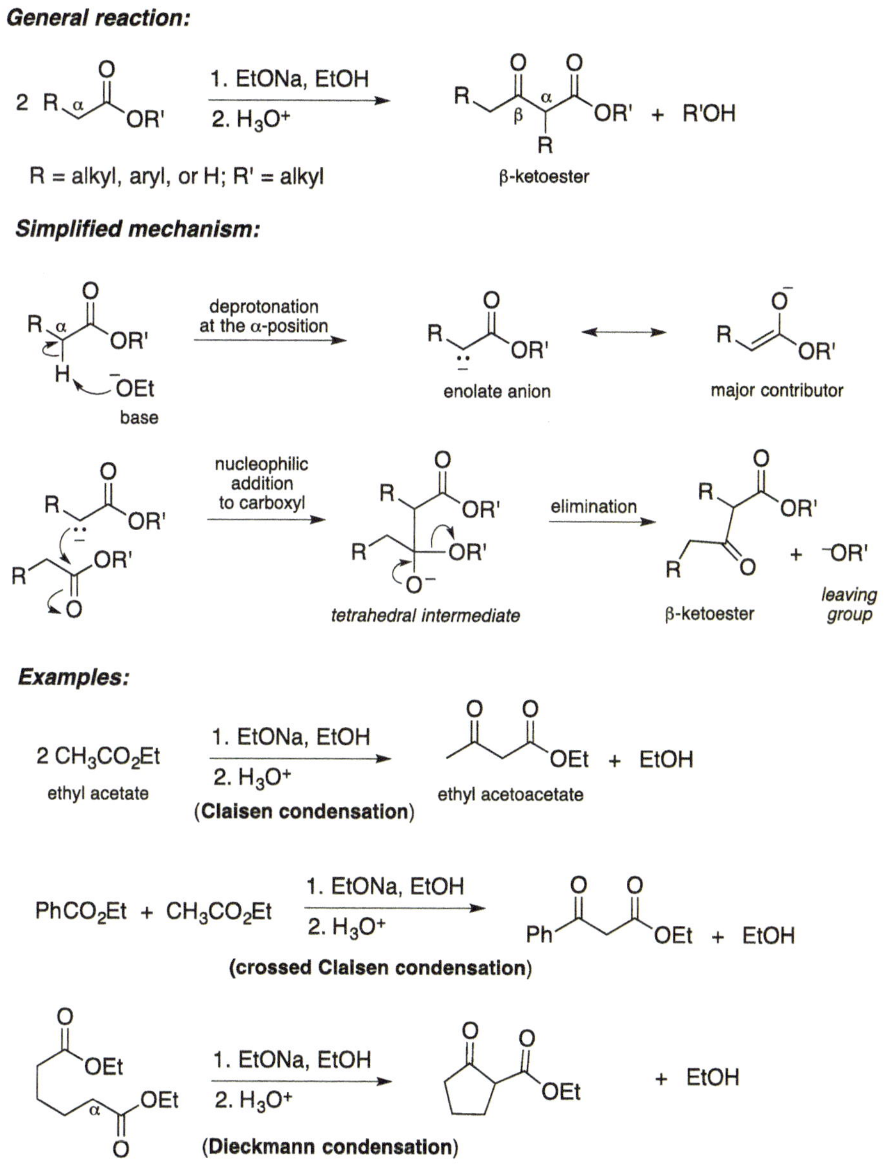

General reaction:

$$2 \text{ R-CH}_2\text{-CO-OR'} \xrightarrow[\text{2. H}_3\text{O}^+]{\text{1. EtONa, EtOH}} \text{R-CH}_2\text{-CO-CHR-CO-OR'} + \text{R'OH}$$

R = alkyl, aryl, or H; R' = alkyl

β-ketoester

Simplified mechanism:

[Mechanism diagram showing: deprotonation at the α-position → enolate anion ↔ major contributor; then nucleophilic addition to carboxyl → tetrahedral intermediate → elimination → β-ketoester + ⁻OR' (leaving group)]

Examples:

$$2 \text{ CH}_3\text{CO}_2\text{Et} \xrightarrow[\text{2. H}_3\text{O}^+]{\text{1. EtONa, EtOH}} \text{CH}_3\text{-CO-CH}_2\text{-CO-OEt} + \text{EtOH}$$

ethyl acetate → ethyl acetoacetate

(Claisen condensation)

$$\text{PhCO}_2\text{Et} + \text{CH}_3\text{CO}_2\text{Et} \xrightarrow[\text{2. H}_3\text{O}^+]{\text{1. EtONa, EtOH}} \text{Ph-CO-CH}_2\text{-CO-OEt} + \text{EtOH}$$

(crossed Claisen condensation)

[Dieckmann condensation: diethyl adipate → ethyl 2-oxocyclopentanecarboxylate + EtOH]

(Dieckmann condensation)

FIGURE 12.20. Condensation reactions of esters.

The product of decarboxylation of a β-ketoacid is the corresponding ketone, while decarboxylation of a β-dicarboxylic acid yields the corresponding monocarboxylic acid. β-ketoacids and β-dicarboxylic acids predominantly exist in cyclic conformation due to intramolecular hydrogen bonding, and the mechanism of decarboxylation involves the concerted rearrangement of bonding electrons, as shown in Figure 12.21.

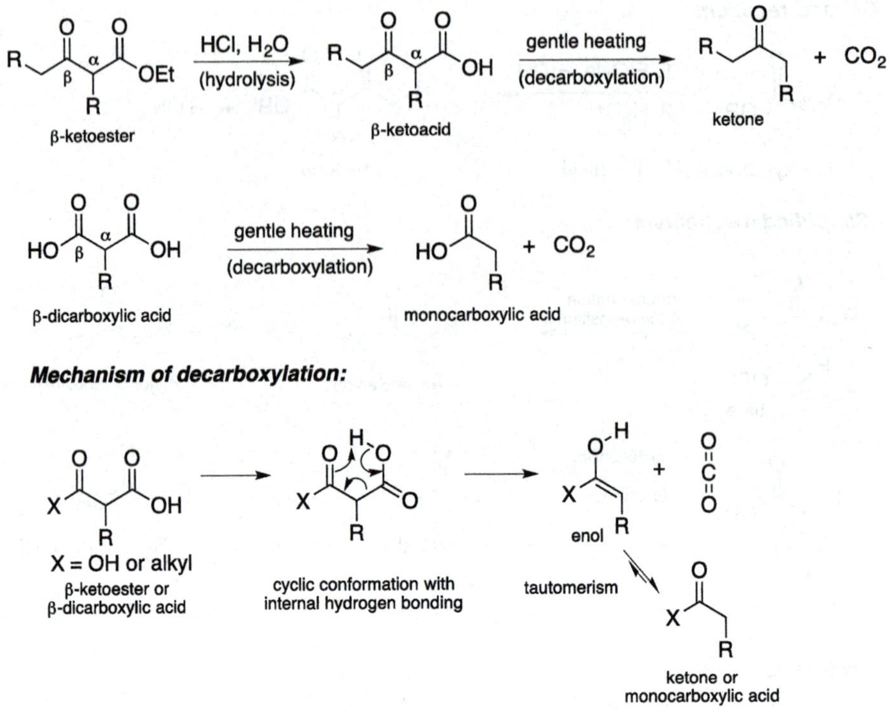

FIGURE 12.21. Decarboxylation of β-ketoacids and β-dicarboxylic acids.

Esters of β-ketoacids and β-dicarboxylic acids are important reagents widely utilized in organic synthesis. The α-hydrogen between two carbonyls in these compounds has an especially high acidity (pK_a about 10) and can be easily removed by treatment with a base, producing highly nucleophilic enolate anions. These enolate anions are then reacted with alkyl halides, carbonyl compounds, or esters, forming the corresponding products of condensation. At the end of reaction, the ester function is hydrolyzed and the resulting β-ketoacid (or β-dicarboxylic acid) is decarboxylated to yield the final product of reaction (Figure 12.22). Such condensation-decarboxylation reaction sequences are commonly known as the **acetoacetic ester synthesis** and the **malonic ester synthesis** (Figure 12.22).

Problems

12.22. Write detailed mechanism for the Dieckman condensation (Figure 12.20).

12.23. Write detailed mechanisms for reactions shown in Figure 12.22.

Acetoacetic ester synthesis:

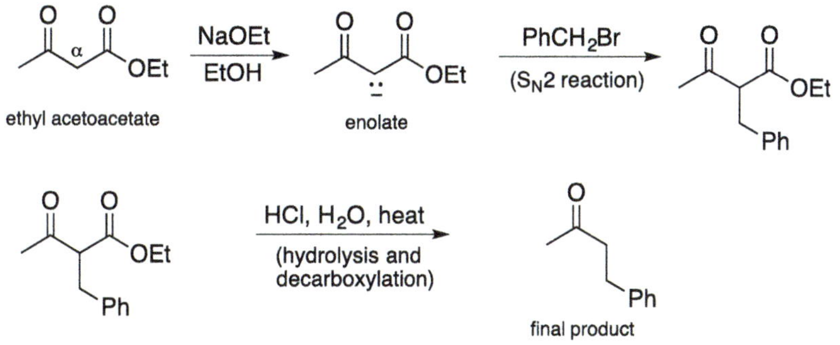

Malonic ester synthesis:

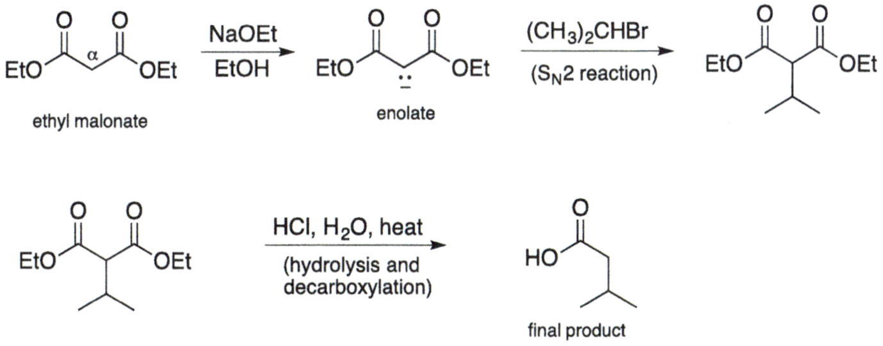

FIGURE 12.22. Examples of acetoacetic ester and malonic ester synthesis.

12.24 Provide the major products of the following reactions:

a)

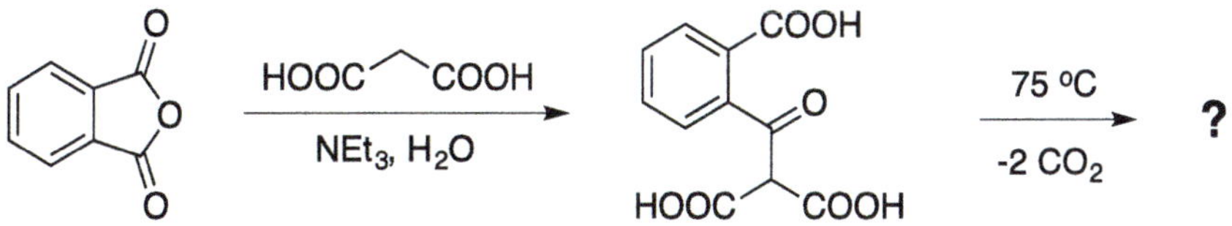

b)

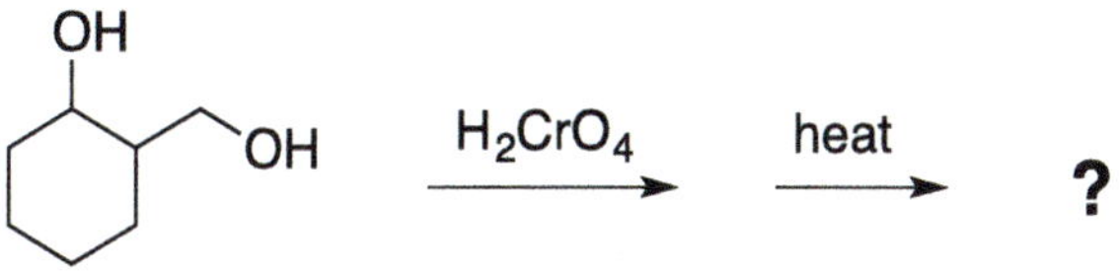

CARBOXYLIC ACIDS AND THEIR DERIVATIVES 259

c)

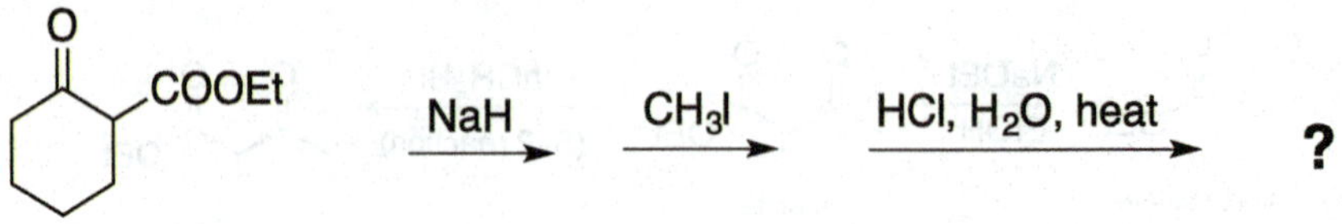

[Structure: 2-carbethoxycyclohexanone] →NaH→ →CH₃I→ →HCl, H₂O, heat→ ?

12.8. POLYESTERS AND POLYAMIDES

Because of chemical inertness and straightforward preparation, esters and amides are widely utilized as structural units of synthetic polymeric materials known under the general names of polyesters and polyamides. Polyesters and polyamides can be formed through a polycondensation reaction, in which molecules of monomers join together, losing water as a by-product. Polyesters can be prepared by the Fischer esterification reaction of dicarboxylic acids and diols. The most important polyester, polyethylene terephthalate (PET) is industrially produced from terephthalic acid and ethylene glycol (Figure 12.23). PET is known under different brand names (such as Dacron, Terylene, Lavsan, and Mylar) and is used in textile fibers for clothing, containers for liquids and foods, and in combination with glass fiber for engineering resins.

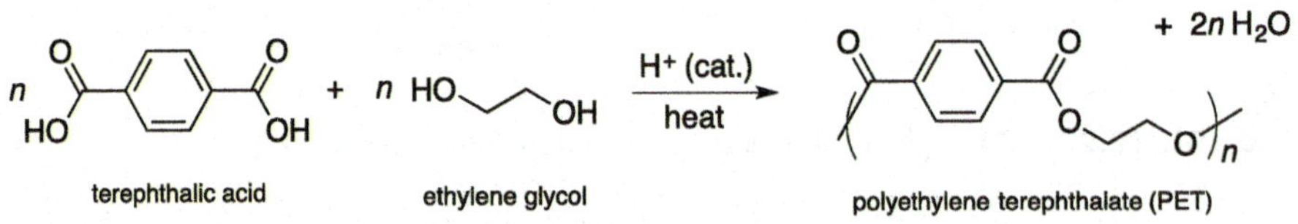

n [terephthalic acid] + n [ethylene glycol] →H⁺ (cat.), heat→ [polyethylene terephthalate (PET)] + $2n$ H₂O

FIGURE 12.23. Preparation of polyethylene terephthalate (PET).

Polyamides commonly occur as natural and synthetic fibers. Natural fibers such as silk and wool are proteins formed from amino acids (see Chapter 16) joined together by the amide links (peptide bonds). Synthetic polyamides are prepared by polycondensation of dicarboxylic acids and diamines, and also have amide links in their structure. For example, the important polymer Nylon 66 is synthesized by polycondensation of hexanedioic acid and hexamethylenediamine (Figure 12.24). Nylon is widely used in synthetic fibers and industrial resins and its annual worldwide production exceeds 2 million tons.

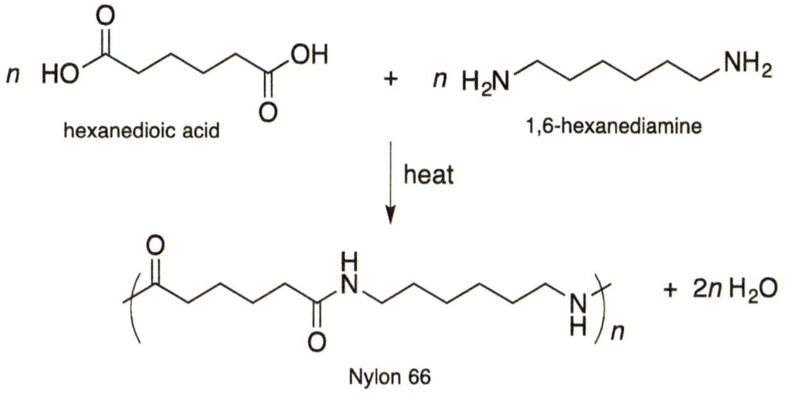

FIGURE 12.24. Preparation of Nylon 66.

Problems

12.25. Battery acid is a 30% solution of H_2SO_4 in water. Why is it not recommended to use a PET bottle for long-term storage of this acid?

12.26. When ε-caprolactam is heated to about 260° C in an inert atmosphere, ring opening occurs and a polymeric material (Nylon 6, also known as Perlon or Capron) is produced. Propose a general structure for the resulting polymer.

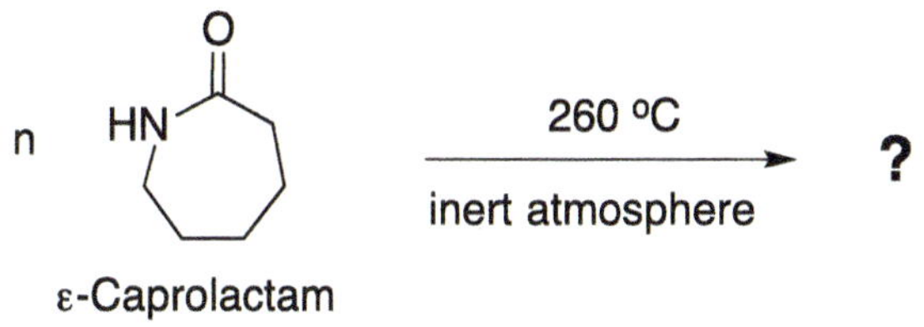

12.27. Kevlar is a polymeric material that is 5-times stronger than steel. Based on its general structure, which compounds would you use to synthesize Kevlar in high yield?

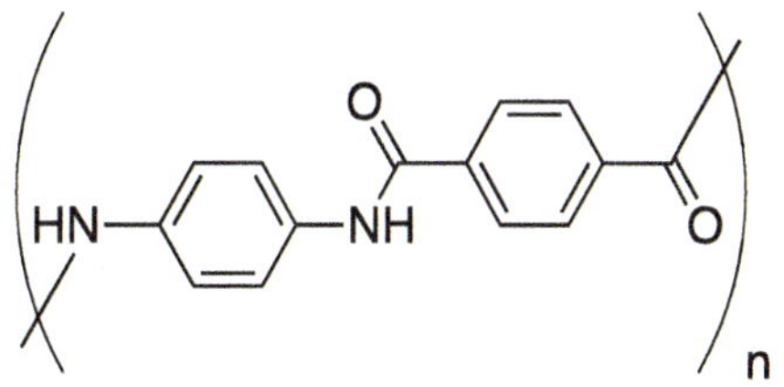

General structure of Kevlar

12.28. When sodium 2-chloroacetate is heated to about 160° C in an inert atmosphere, a polymer called polyglucolide and sodium chloride are produced. What is the general structure of this polymer?

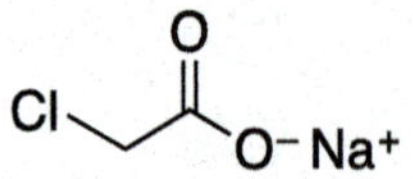

Cl–CH$_2$–C(=O)–O$^-$ Na$^+$

sodium 2-chloroacetate

CHAPTER 13

Non-cyclic Conjugated Systems

A **conjugated system** is a system of connected sp^2 or sp hybridized atoms of carbon, nitrogen, or oxygen, bearing p-orbitals on each atom. Organic molecules containing conjugated systems of atoms are characterized by delocalization of electrons, which in general lowers the overall energy of the molecule and increases stability. The molecules with conjugation often have alternating single and multiple bonds in their structures and can be represented by several resonance contributors involving each atom of the conjugated system. Cyclic and non-cyclic conjugated systems of atoms are very common in natural compounds. The present chapter provides an overview of the structural features and chemical reactions of non-cyclic conjugated systems. The completely conjugated cyclic systems represent another extremely important class of aromatic compounds that are the focus of Chapter 14.

13.1. STRUCTURE AND PROPERTIES OF CONJUGATED SYSTEMS

Common non-cyclic conjugated systems include conjugated dienes, such as 1,3-butadiene, and α,β-unsaturated carbonyl compounds, which are the final products of aldol condensation (Section 11.10). These compounds are characterized by special physical and chemical properties due to the interaction of adjacent p orbitals on each atom of the conjugated system.

Dienes are hydrocarbons with two double bonds; hydrocarbons with three or more double bonds are called polyenes (Section 6.1). **Conjugated dienes and polyenes** have alternating single and multiple bonds in their structures. In contrast, the unconjugated dienes have two or more single bonds separating the double bonds. A third class of dienes, called **cumulated dienes** (or cumulenes) is characterized by the presence of adjacent double bonds sharing an sp hybridized carbon atom (Figure 13.1).

Conjugated dienes occur widely in nature and play an important role in technology. Isoprene (2-methyl-1,3-butadiene) is produced by many plants and its polymers are the main component of natural rubber. Isoprene and 1,3-butadiene are also important industrial products used as monomers in the production of synthetic rubber.

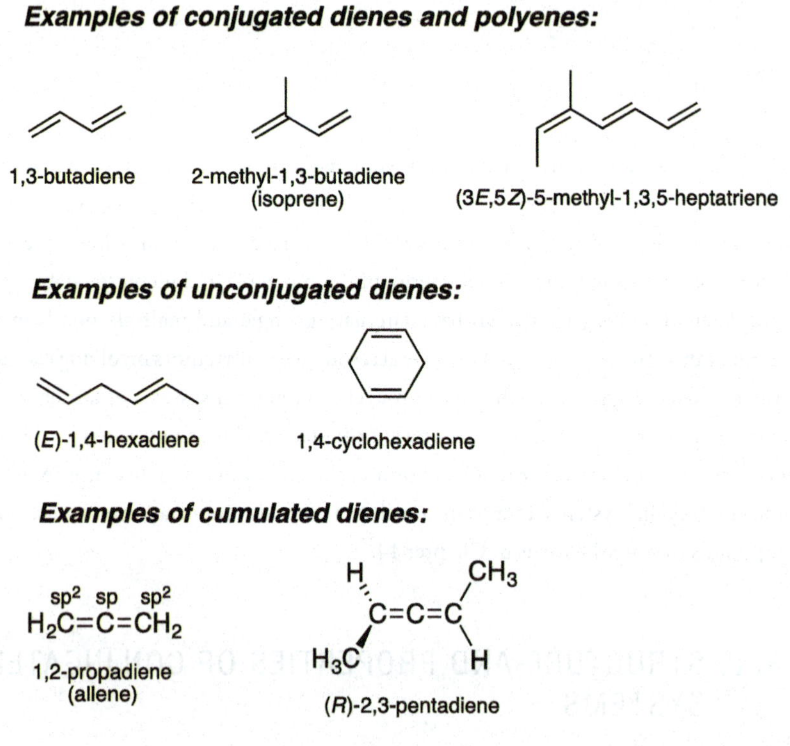

Examples of conjugated dienes and polyenes:

1,3-butadiene

2-methyl-1,3-butadiene (isoprene)

(3E,5Z)-5-methyl-1,3,5-heptatriene

Examples of unconjugated dienes:

(E)-1,4-hexadiene

1,4-cyclohexadiene

Examples of cumulated dienes:

sp² sp sp²
H₂C=C=CH₂

1,2-propadiene (allene)

(R)-2,3-pentadiene

FIGURE 13.1. Classification of dienes and polyenes.

According to the heat of hydrogenation values (Section 6.5.4), conjugated dienes are more stable (i.e., have lower energy) than unconjugated dienes. The **molecular orbital theory** (MO) offers a general explanation for the higher stability of conjugated dienes (Section 1.7). Specifically, the bonding molecular orbitals in conjugated dienes have lower energy because of the additional interaction of atomic p orbitals, as illustrated in Figure 13.2. The structural fragment of conjugated diene has four adjacent sp² hybridized carbon atoms, with one p orbital on each atom. According to the basic rules of MO theory (Section 1.7), the combination of four atomic p orbitals should give four molecular π orbitals. Overlap of p orbitals is possible only when the adjacent orbitals have wave functions with matching signs, indicating that they are in phase (the sign of a wave function is indicated by p orbitals with black or white lobes in Figure 13.2). When all four p-orbitals are in phase they combine together to form the lowest energy bonding MO (i.e., $π_1$), occupied by a pair of bonding electrons shared by all four atoms. The second bonding MO ($π_2$) has one node, in which the signs of the wave function do not match. The unoccupied (antibonding) MOs ($π_3^*$ and $π_4^*$) have two and three nodes respectively, and are higher in energy than the initial atomic p-orbitals. The highest occupied molecular orbital, i.e., $π_2$ (standard abbreviation **HOMO**) and the lowest unoccupied molecular orbital, i.e., $π_3$ (standard abbreviation **LUMO**) are especially important in the theoretical explanation of the reactivity of conjugated dienes (Section 13.4). Overall, a conjugated system is characterized by lower energy, and hence greater stability, of bonding MOs and by the delocalization of bonding π electrons over four atoms.

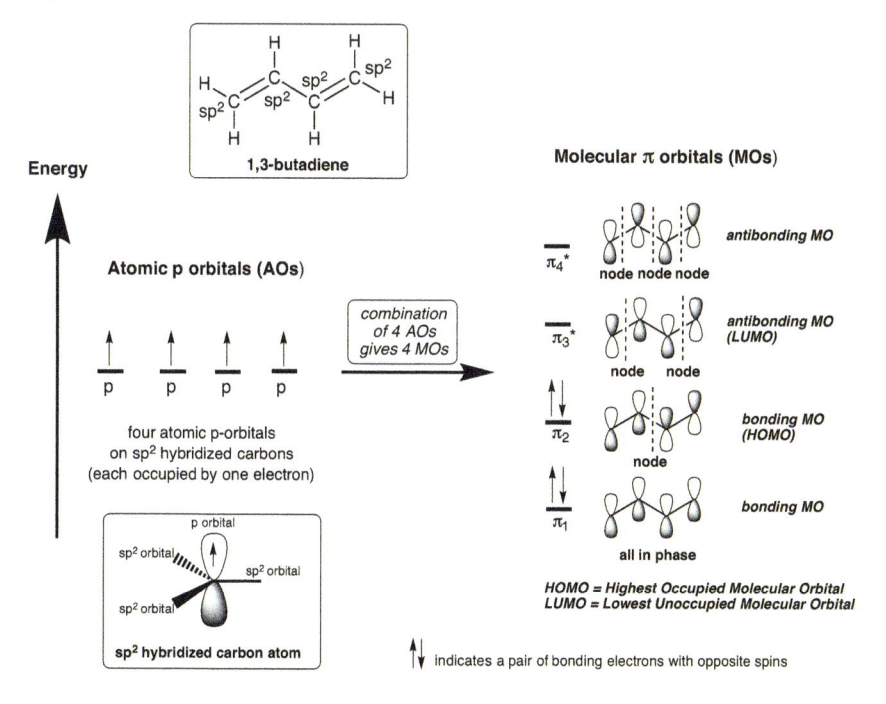

FIGURE 13.2. Molecular π orbitals in 1,3-butadiene (σ orbitals are not shown).

The electronic structure of α,β-unsaturated carbonyl compounds is similar to that of conjugated dienes. It is formed by three sp² hybridized carbons and one sp² hybridized oxygen. Because of the higher electronegativity of oxygen, the molecule has additional resonance contributors with a negative charge on oxygen and a positive charge on carbon (Figure 13.3). The presence of these charged resonance contributors explain the high electrophilic character of the β-carbon in reactions with nucleophilic reagents (Section 13.3).

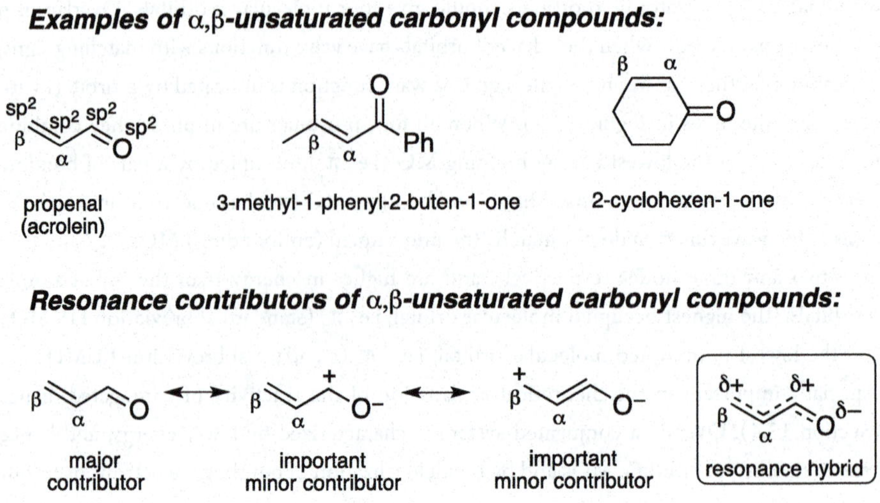

Examples of α,β-unsaturated carbonyl compounds:

propenal (acrolein)

3-methyl-1-phenyl-2-buten-1-one

2-cyclohexen-1-one

Resonance contributors of α,β-unsaturated carbonyl compounds:

major contributor ↔ important minor contributor ↔ important minor contributor   resonance hybrid

FIGURE 13.3. α, β-Unsaturated carbonyl compounds.

Problems

13.1. The selected spectroscopic data of propenal (acrolein) are compared to compounds with similar structural features.

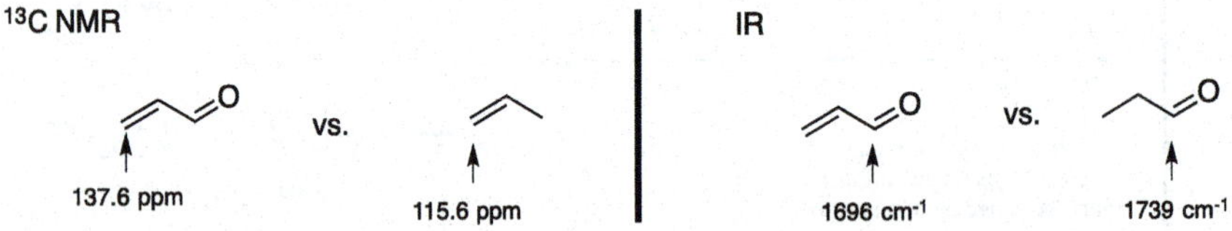

^{13}C NMR

137.6 ppm   vs.   115.6 ppm

IR

1696 cm^{-1} vs.   1739 cm^{-1}

Explain these data. Why is the terminal alkene carbon in propenal in ^{13}C NMR shifted downfield by 22 ppm, compared to the corresponding carbon in propene? Why does the carbonyl group of propenal absorb IR irradiation at lower wavenumber compared to propanal?

13.2. CONJUGATE ELECTROPHILIC ADDITION REACTIONS OF DIENES

Reactions of conjugated dienes with one molar equivalent of an electrophilic reagent, such as chlorine or hydrochloric acid (see Chapter 6), usually give a mixture of two products: (1) the normal product of electrophilic addition to one double bond; and (2) the product of **conjugate addition** to carbons *1* and *4* of the conjugated system of four carbons (Figure 13.4).

The first product is commonly named as the **1,2-addition product** and the second product is called the **1,4-addition product**. This reaction proceeds according to the typical mechanism of electrophilic addition (see Figure 6.7 and the related text in Chapter 6). Initial addition of the electrophile produces an allylic carbocationic intermediate, which has two resonance contributors with positive charges at carbons **2** and **4** of the conjugated system. The reaction of this carbocation with the nucleophile (Cl⁻) in the second step yields the final products of 1,2- and 1,4-addition (Figure 13.4).

Examples of reactions:

Mechanism:

FIGURE 13.4. Conjugate electrophilic addition.

Problems

13.2. Provide the products of the following conjugate electrophilic addition reactions:

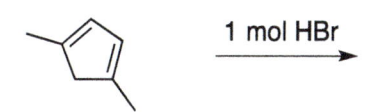

a)

1 mol HBr

b)

[structure: bicyclic diene] → 1 mol Br₂

13.3. Quite often the outcome of the conjugate electrophilic addition is temperature dependent, due to both the kinetics and the relative stability of the products. In the case of the addition of HCl to 1,3-butadiene (Figure 13.4) at -78 °C, the least stable isomer is formed as a major product (chemists say the reaction is kinetically controlled), while at room temperature, the most stable isomer is formed as a major product (thermodynamic control). Based on this information, which isomer is the major product at −78 °C and room temperature?

13.3. CONJUGATE NUCLEOPHILIC ADDITION REACTIONS OF α,β-UNSATURATED CARBONYL COMPOUNDS

The β-carbon atom in α,β-unsaturated carbonyl compounds has a partial positive charge due to the resonance conjugation (see Figure 13.3 and the related text) and can react with nucleophilic reagents (Nu:⁻) according to the general scheme shown in Figure 13.5. The initial

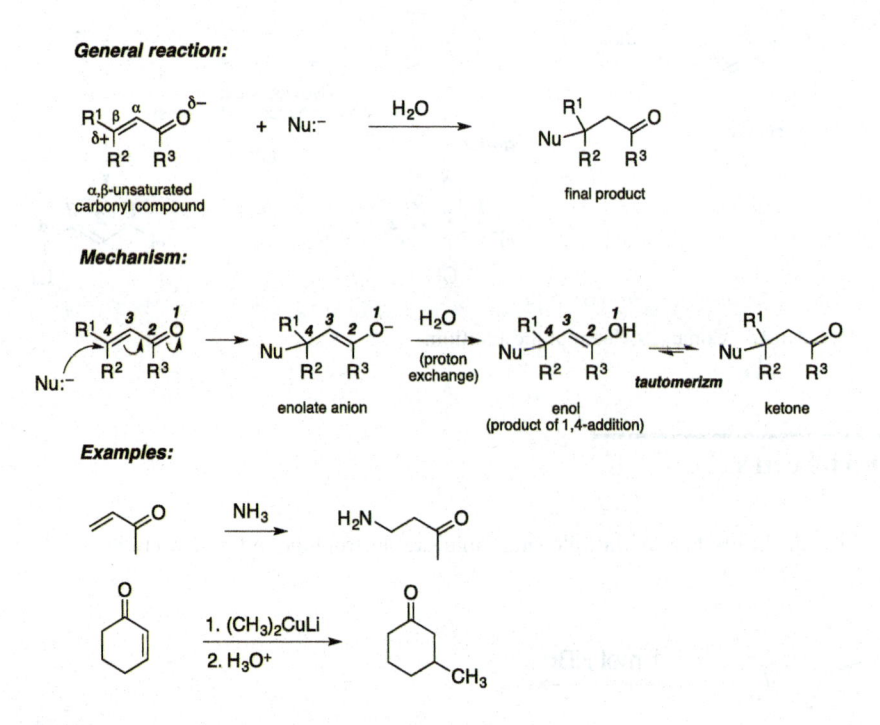

FIGURE 13.5. Conjugate nucleophilic addition.

product of this reaction is an enol, formed by the addition of a nucleophile and a proton to atoms 1 and 4 of the conjugated system (Figure 13.5). Therefore this reaction is also classified as the 1,4-conjugate addition but 1,2-addition products are also possible in this reaction. For example, reactions of α,β-unsaturated carbonyl compounds with LiAlH$_4$ or RMgBr involve the addition of a hydride anion or alkyl anion to the carbonyl carbon (atom 2 of the conjugated system), producing the corresponding alcohols (see Sections 11.3 and 11.4).

Particularly important is the conjugate addition reaction of α,β-unsaturated carbonyl compounds with enolate anions as nucleophiles. This reaction is known as the **Michael addition** reaction, named after the American organic chemist Arthur Michael. Figure 13.6 shows an example of a Michael addition reaction, of the enolate anion generated from acetoacetic ester (Section 12.7) with α,β-unsaturated carbonyl compound.

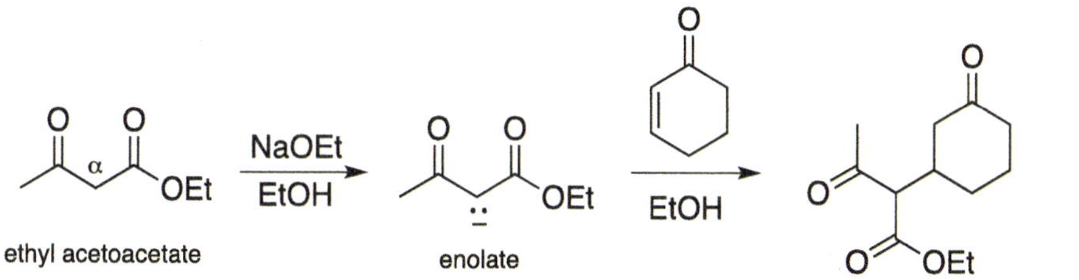

FIGURE 13.6. Example of a Michael addition reaction.

Problems

13.4. What are products of the following transformations?

a)

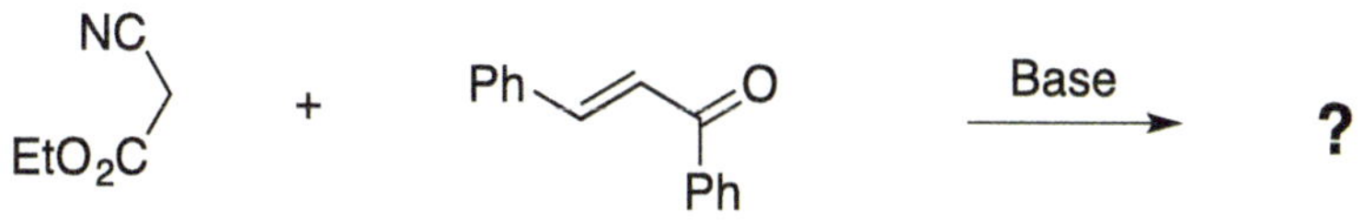

b)

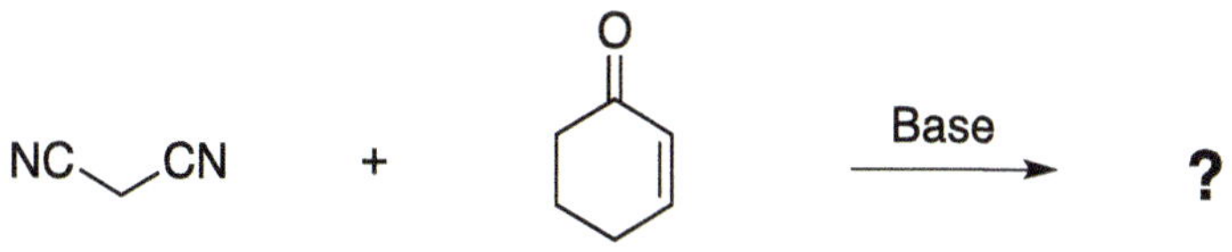

c)

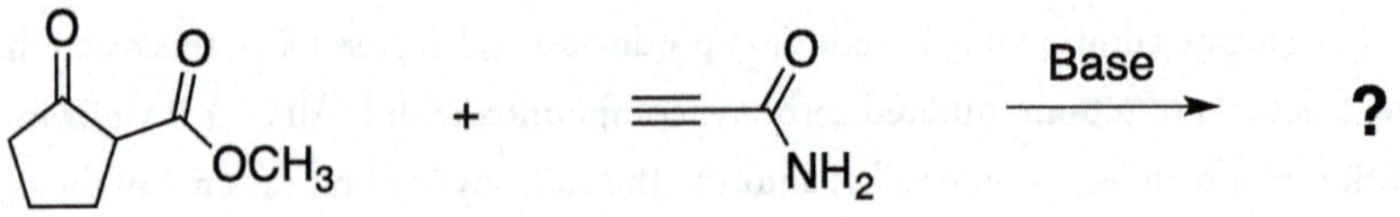

[structure: methyl 2-oxocyclopentanecarboxylate] + [structure: propiolamide, HC≡C-C(=O)-NH₂] →(Base) ?

13.5. For each of the following molecules, propose a synthesis that involves the reaction of an α,β-unsaturated ketone with a Gilman reagent.

a)

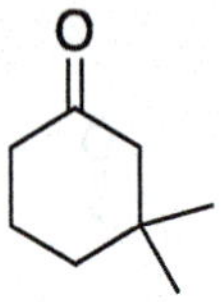

[structure: 4,4-dimethylcyclohexanone]

b)

[structure: 3-allylcyclohexanone]

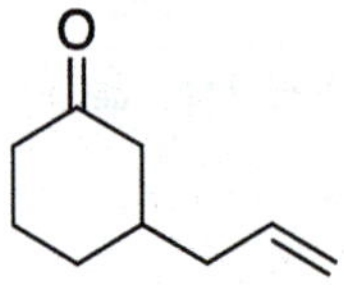

13.6. The formation of a bicyclic product involves a stepwise sequences of reactions. The first step is a Michael addition, which is followed by an aldol reaction, and finally an elimination step. Draw the structures of the intermediates.

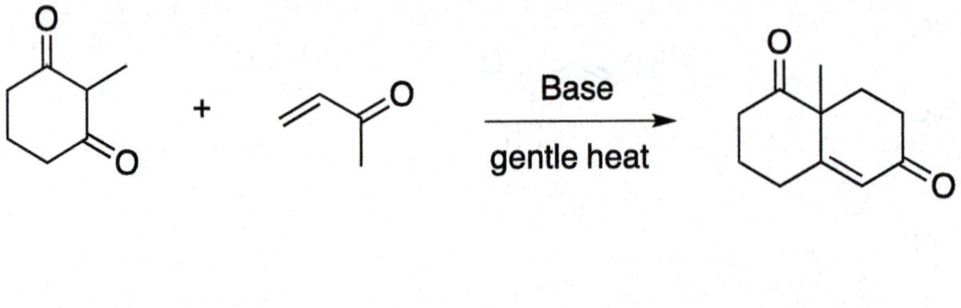

[structure: 2-methyl-1,3-cyclohexanedione] + [structure: methyl vinyl ketone] →(Base, gentle heat) [structure: bicyclic enedione product]

13.4. DIELS-ALDER CYCLOADDITION REACTION

The **Diels-Alder reaction** (also known as [4 + 2] cycloaddition reaction) is a reaction between a conjugated diene and a substituted alkene or alkyne (termed the dienophile), producing substituted cyclohexene (i.e., 1,4-cyclohexadiene; see Figure 13.7). This reaction was originally described in 1928 by the German chemists Otto Diels and Kurt Alder in 1928, who were awarded the Nobel Prize in Chemistry in 1950. The Diels-Alder reaction provides a straightforward and highly stereoselective approach to the synthesis of various important cyclic molecules.

A diene (4 atoms) combines with a dienophile (2 atoms) forming a six-membered ring. The reaction occurs via concerted movement of electrons in a cyclic transition state.

FIGURE 13.7. Diels-Alder Reactions ([4 + 2] cycloaddition).

Diels-Alder reactions are driven by the interaction of π molecular orbitals of the diene and dienophile. During the course of the reaction, the highest occupied molecular orbital of diene (i.e., HOMO, π_2 in Figure 13.2) combines in matching phase with the unoccupied π^* orbital of alkene, resulting in the formation of two new σ-bonds (Figure 13.8).

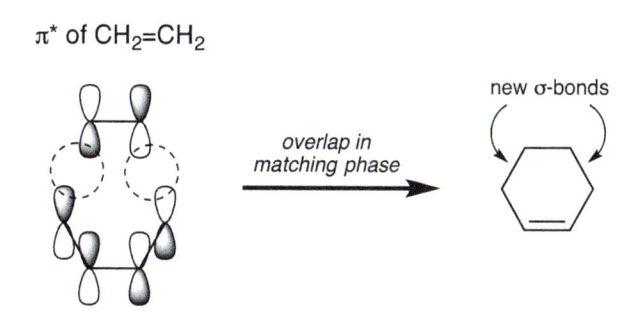

FIGURE 13.8. Orbital interaction in the [4 + 2] cycloaddition reaction of 1,3-butadiene with ethylene.

NON-CYCLIC CONJUGATED SYSTEMS   271

Diels-Alder reactions occur in one step as a concerted movement of electrons in a cyclic transition state, formed by the six atoms C1 - C6 (Figure 13.7). Two new single bonds (C1–C6 and C4–C5) and a double bond (C2=C3) are formed as a result of [4+2] cycloaddition. The diene molecule must participate in the reaction in the appropriate conformation (the so-called **s-cis** conformation; s refers to the single bond between the two double bonds) to assure the best orbital overlap with the dienophile molecule. The configuration of substituents in the dienophile (carbons C5 and C6) is preserved in the product; the reactions of *cis*- or *trans*-alkenes produce the corresponding *cis*- or *trans*-substituted cyclohexene products (Figure 13.9). The most reactive dienophiles usually have electron-withdrawing substituents (i.e., carbonyl, carboxyl or nitrile groups) while typical dienes have electron-donating substituents (i.e., alkyl or aryl).

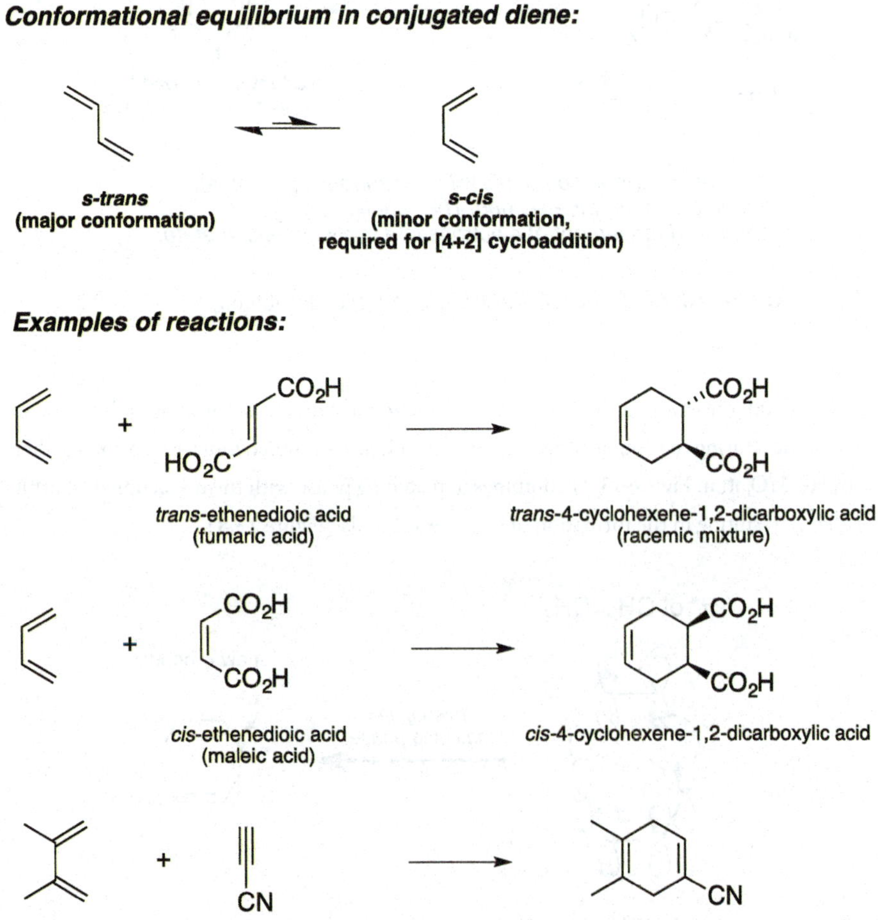

Conformational equilibrium in conjugated diene:

s-trans
(major conformation)

s-cis
(minor conformation, required for [4+2] cycloaddition)

Examples of reactions:

trans-ethenedioic acid
(fumaric acid)

trans-4-cyclohexene-1,2-dicarboxylic acid
(racemic mixture)

cis-ethenedioic acid
(maleic acid)

cis-4-cyclohexene-1,2-dicarboxylic acid

FIGURE 13.9. Stereochemistry of Diels-Alder reaction.

Cyclic dienes, such as cyclopentadiene, are particularly reactive in Diels-Alder reactions, producing bicyclic products of cycloaddition. Figure 13.10 shows several examples of cycloaddition reactions of cyclic dienes.

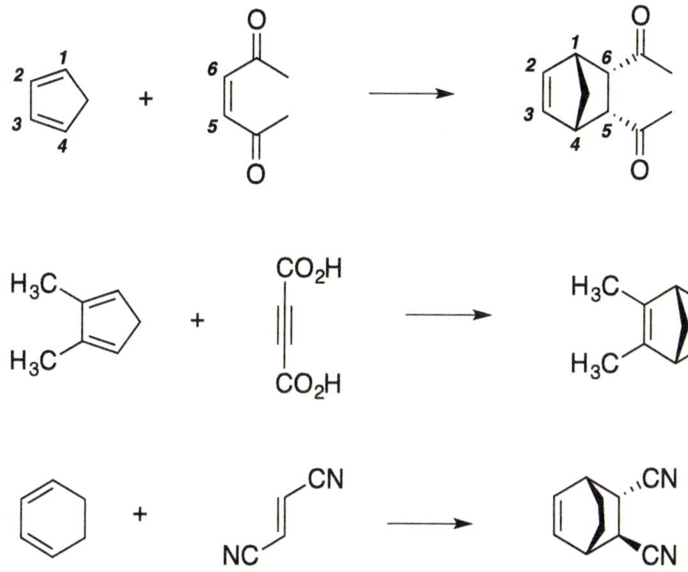

FIGURE 13.10. Diels-Alder reactions of cyclic dienes.

Problems

13.7. What are the products of the following transformations?

a)

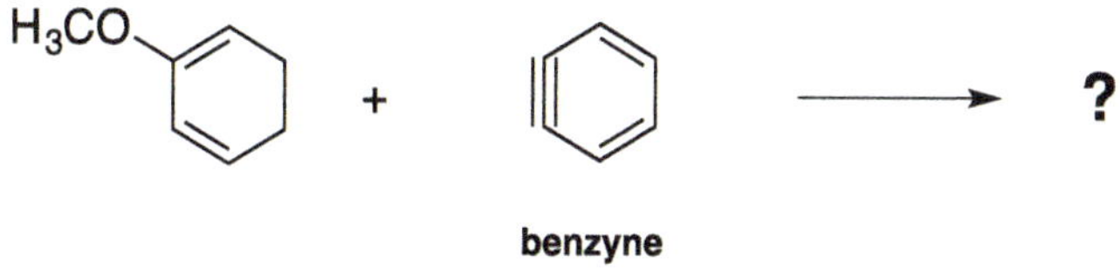

H_3CO-diene + benzyne ⟶ ?

b)

(2,3-dimethylbutadiene) + (methyl vinyl ketone) ⟶ ?

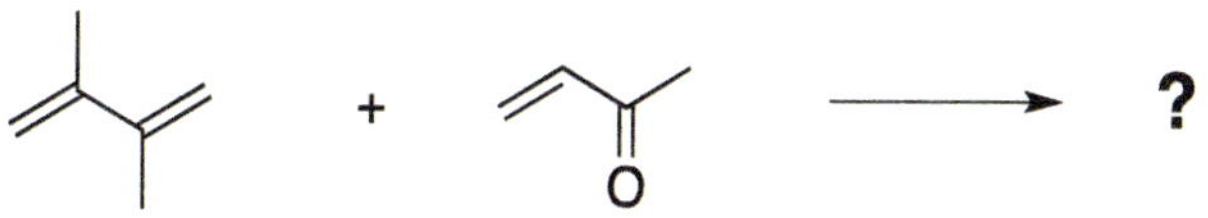

c)

$\diagup\!\!\diagdown\!\!\diagup$ + $\diagdown\!\!\diagup\!\!\diagdown$CHO ⟶ ?

13.8. Suggest dienes and dienophiles required for the synthesis of the following compounds via Diels-Alder reaction:

CHAPTER 14

Benzene and Aromatic Compounds

Aromatic compounds are conjugated cyclic systems of sp² hybridized atoms with completely filled bonding π orbitals. Because of the conjugation and delocalization of π-electrons around the ring, aromatic compounds are very stable (i.e., they have low overall energy). The term "aromatic" was initially introduced in the mid 1850's for compounds related to benzene, many of which have special odors (aromas) distinctly different from saturated hydrocarbons. In modern chemistry the word aromatic often refers to benzene derivatives; however, various non-benzene aromatic compounds also exist. Aromatic compounds are extremely important in nature and in our lives. In living organisms, for example, six- and five-membered nitrogen-containing aromatic rings are present in nucleobases, which are the basic building blocks of deoxyribonucleic acid (DNA) and ribonucleic acid (RNA). Large aromatic rings (porphyrins) are essential components of hemoglobin and chlorophyll. Benzene and its derivatives are important industrial chemicals with many practical applications. This chapter provides an overview of the structural features of aromatic compounds and summarizes chemical reactions of benzene and substituted benzenes.

14.1. STRUCTURE OF BENZENE AND THE CONCEPT OF AROMATICITY

Benzene is an important organic compound with molecular formula C_6H_6. Pure benzene was first isolated by the famous English scientist Michael Faraday in 1825, from the oily residue derived from the production of illuminating gas. Benzene's unusual properties have been recognized since its first isolation. It is a highly unsaturated compound with four units of hydrogen deficiency (Section 1.3). However, in contrast to alkenes and alkynes, benzene does not participate in electrophilic addition reactions with common electrophiles such as HCl or HBr. In general, it is a very stable compound with exceptionally low chemical reactivity. In 1865, the German chemist August Kekulé proposed the cyclic structure of benzene with alternating single and double bonds. Kekulé also suggested a fast equilibrium between the two forms of the ring, which in modern interpretation can be shown as two resonance contributors (Figure 14.1). The real structure of benzene (the resonance hybrid) can be represented as a planar hexagon with internal bond angles of 120°, and six C–C bonds of identical length at 1.40 Å. The C–C bonds in benzene are longer than a typical double bond (1.35 Å) but shorter than a single bond (1.47 Å). This structure has been confirmed by experimental studies and is consistent with a completely conjugated cyclic system of six sp^2 hybridized carbons, characterized by delocalization of π-electrons. According to experimental studies of heats of hydrogenation, a molecule of benzene has much lower energy (about 36 kcal/mol) compared to the three isolated double bonds.

resonance contributors of benzene **resonance hybrid**

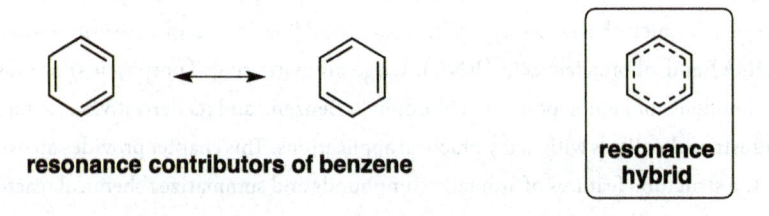

FIGURE 14.1. Resonance description of benzene.

The modern explanation of the exceptional stability of benzene and other aromatic compounds was developed in the 1930's and is based on molecular orbital (MO) theory. The six-membered ring of benzene is formed by six adjacent sp^2 hybridized carbon atoms, with one p orbital on each atom. According to the basic rules of MO theory, the combination of six atomic p orbitals should give six molecular π orbitals. Molecular orbitals in benzene can be constructed analogous to the conjugated dienes (see Figure 13.2 and the related text in Section 13.1). Figure 14.2 shows energy levels of six molecular orbitals of benzene in comparison with the initial energy of six atomic p orbitals. The lowest energy bonding

MO (π_1) is formed by a combination of all six atomic p orbitals that are in phase, while the highest energy antibonding MO (π_6^*) has all six p-orbitals out of phase. It is essential that all three of benzene's bonding MO's are completely filled with electron pairs, which provides the molecule's exceptional stability.

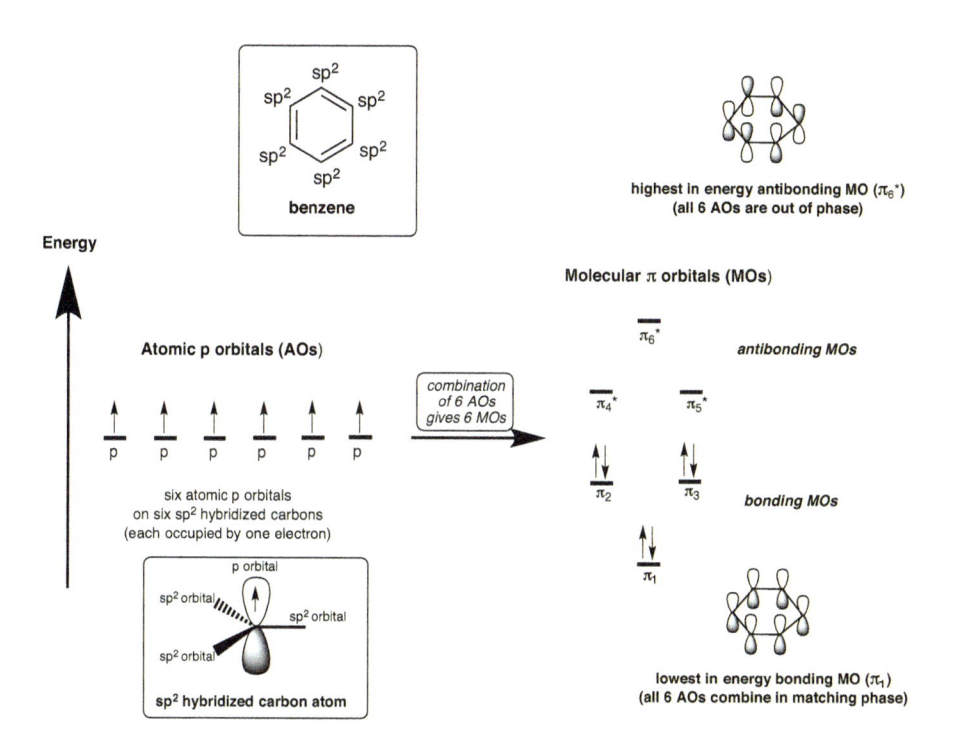

FIGURE 14.2. Molecular π orbitals in benzene.

In 1931, the German physical chemist Erich Hückel first recognized that the number of bonding electrons filling π orbitals in completely conjugated cyclic systems is an important feature that helps explain the exceptional stability (**aromaticity**) of aromatic compounds. In aromatic systems, all bonding π MO's must be completely filled with electron pairs. Theoretical analysis of molecular orbitals in various conjugated cyclic systems has led to the development of **Hückel's Rule** for aromaticity: A cyclic, planar, completely conjugated system of sp² hybridized atoms is aromatic if it is occupied by 2, 6, 10, 14, 18, ... (4n + 2) π electrons, where "n" is zero or any positive integer. The mathematical formula $4n + 2$ identifies the number of electrons required to completely fill bonding π orbitals of conjugated cyclic systems. For example, n = 0 produces the number 2 ($4 \times 0 + 2 = 2$); when n = 1 6 bonding π electrons are required in benzene.

The simplest aromatic system (2 π electrons) is represented by the cyclopropenyl cation (Figure 14.3), which is formed by three sp² hybridized carbons, one of which has an empty p orbital (for electronic structure of a carbocation, see Figure 5.8 in Chapter 5).

Cyclopropenyl cations are completely conjugated systems (as illustrated by a set of resonance contributors involving all three atoms; Figure 13.3) with one π bond occupied by two electrons. Additional examples of charged aromatic systems include cyclopentadienyl anions, cycloheptatrienyl cations, and cyclooctatrienyl dianions (Figure 14.3). Note that the anionic aromatic molecules (cyclopentadienyl anion and cyclooctatrienyl dianion) include sp² hybridized carbon with a pair of electrons occupying a p orbital. All charged aromatic systems are characterized by increased thermal stability and can be isolated as the salts with appropriate counter ions.

2 π electron aromatic system:

cyclopropenyl cation (one π bond provides 2 π electrons in a completely conjugated system)

resonance hybrid (1/3+, 1/3+, 1/3+)

6 π electron aromatic systems:

cyclopentadienyl anion (2 π bonds and one lone pair provide 6 π electrons in a completely conjugated system)

resonance hybrid (1/5−, 1/5−, 1/5−, 1/5−, 1/5−)

cycloheptatrienyl cation
(one of the 7 resonance contributors)
3 π bonds provide 6 π electrons in a completely conjugated system

resonance hybrid (1/7+, 1/7+, 1/7+, 1/7+, 1/7+, 1/7+, 1/7+)

10 π electron aromatic system:

cyclooctatrienyl dianion
(one of the resonance contributors)
3 π bonds and 2 lone pairs provide 10 π electrons in a completely conjugated system

resonance hybrid (2−)

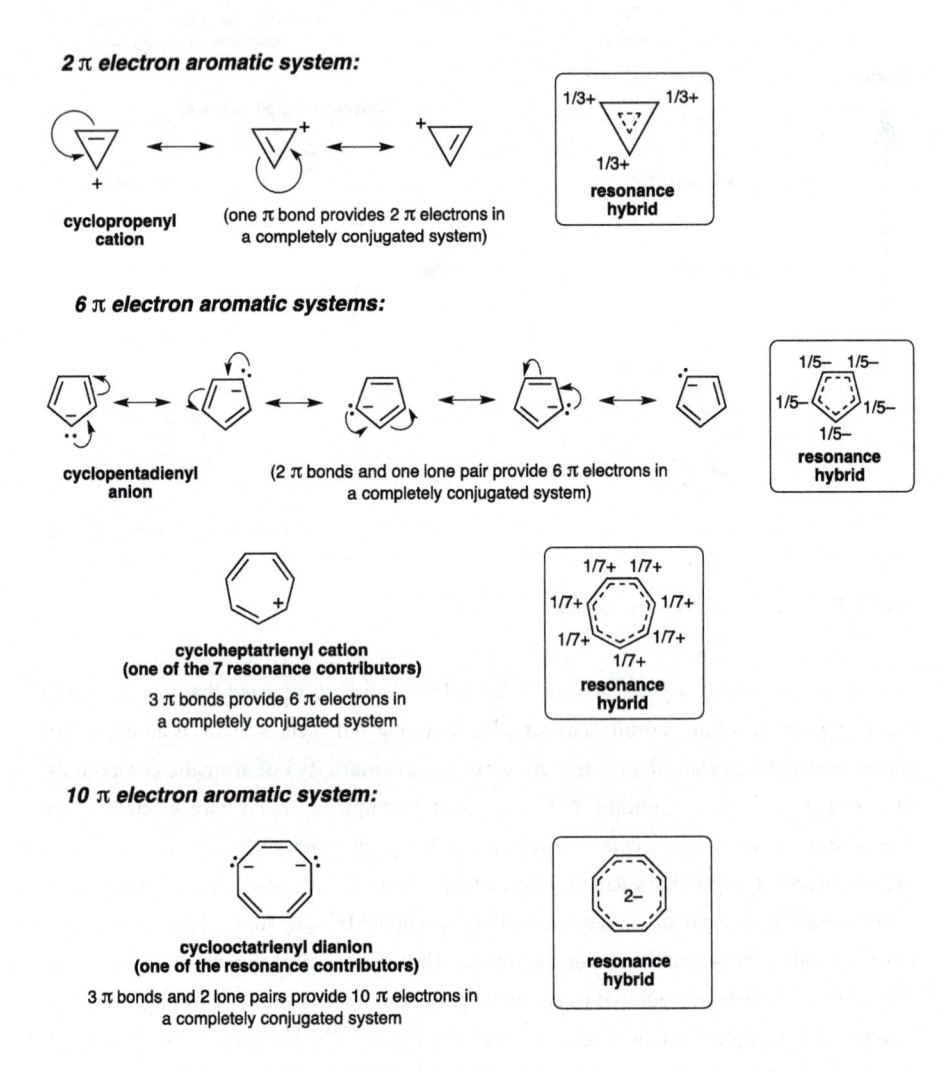

FIGURE 14.3. Examples of charged aromatic systems.

Hückel's theory explains why some simple cyclic conjugated systems are extremely unstable. For example, cyclobutadiene is a completely conjugated system with two π bonds occupied by four electrons. Figure 14.4 shows the energy levels of four π molecular orbitals for this system. Recall that energy levels of molecular orbitals can be both predicted by theoretical calculations, and measured by physical experimental methods. According to the basic principles of quantum mechanics (Section 1.1), placement of four π electrons in four π orbitals results in one bonding, one antibonding, and two non-bonding orbitals of the same energy each occupied by single electrons (Figure 14.4). This is a highly unfavorable electronic system (called **antiaromatic**), making cyclobutadiene an extremely unstable compound (lifetime at room temperature, shorter than five seconds). Another example of a non-aromatic fully conjugated cyclic system is cyclooctatetraene, an unstable compound with an eight-membered ring made of alternating single and double bonds that contains 8 π electrons.

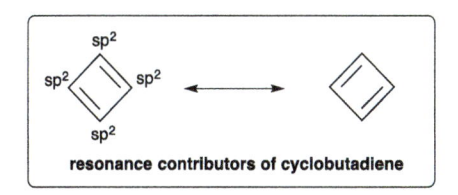

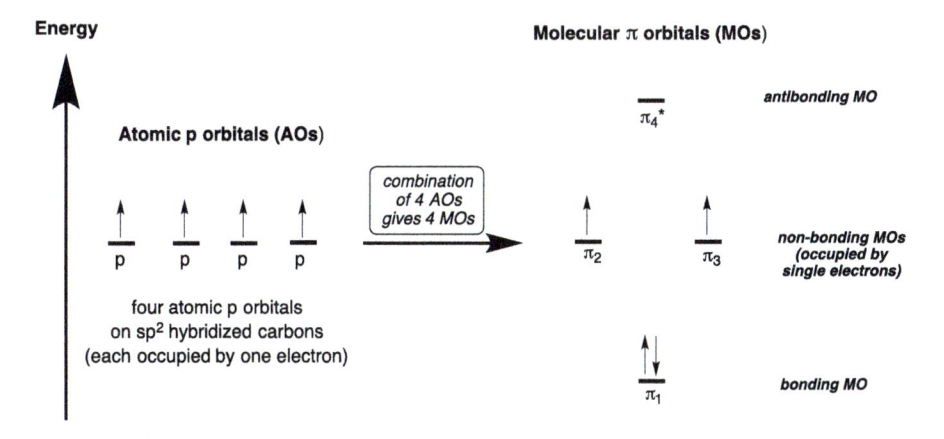

FIGURE 14.4. Electronic structure of cyclobutadiene.

In **heterocyclic aromatic compounds** (or **aromatic heterocycles**), the completely conjugated, aromatic cyclic system of sp^2 hybridized atoms includes nitrogen, oxygen, or sulfur atoms, along with carbons. Several common aromatic heterocycles (pyridine, pyrrole, furan, and thiophene) are shown in Figure 14.5. It should be noted that only one lone pair of electrons occupies the p orbital of the sp^2 hybridized oxygen or sulfur atom in furan and thiophene. The second lone pair is placed in the sp² orbital and is not involved in conjugation (i.e., not counted as π electrons). Likewise, the lone pair occupies the sp² orbital on nitrogen and is not counted as π electrons in pyridine. All these heterocycles are very stable aromatic systems containing 6 π electrons. Aromatic heterocycles occur widely in nature and play an important role in our lives.

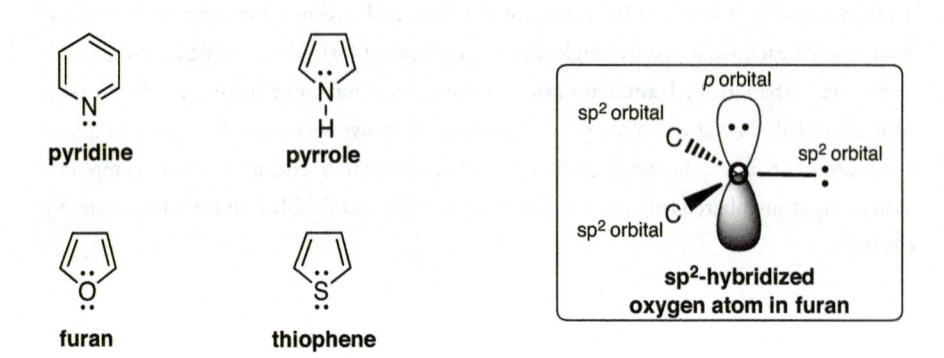

FIGURE 14.5. Heterocyclic aromatic compounds.

Problems

14.1. Draw all possible resonance contributors for a cycloheptatrienyl cation, pyridine, pyrrole, furan, and thiophene.

14.2. In ground state, the central double bond of pentaheptafulvalene is polarized with a partial negative charge on the carbon atom of the pentagon carbon ring, and a partial positive charge on the carbon atom of the heptagon carbon ring. Explain this using resonance structures.

pentaheptafulvalene

14.3. Pyridine is commonly used as a base in organic synthesis. Explain why pyrrole cannot be used for the same purpose.

pyridine   pyrrole

280   ORGANIC CHEMISTRY

14.4. Explain: 2,4-cyclopentadienone is unstable and cannot be isolated; 2,4,6-cycloheptatrienone, on the other hand, is quite stable and can readily be isolated.

2,4-cyclopentadienone   2,4,6-cycloheptatrienone

14.2. NOMENCLATURE OF BENZENE DERIVATIVES

Derivatives of benzene are very common naturally occurring compounds and important industrial products with many practical applications. Many simple benzene derivatives are known by their common names, which are retained in the IUPAC systematic nomenclature. Several important benzene derivatives and their names are shown in Figure 14.6.

toluene   benzoic acid   benzaldehyde   phenol   aniline

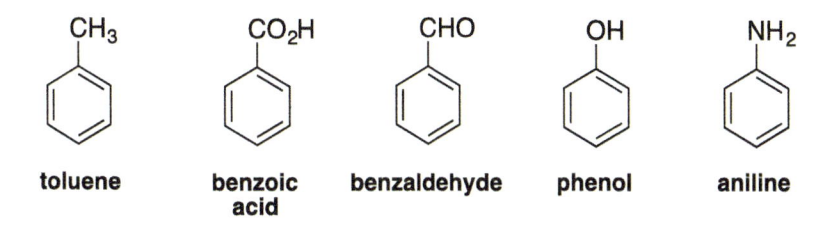

FIGURE 14.6. Important derivatives of benzene named exclusively by common names.

Other monosubstituted benzenes can be systematically named by using the corresponding prefix for the substituent and the word benzene as the parent name (e.g., ethylbenzene, chlorobenzene, and nitrobenzene). The benzene ring attached to a carbon chain bearing additional functional groups or substituents is called a **phenyl** substituent (Ph or C_6H_5). The general name of a phenyl group with additional substituents in the benzene ring is **aryl**, abbreviated as Ar. A **benzyl** group, abbreviated as Bn has the formula $C_6H_5CH_2$ (or $PhCH_2$). Figure 14.7 shows examples of monosubstituted benzene derivatives and their corresponding names.

The relative position of substituents in disubstituted benzenes is commonly indicated by the designators *ortho-* (*o-*, for substituents at carbons 1 and 2), *meta-* (*m-*, for substituents at carbons 1 and 3), and *para-* (*p-*, for substituents at carbons 1 and 4). The names of important benzene derivatives shown in Figure 14.6 are used as the parent names, and the second substituent (i.e., the lower priority functional group; see Section 1.3) is listed as the prefix with

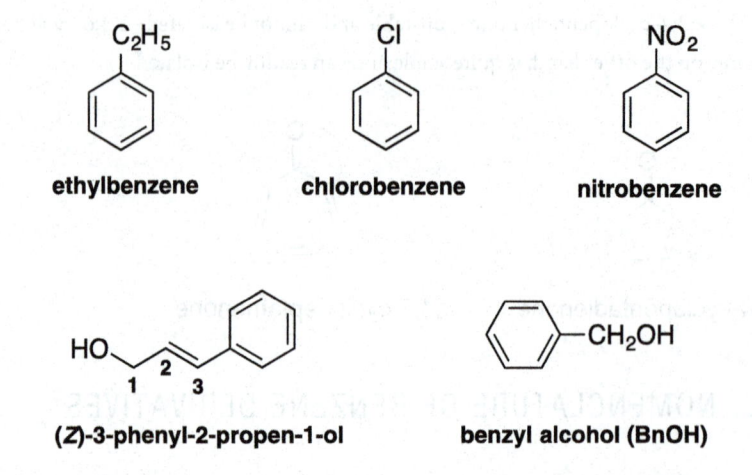

FIGURE 14.7. Names of monosubstituted benzene derivatives.

the corresponding designator (e.g., *o*-chlorotoluene, *m*-hydroxybenzaldehyde, *p*-nitrophenol, and *p*-aminobenzoic acid; Figure 14.8). The *o*-, *m*-, *p*- nomenclature is also commonly used in describing the mechanism of reactions of aromatic compounds and the directing effects of substituents (Section 14.4).

o-chlorotoluene
(2-chlorotoluene)

m-hydroxybenzaldehyde
(3-hydroxybenzaldehyde)

p-nitrophenol
(4-nitrophenol)

p-aminobenzoic acid
(4-aminobenzoic acid)

FIGURE 14.8. Names of disubstituted benzene derivatives.

In benzene derivatives with three or more substituents, position is indicated by numbers (Figure 14.9). In the compound name, the substituents names with their corresponding numbers are listed in alphabetical order as prefixes to the parent name. The position numbers must be as small as possible and the position number 1 is assigned to the parent functional group, but is not listed in the name.

4-bromo-2-chlorotoluene

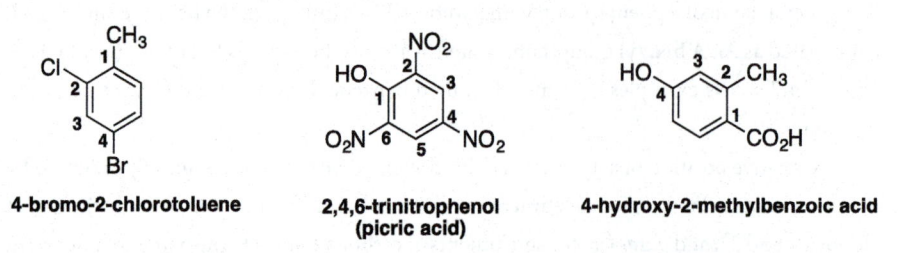

2,4,6-trinitrophenol
(picric acid)

4-hydroxy-2-methylbenzoic acid

FIGURE 14.9. Names of benzene derivatives with more than two substituents.

282 ORGANIC CHEMISTRY

Problems

14.5. What are the IUPAC names of the following compounds?

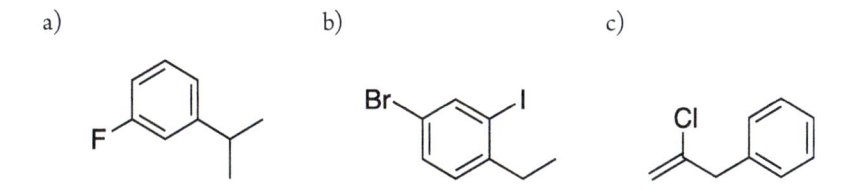

a)   b)   c)

14.6. What are the IUPAC names of the major products of the following reactions?

a)

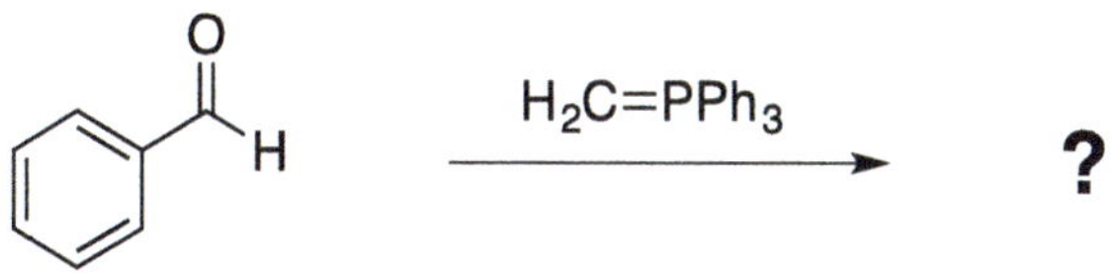

$$\text{PhCHO} \xrightarrow{H_2C=PPh_3} ?$$

b)

$$\xrightarrow{H_2CrO_4} ?$$

14.7. Provide the IUPAC names of the following compounds on the basis of their NMR spectra:

a) Molecular formula: C_8H_7BrO. Compound contains ketone functional group.

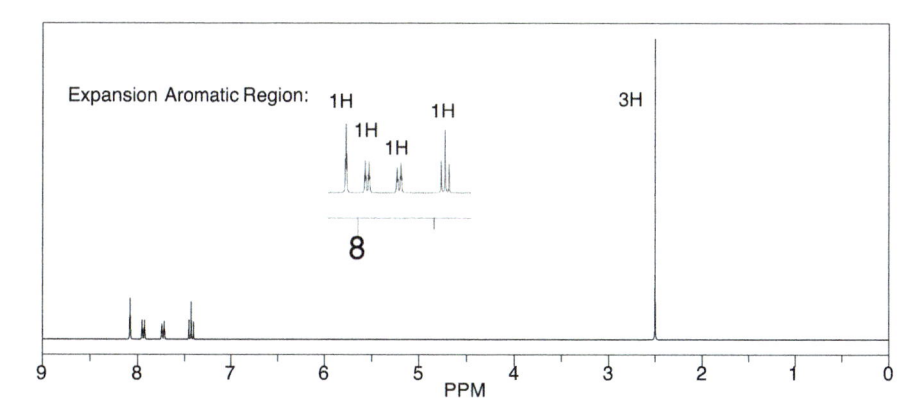

Expansion Aromatic Region: 1H, 1H, 1H, 1H — 8

3H

b) Molecular formula: $C_7H_5ClO_2$. The compound contains a carboxylic acid functional group.

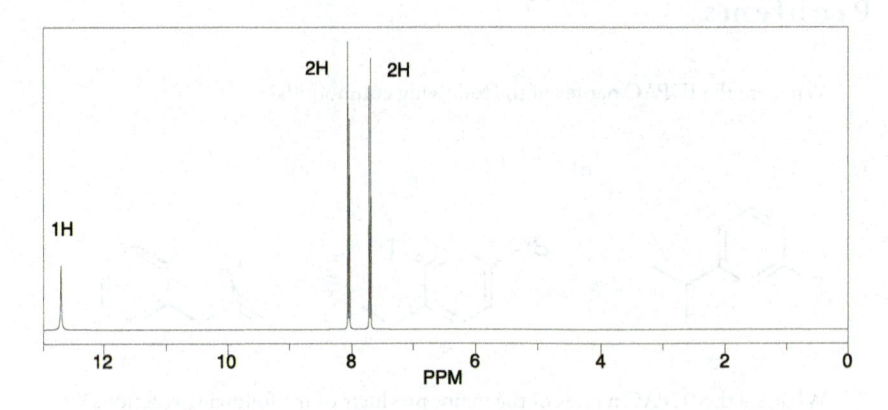

c) Molecular formula: C_8H_8O. The compound contains an aldehyde functional group.

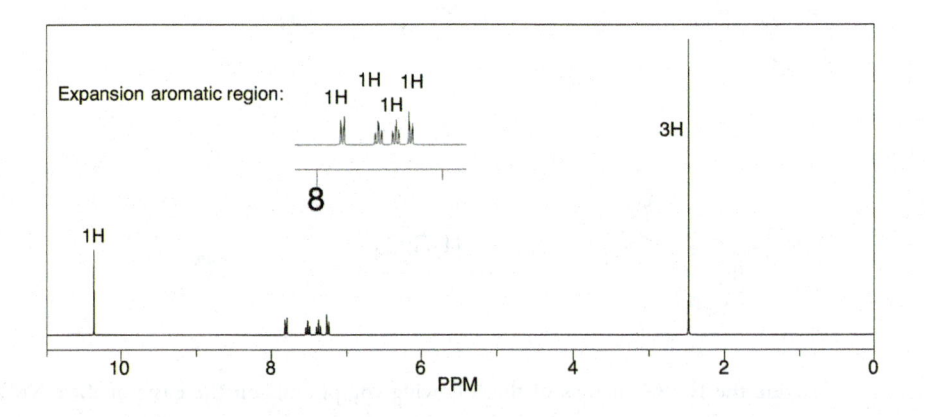

14.3. ELECTROPHILIC AROMATIC SUBSTITUTION REACTIONS OF BENZENE

Benzene does not undergo electrophilic addition reactions, which is the characteristic mode of reaction for alkenes and alkynes, because such additions will destroy aromatic conjugation in benzene rings. However, benzene can react with strong electrophiles, forming products by replacing a hydrogen atom with the electrophile (Figure 14.10). This type of reaction is called **electrophilic aromatic substitution** and its mechanism includes two steps: 1) addition of the electrophile (E^+), producing a resonance-stabilized carbocationic intermediate; and 2) elimination of a proton from the carbocationic intermediate, resulting in the formation of a substituted benzene product. The stable aromatic ring is retained in the product, making this reaction thermodynamically possible. The first step in the reaction is the slow, rate-limiting step, requiring high activation energy to produce the non-aromatic carbocation from the very stable aromatic benzene ring. Participation of highly reactive, strong electrophiles (e.g., Cl^+, Br^+, O_2N^+, HO_3S^+, R^+, or RCO^+) is necessary in the first step. The second, faster step re-forms the aromatic system, releasing energy. The products of these reactions, chlorobenzene, bromobenzene, nitrobenzene, benzenesulfonic acid, alkylbenzenes, and acylbenzenes, are manufactured at industrial scale and have numerous applications in chemical technology.

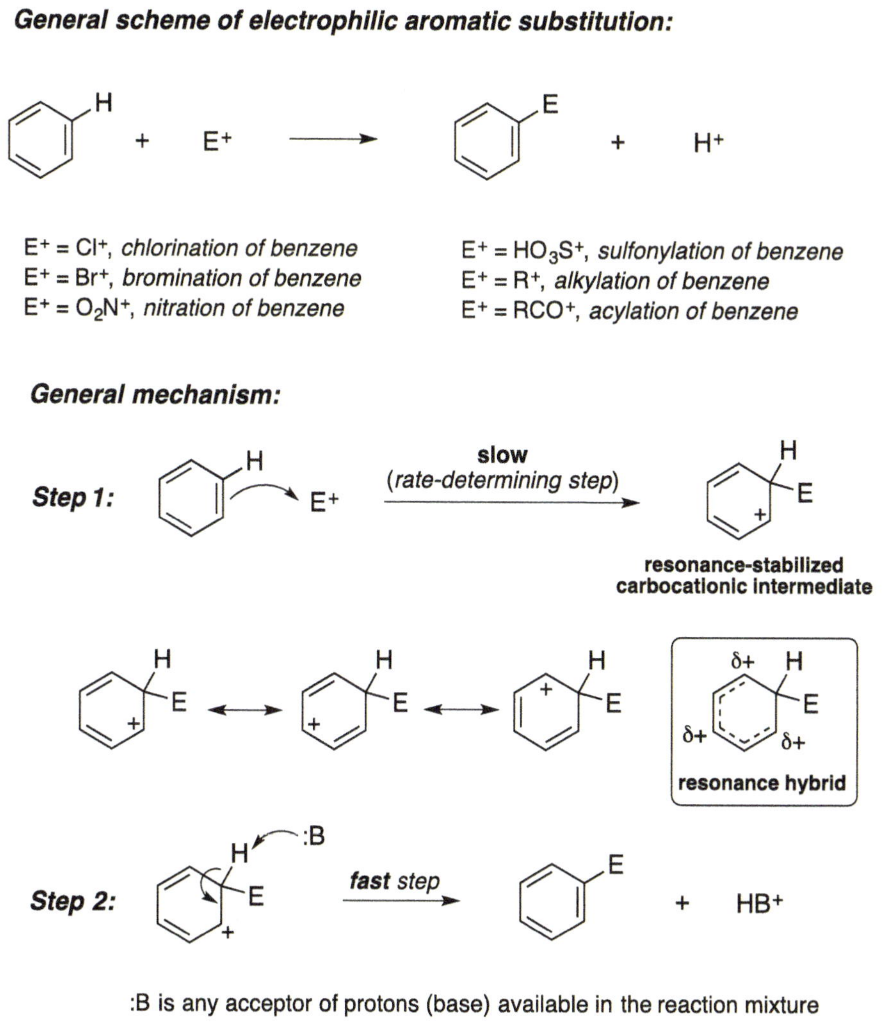

General scheme of electrophilic aromatic substitution:

$E^+ = Cl^+$, *chlorination of benzene*
$E^+ = Br^+$, *bromination of benzene*
$E^+ = O_2N^+$, *nitration of benzene*

$E^+ = HO_3S^+$, *sulfonylation of benzene*
$E^+ = R^+$, *alkylation of benzene*
$E^+ = RCO^+$, *acylation of benzene*

General mechanism:

Step 1: slow (rate-determining step) → resonance-stabilized carbocationic intermediate

resonance hybrid

Step 2: fast step → + HB^+

:B is any acceptor of protons (base) available in the reaction mixture

FIGURE 14.10. Electrophilic aromatic substitution reaction of benzene.

14.3.1. Halogenation

Benzene reacts with Cl_2 or Br_2 in the presence of a Lewis acid (usually, the salts of irons such as $FeCl_3$ or $FeBr_3$), producing chlorobenzene or bromobenzene, respectively (Figure 14.11). The Lewis acid is required to polarize the molecule of halogen, generating the strongly electrophilic species Cl^+ or Br^+, which further reacts with benzene, according to the general mechanism shown in Figure 14.10. In the second step, the proton is eliminated from the carbocationic intermediate with the assistance of a halide anion, forming the final products of the reaction. It should be noted that only the reactions of Cl_2 and Br_2 are practically useful; fluorine reacts with benzene violently, while iodine is not sufficiently reactive, even in the presence of a Lewis acid.

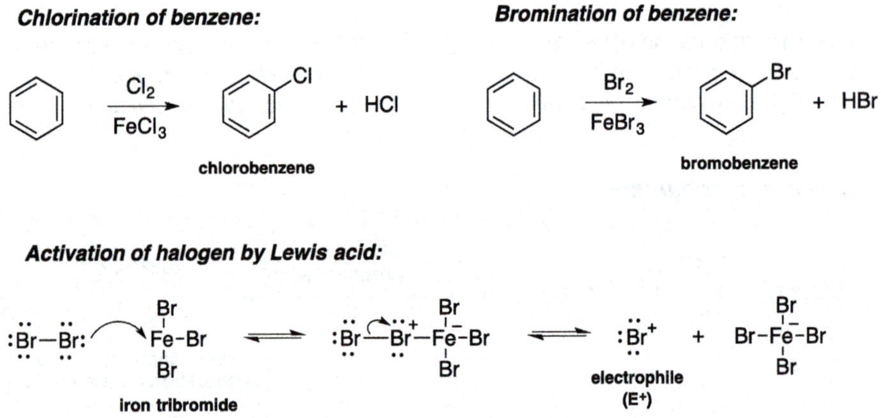

FIGURE 14.11. Chlorination and bromination of benzene.

14.3.2. Nitration

Benzene reacts with nitric acid (HNO_3) in the presence of sulfuric acid (H_2SO_4), producing nitrobenzene (Figure 14.12). The presence of a strong acid (e.g., H_2SO_4) is required to convert nitric acid to a strong electrophile, i.e., nitronium cation (O_2N^+), which further reacts with benzene according to the general mechanism shown in Figure 14.10. In the second step of the reaction mechanism, the proton is eliminated from the carbocationic intermediate with the assistance of any acceptor of protons available in the reaction mixture (most likely, a molecule of H_2O). Nitrobenzene is an important industrial compound used as a feedstock in the production of aniline (Chapter 15).

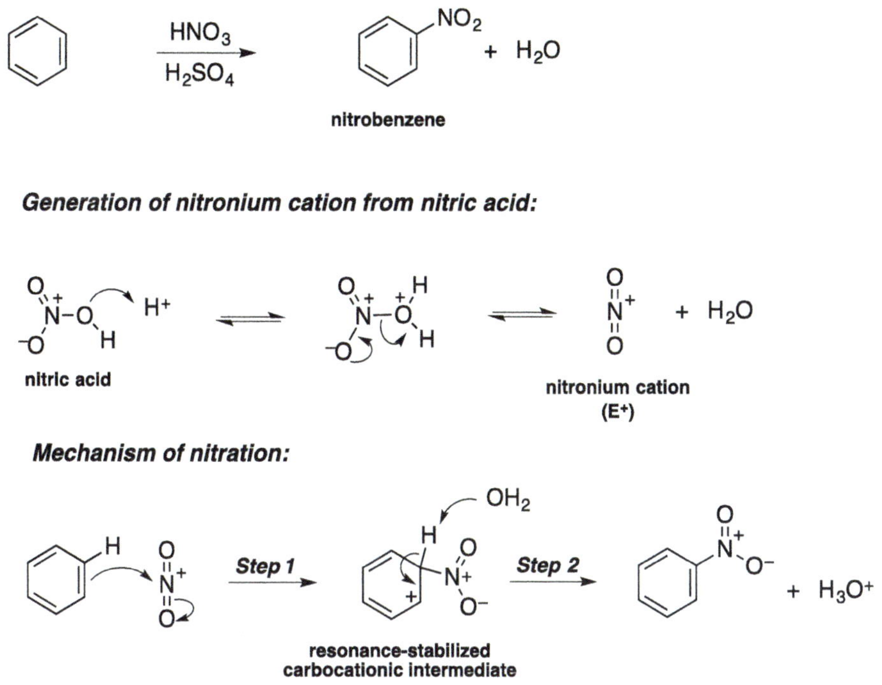

FIGURE 14.12. Nitration of benzene.

14.3.3. Sulfonation

Benzenesulfonic acid, another important industrial product, is prepared by the reaction of benzene with a mixture of concentrated sulfuric acid and sulfur trioxide (Figure 14.13).

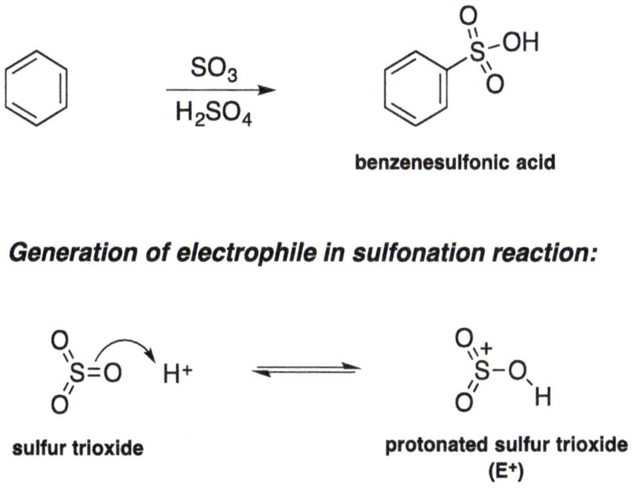

FIGURE 14.13. Sulfonation of benzene.

BENZENE AND AROMATIC COMPOUNDS 287

In this reaction, sulfur trioxide is protonated by sulfuric acid, creating the cationic species HO_3S^+, which then reacts as a strong electrophile during the continued course of the reaction.

14.3.4. Friedel-Crafts alkylation and acylation

The **Friedel-Crafts reaction** refers to the electrophilic alkylation or acylation of benzene. It was developed by Charles Friedel and James Crafts in 1877. In these reactions, the strongly electrophilic alkyl cations or acyl cations are generated from alkyl chlorides or acyl chlorides and a Lewis acid, such as $AlCl_3$, in the presence of benzene. Electrophilic aromatic substitution reactions of benzene with these strong electrophiles produce alkylbenzenes or acylbenzenes as the final products (Figure 14.14).

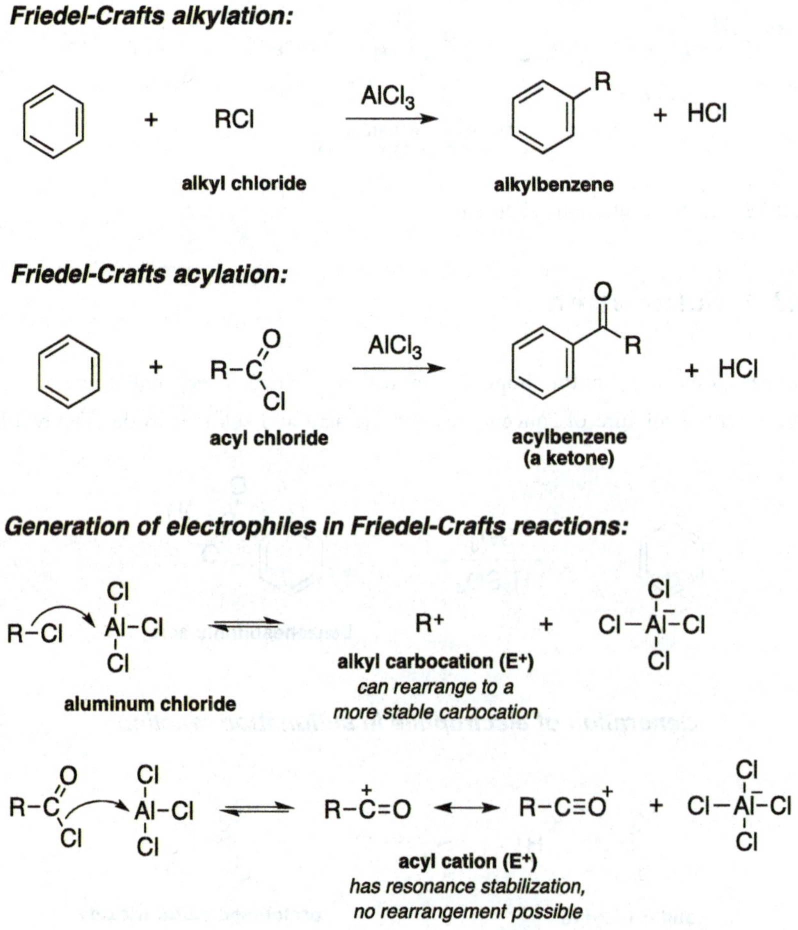

FIGURE 14.14. Friedel-Crafts alkylation and acylation of benzene.

Friedel-Crafts alkylation reactions may proceed with rearrangement of a less stable carbocation to a more stable carbocation by a 1,2-hydride shift or a 1,2-alkyl shift, as explained in Section 5.2.2 (see Figure 5.9). For example, the reaction of 1-chloropropane with AlCl$_3$ and benzene gives the rearranged product (i.e., 2-propylbenzene) as a result of a 1,2-hydride shift in the initial 1° carbocation, leading to the more stable 2° carbocation (Figure 14.15). In contrast, Friedel-Crafts acylation reactions never produce products of rearrangement. The intermediate acyl cation has enhanced stability due to resonance delocalization (Figure 14.14), and its rearrangement to the less stable 2° or 3° alkyl carbocation is impossible. The product of Friedel-Crafts acylation, an aromatic ketone, can be reduced by Wolff-Kishner reduction or Clemmensen reduction (Section 11.6) to give unrearranged, normal alkylbenzene, such as 1-propylbenzene, which is not available via the Friedel-Crafts alkylation (Figure 14.15).

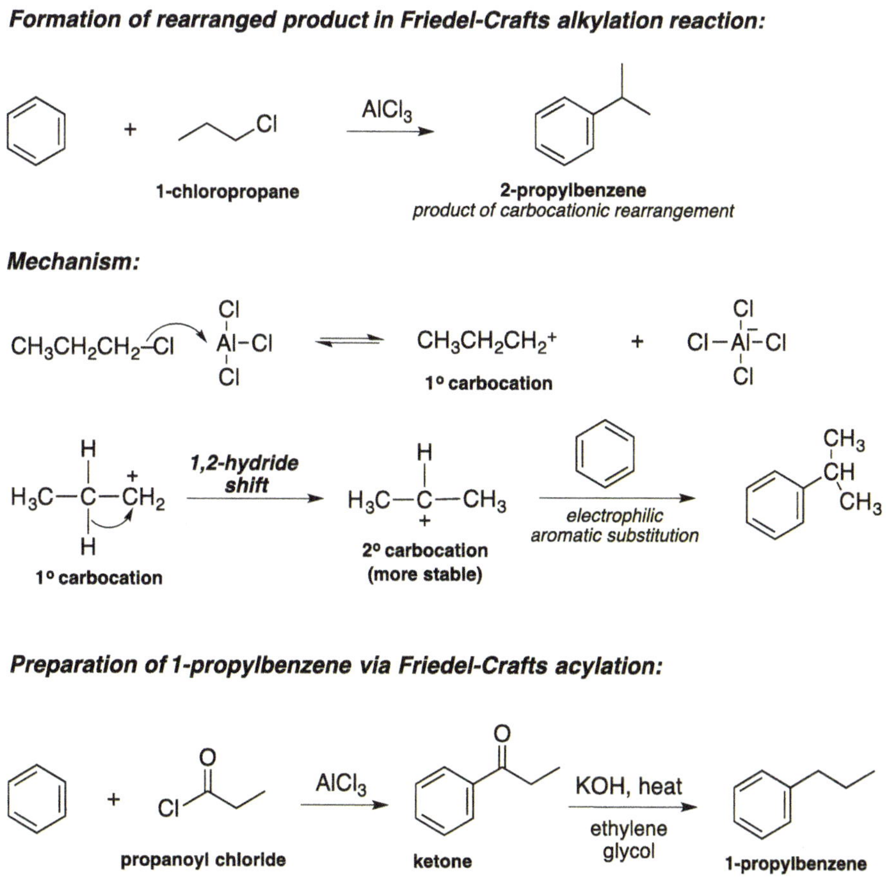

Formation of rearranged product in Friedel-Crafts alkylation reaction:

Mechanism:

Preparation of 1-propylbenzene via Friedel-Crafts acylation:

FIGURE 14.15. Rearrangement in Friedel-Crafts alkylation and preparation of unrearranged alkylbenzene via a Friedel-Crafts acylation-reduction sequence.

14.4. ELECTROPHILIC AROMATIC SUBSTITUTION REACTIONS OF SUBSTITUTED BENZENES

The presence of a substituent on a benzene ring has a strong effect on electrophilic aromatic substitution on the ring, leading to disubstituted benzene derivatives. Any electron-donating substituent will make a phenyl ring more active when it reacts with an electrophile (E⁺) because of the electrostatic attraction between the partially negatively charged ring and the positively charged E⁺ species. The presence of activating substituent increases the reaction rate and makes reaction possible at a lower temperature. The activating substituents can donate electronic density to the benzene ring by inductive effect (alkyl or aryl groups, see Section 2.3) or by resonance donation of the lone electron pair from oxygen or nitrogen atoms in substituents, such as OH, OR, NH_2, and NR_2 (Figure 14.16). In addition to increasing electron density in the benzene ring, the electron-donating substituents provide additional stabilization to the carbocationic intermediate and thus, lower the activation

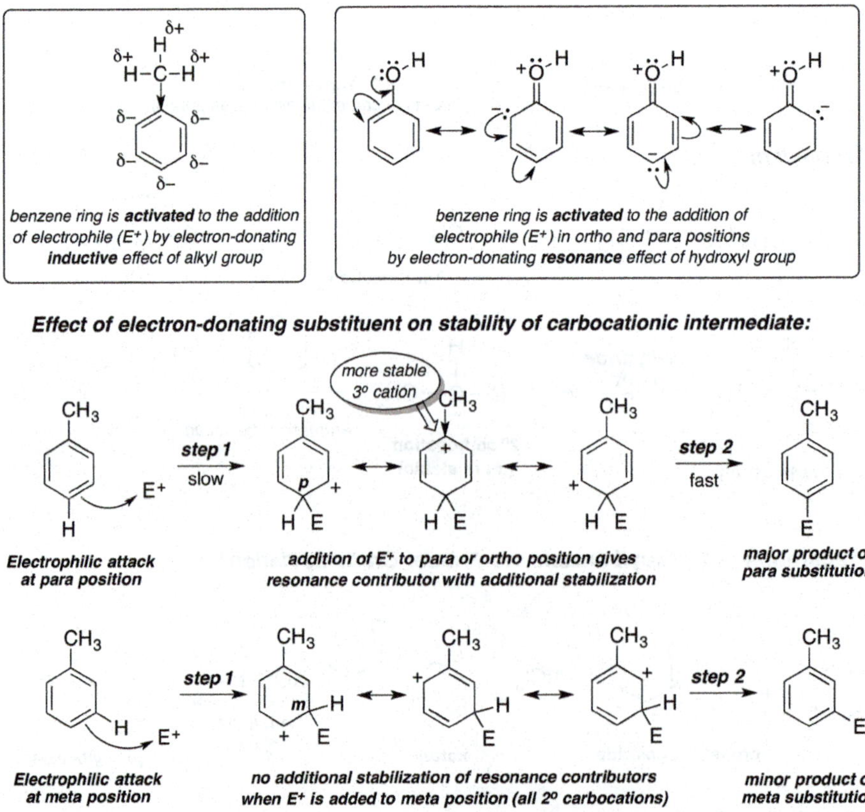

FIGURE 14.16. Electron-donating groups (alkyl, OH, OR, NR_2) activating the phenyl ring and directing electrophilic substitution to ortho and para positions.

290   ORGANIC CHEMISTRY

energy of the first step of the reaction. Importantly, the activating effect of electron-donating substituents works exclusively in ortho and para positions of the benzene ring because of charge distribution in resonance contributors (Figure 14.16). Therefore, all ring activating (i.e., electron-donating) substituents are also **ortho- and para-directing substituents** (or **o-, p-directors**).

Figure 14.17 shows several examples of electrophilic substitution reactions of benzene derivatives bearing activating electron-donating groups. It should be noted that these reactions occur faster and at a lower temperature due to the presence of electron-donating groups. In general, alkyl substituents have a weak activating effect on the reaction, while the resonance electron donors (i.e., OH, OR, NH_2, and NR_2) are strong activators. Electrophilic substitution reactions of strongly activated benzene derivatives usually do not require the presence of iron salts as catalysts, and proceed readily even below room temperature.

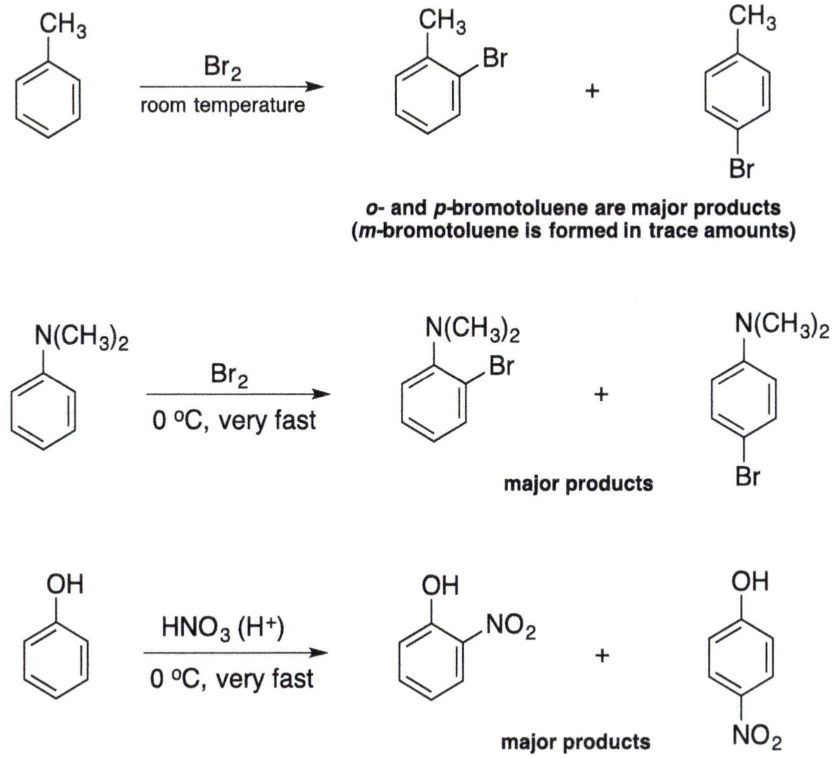

FIGURE 14.17. Electrophilic substitution reactions of benzene derivatives bearing activating electron-donating groups.

The presence of electron-withdrawing substituents on benzene rings has the opposite, deactivating effect on the reactivity in electrophilic aromatic substitution. The deactivating substituents can withdraw electronic density from the benzene ring by inductive effect (CF_3, CCl_3, and R_3N^+) or by resonance interaction (NO_2, CHO, CO_2H, CO_2R, SO_2OH, etc.). In addition to decreasing electron density in the benzene ring, the electron-withdrawing substituents destabilize the carbocationic intermediates and thus slow down the reaction. Importantly, the deactivating effect of electron-withdrawing substituents works in ortho and para positions of the benzene ring because of charge distribution in resonance contributors (Figure 14.18). Only meta position on the ring remains relatively active but still less reactive compared to benzene itself. Therefore, the majority of ring deactivating (i.e., electron-withdrawing) substituents are also **meta directing substituents** (**m-directors**). The halogen substituents (e.g., F, Cl, or Br), which are o-, p-directors, represent the only exception from this rule. Halogen atoms have strong electron-withdrawing inductive effect due to their electronegativity. This is why fluoro-, chloro-, or bromobenzene have relatively

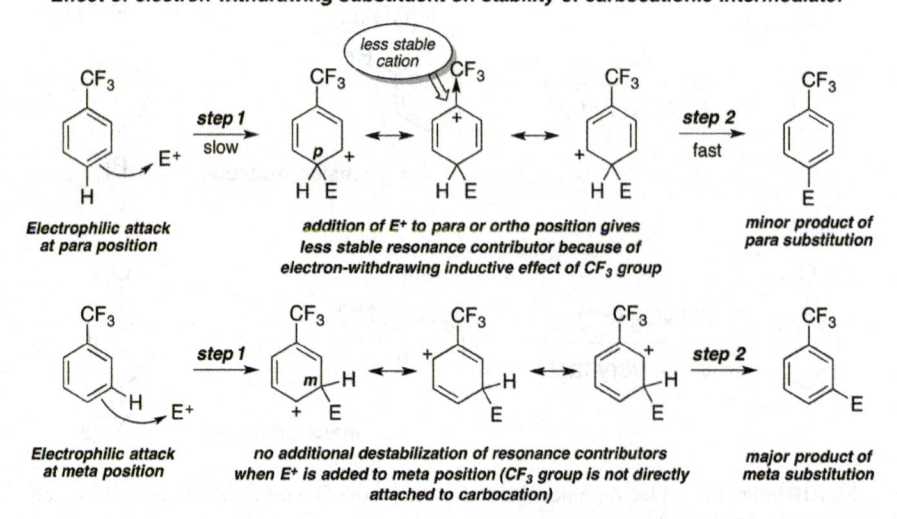

FIGURE 14.18. Electron-withdrawing groups (CF_3, CCl_3, and R_3N^+, NO_2, CHO, CO_2H, CO_2R, SO_2OH, etc) deactivate phenyl ring and in general direct electrophilic substitution to meta position.

292   ORGANIC CHEMISTRY

low reactivity compared to benzene. However, because of the presence of lone electron pairs, halogen atoms can donate electronic density by resonance to ortho and para positions during the course of reaction, and thus stabilize carbocationic intermediates. Overall, halogen substituents deactivate benzene rings and direct electrophilic substitution to ortho and para positions.

Figure 14.19 shows examples of electrophilic substitution reactions of benzene derivatives bearing deactivating electron-withdrawing groups. Note that these reactions occur slow even at higher temperature due to the presence of deactivating groups. The strongest deactivating effect is due to a nitro group (NO_2) and positively charged trialkylammonium substituents (R_3N^+).

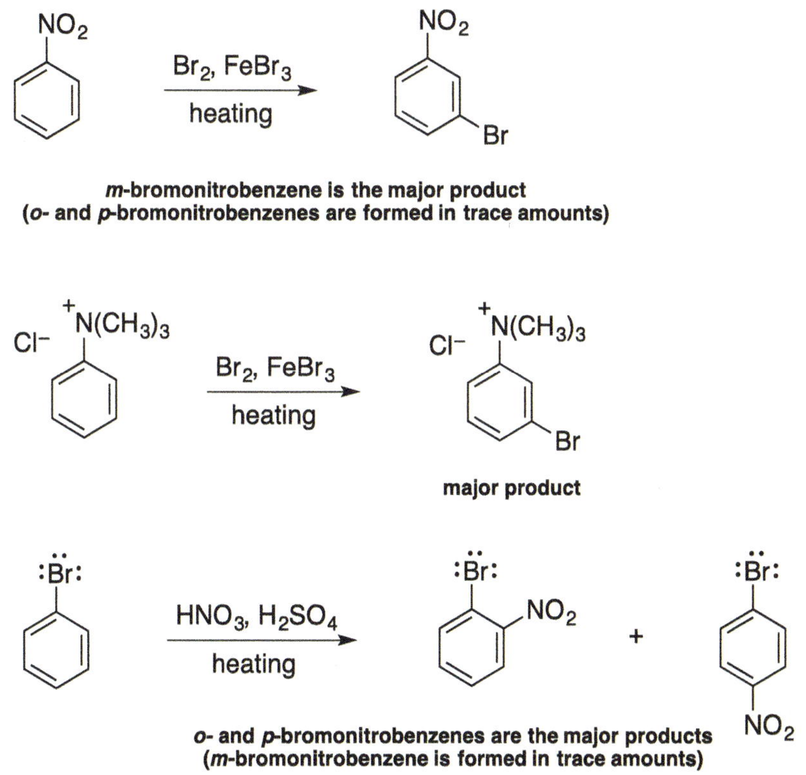

FIGURE 14.19. Electrophilic substitution reactions of benzene derivatives bearing deactivating electron-withdrawing groups.

In the reactions of benzene derivatives bearing two or more substituents, the activating and directing effect of each substituent should be evaluated and compared in order to predict the major product. In general, stronger activators have the strongest directing effects. Several examples of such reactions are shown in Figure 14.20.

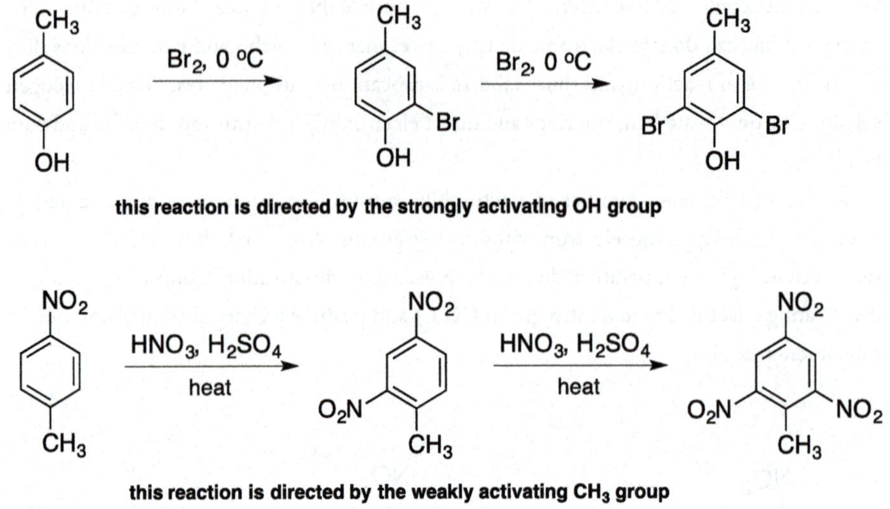

FIGURE 14.20. Electrophilic substitution reactions of benzene derivatives bearing two or more substituents.

Problems

14.8. Propose mechanisms for the reactions shown in Figures 14.17 and 14.19.

14.9. Provide the major products of the following transformations:

a)

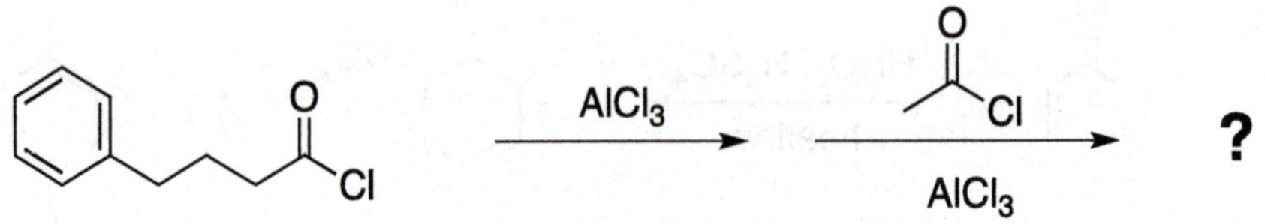

PhCH₂CH₂CH₂C(O)Cl →(AlCl₃) →(CH₃C(O)Cl, AlCl₃) ?

b)

HO-C₆H₄-CH₂-CH(⁺NH₃)-C(O)OH →(HNO₃, H₂SO₄) ?

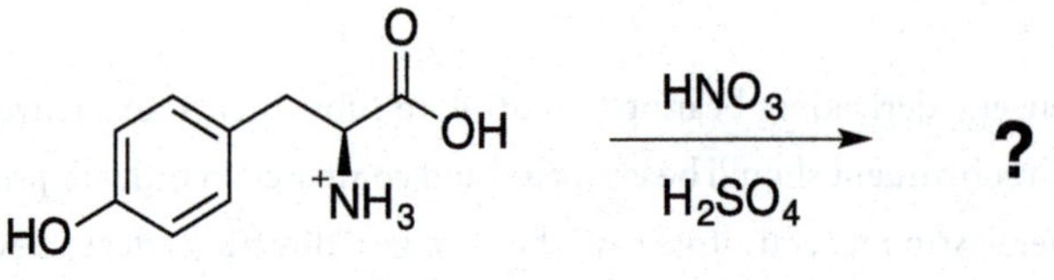

c)

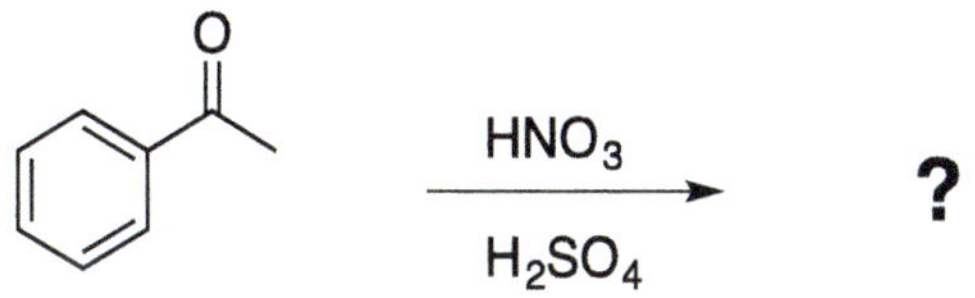

[acetophenone] —HNO₃/H₂SO₄→ ?

d)

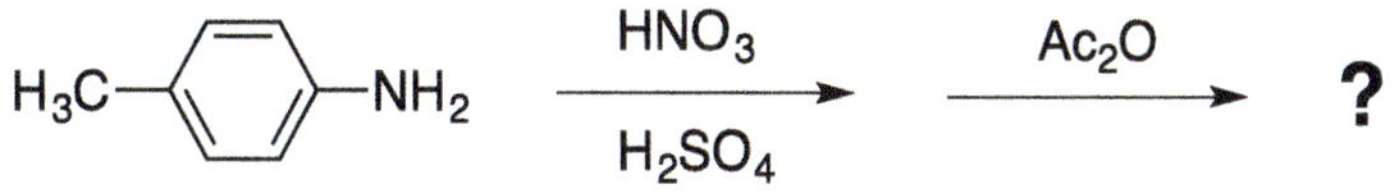

H₃C—⌬—NH₂ —HNO₃/H₂SO₄→ —Ac₂O→ ?

e)

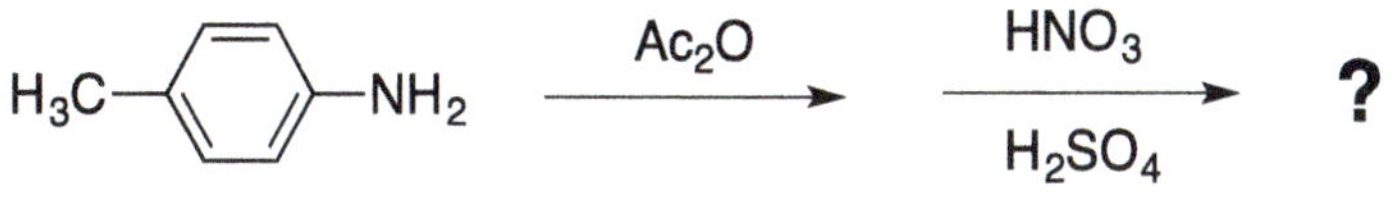

H₃C—⌬—NH₂ —Ac₂O→ —HNO₃/H₂SO₄→ ?

14.10 The following synthesis involves three elementary steps: 1. a transesterfication; 2. an electrophilic aromatic substitution; 3. a dehydration. Draw the intermediates.

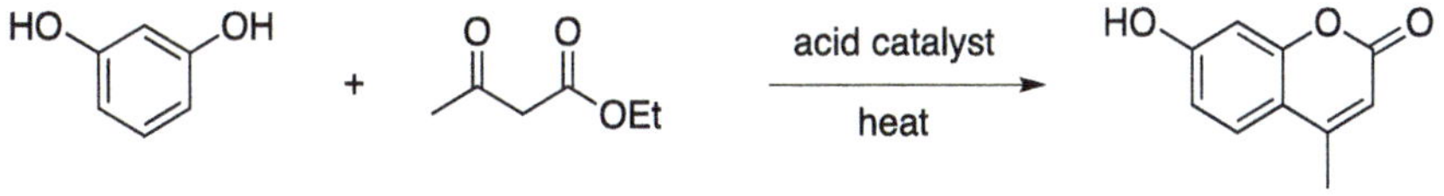

HO—⌬—OH + [ethyl acetoacetate, CH₃COCH₂COOEt] —acid catalyst, heat→ [4-methylumbelliferone structure]

14.11 The following two reactions result in the same product. Explain.

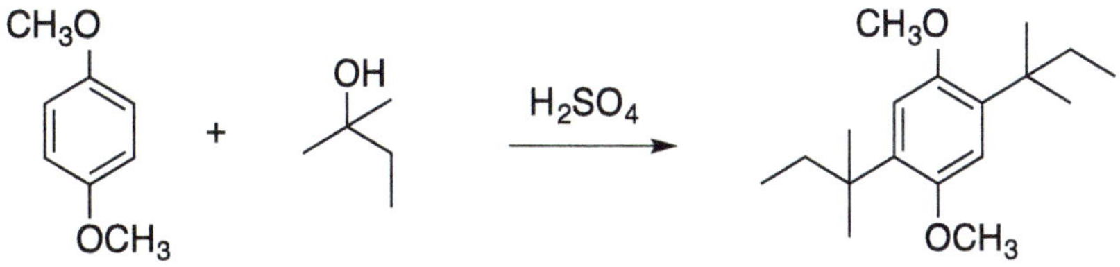

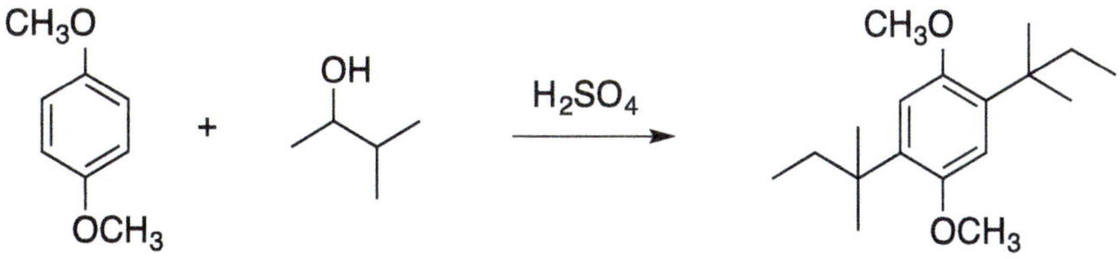

[1,4-dimethoxybenzene] + [2-methyl-2-butanol, (CH₃)₂C(OH)CH₂CH₃] —H₂SO₄→ [2,5-di-tert-amyl-1,4-dimethoxybenzene]

[1,4-dimethoxybenzene] + [3-methyl-2-butanol] —H₂SO₄→ [same product]

BENZENE AND AROMATIC COMPOUNDS   295

14.5. NUCLEOPHILIC AROMATIC SUBSTITUTION REACTIONS

Nucleophilic aromatic substitution reactions involve replacement of a leaving group (usually, a halogen atom) in the aromatic substrate with a nucleophile. Unlike alkyl halides (Chapter 5), aryl halides do not undergo nucleophilic substitution by S_N2 or S_N1 mechanisms. In particular, the S_N2 mechanism requires the approach of a nucleophile from the side opposite the leaving group (Figure 5.5), which is impossible in the planar molecule of aryl halide (such as bromobenzene). Nucleophilic substitution in aryl halides by S_N1 mechanisms requires the initial formation of the phenyl carbocation (i.e., Ph^+), which is an extremely high energy, unstable intermediate. Formation of phenyl carbocations is possible only under very special conditions, such as the decomposition of aryldiazonium salts (Chapter 15). In general, nucleophilic substitution reactions in aromatic compounds are less common than electrophilic substitution reactions, and proceed by two special mechanisms involving addition and elimination steps in different sequences.

14.5.1. Addition/elimination mechanism

Chlorobenzene or bromobenzene do not react with common nucleophiles, such as HO^-, RO^-, or RNH_2 under normal conditions. However, the presence of a strong electron-withdrawing substituent, such as the nitro group, activates the benzene ring towards nucleophilic attack. In particular, p-chloronitrobenzene has resonance contributors with positively charged carbons (Figure 14.21), making the benzene ring more active in reactions with nucleophiles (i.e., Nu:⁻), because of the electrostatic attraction between the partially positively charged ring and the negatively charged Nu:⁻ species.

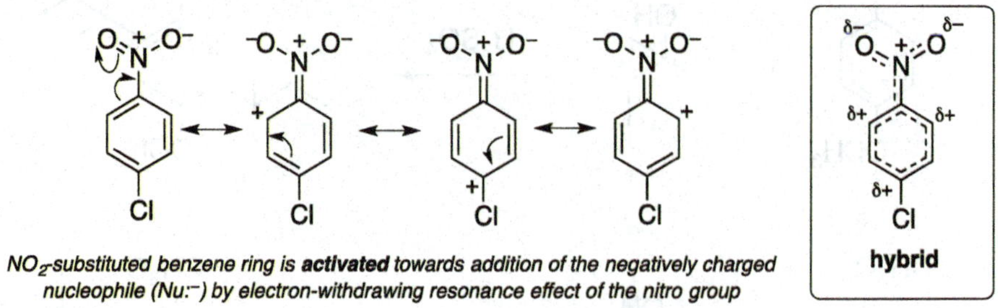

NO_2-substituted benzene ring is **activated** towards addition of the negatively charged nucleophile (Nu:⁻) by electron-withdrawing resonance effect of the nitro group

FIGURE 14.21. Resonance contributors of p-chloronitrobenzene.

A gentle heating of *p*-chloronitrobenzene with aqueous sodium hydroxide followed by neutralization of the base, yields *p*-nitrophenol, resulting from nucleophilic substitution of a chlorine atom with a hydroxide anion (Figure 14.22). This reaction proceeds via the initial addition of HO⁻ to the partially positively charged carbon bearing the halogen atom. The resulting negatively charged intermediate (the Meisenheimer complex) is stabilized by resonance participation of the nitro group in para position. Elimination of the chloride anion in the second step results in the formation of the final product. It should be noted that resonance stabilization of the intermediate is possible only when the NO₂ group is located in para or ortho position, relative to the leaving group. The presence of additional nitro group(s) in the benzene ring further facilitates the reaction.

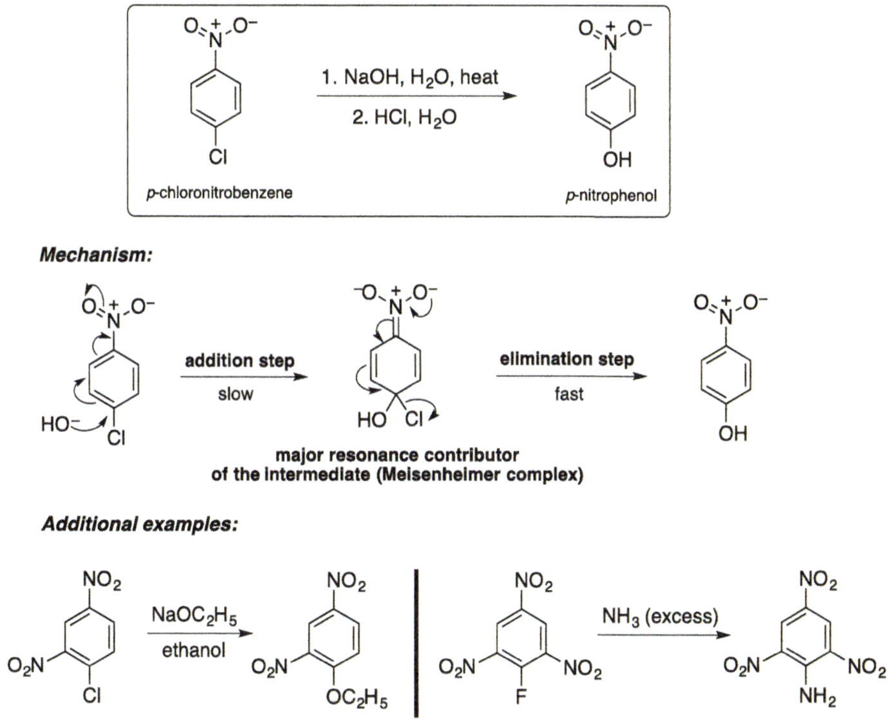

FIGURE 14.22. Nucleophilic aromatic substitution via addition/elimination mechanism.

14.5.2. Elimination/addition mechanism via benzyne intermediate

Chlorobenzene or aryl chlorides bearing electron-donating substituents can be converted to the corresponding phenols by basic hydrolysis at very high temperatures and pressure (about 350 °C and 300 bar pressure). This is an important industrial process (the **Dow process**) for the large-scale preparation of phenol from chlorobenzene. Advanced experimental studies have revealed the mechanism of this reaction. The initial β-elimination step (Chapter 5) produces an extremely unstable and highly reactive benzyne intermediate (Figure 14.23). Addition of hydroxide anion to benzyne generates carbanion, which is finally protonated via a proton transfer reaction with water. In the end, the reaction mixture is neutralized by addition of aqueous acid in order to avoid deprotonation of phenol by excessive base. The analogous reaction of *p*-chlorotoluene produces a mixture of para- and meta-methylphenols (*p*-cresol and *m*-cresol) via nucleophilic addition of the hydroxide anion to each of the two alkyne carbons of the 4-methylbenzyne intermediate

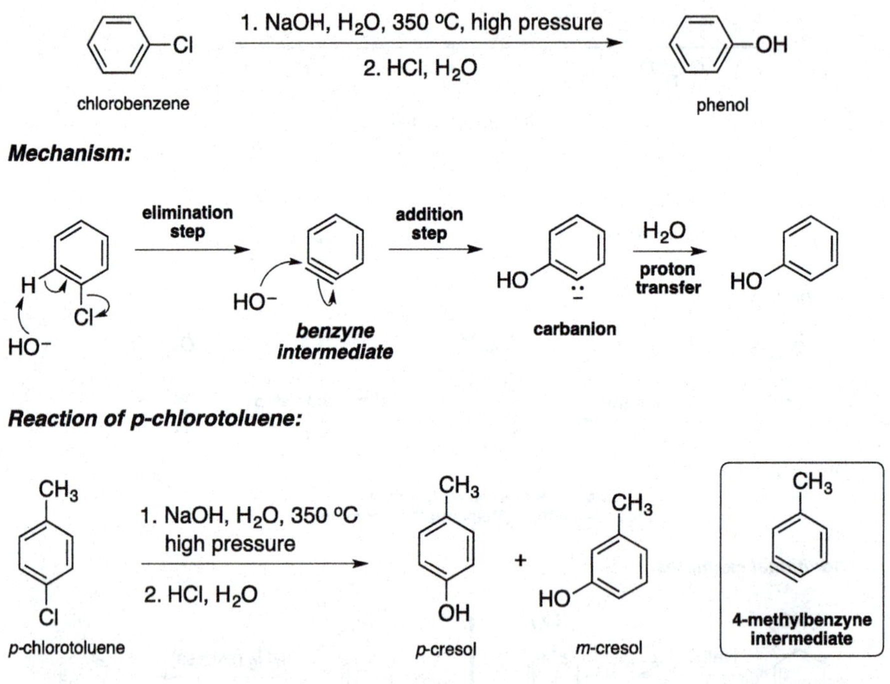

FIGURE 14.23. Preparation of phenols from aryl chlorides via benzyne intermediate.

The use of a stronger base, such as amide anions instead of hydroxide anions, facilitates the β-elimination step, leading to the benzyne intermediate. The reaction of chlorobenzene or aryl halides bearing electron-donating substituents with sodium amide, proceeds under very mild conditions, yielding the corresponding anilines via the elimination/addition mechanism involving benzyne or 3-methylbenzyne intermediates (Figure 14.24).

FIGURE 14.24. Preparation of anilines from aryl chlorides by reaction with sodium amide.

Problems

14.12 Propose mechanisms for the reactions shown in Figure 14.24.

14.13 What are the major products of the following reactions?

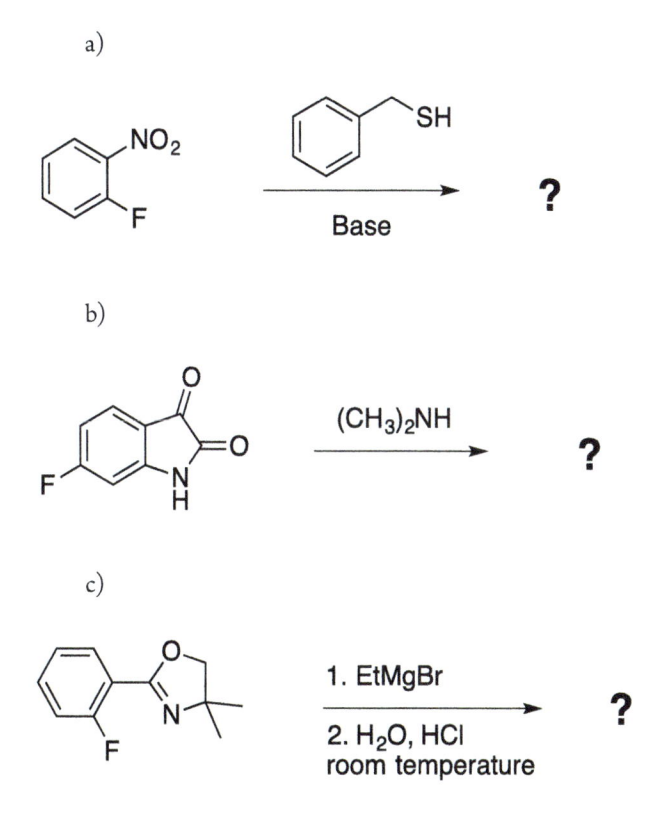

a) 2-fluoronitrobenzene + benzyl thiol (PhCH₂SH), Base → ?

b) 6-fluoroisatin + (CH₃)₂NH → ?

c) 2-(2-fluorophenyl)-4,4-dimethyl-4,5-dihydrooxazoline
1. EtMgBr
2. H₂O, HCl, room temperature
→ ?

d)

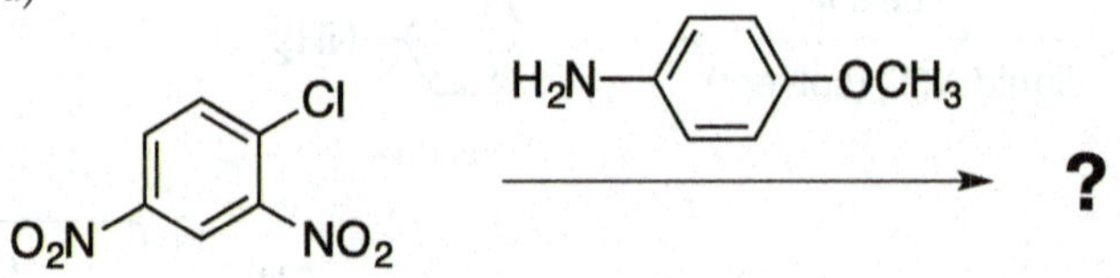

[Reaction: 1-chloro-2,4-dinitrobenzene type structure (Cl, NO₂, O₂N on benzene) + H₂N—C₆H₄—OCH₃ → ?]

e)

[Reaction: 4-fluorobenzaldehyde (OHC—C₆H₄—F) + HO—C₆H₄—OCH₃, K₂CO₃ → ?]

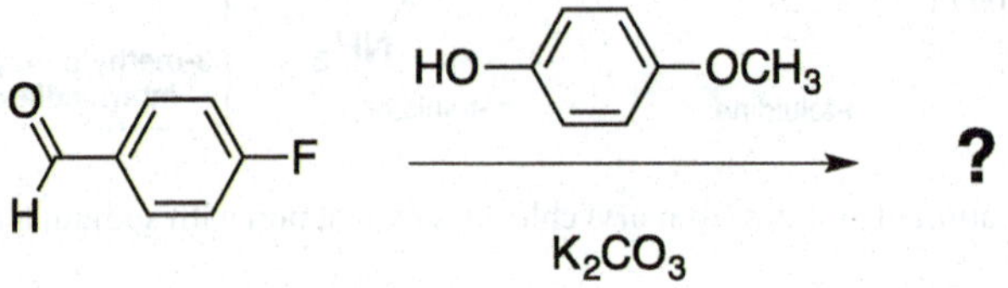

14.6. REACTIONS OF ALKYLBENZENES AT THE BENZYLIC POSITION

Because of the high stability of benzene rings, the reactions of alkylbenzenes with radical reagents (see Sections 3.6 and 6.6.1) and oxidants (Sections 6.5 and 8.7) generally occur at the alkyl substituent, keeping the aromatic ring unchanged. The hydrogen atoms at the alkyl carbon adjacent to the aromatic ring (called **benzylic carbon,** or the **benzylic position**) are particularly reactive in radical halogenation reactions (Section 3.6). For example, ethylbenzene reacts with Cl_2 in the presence of light, forming 1-chloro-1-phenylethane as the major product (Figure 14.25). This reaction has the same radical substitution chain

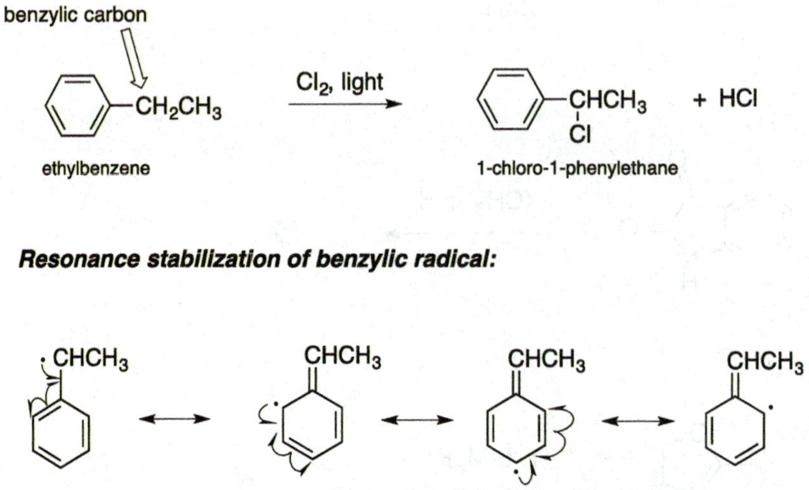

benzylic carbon

C₆H₅—CH₂CH₃ →[Cl_2, light] C₆H₅—CHCH₃ + HCl
                                       |
                                       Cl

ethylbenzene   1-chloro-1-phenylethane

Resonance stabilization of benzylic radical:

[Four resonance structures of the benzylic radical ·CHCH₃ attached to benzene ring, showing delocalization of the radical]

FIGURE 14.25. Chlorination of alkylbenzene at the benzylic position.

300 ORGANIC CHEMISTRY

mechanism as shown for alkanes in Figure 3.13 (Section 3.6), and involves the formation of benzylic radicals as reaction intermediates. Benzylic radicals have about the same stability as allylic radicals (see Section 6.6.1) because of the resonance delocalization of the single electron, as shown in Figure 14.25. The resonance stabilization of benzylic radical explains the high reactivity of alkylbenzenes at the benzylic position.

Radical bromination at the benzylic position can be accomplished using N-bromosuccinimide (NBS), analogous to allylic bromination (Section 6.6.1). Several examples of benzylic bromination are shown in Figure 14.26.

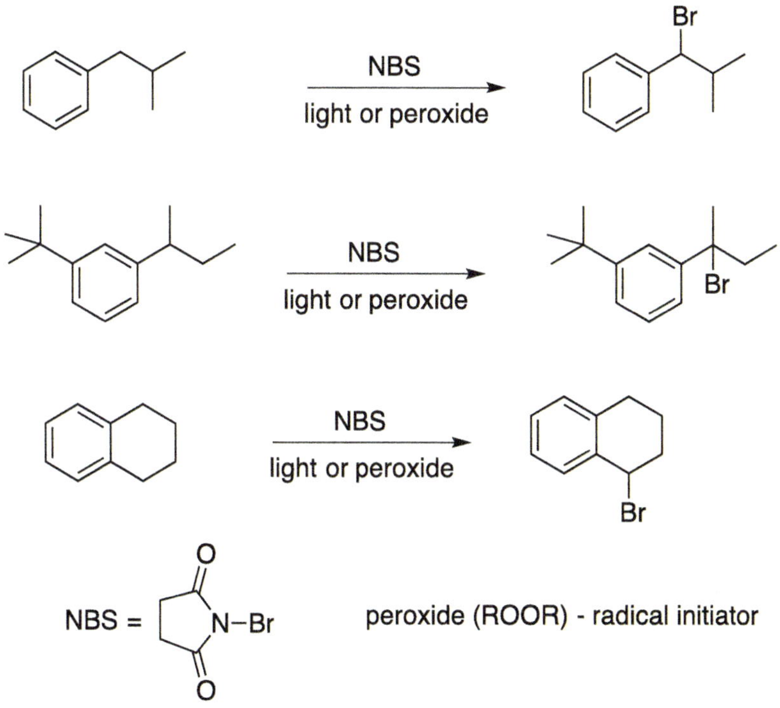

FIGURE 14.26. Bromination of alkylbenzenes at the benzylic position.

The alkyl group in alkylbenzenes can be oxidized at the benzylic position using common oxidants, such as IBX and chromic acid (Section 8.7). The milder oxidant, IBX, oxidizes toluene to benzaldehyde, and unbranched alkylbenzenes are oxidized to the corresponding ketones (Figure 14.27). The mechanism of these reactions involves initial generation of the resonance stabilized benzylic radicals and cations via single electron transfer processes. The initial products are benzylic alcohols, which are immediately oxidized by IBX to the final carbonyl compounds. The stronger oxidant, chromic acid, completely oxidizes any alkyl chain containing benzylic hydrogens to the carboxylic group via a complex, multistep mechanism.

The 3° alkyl substituents are not oxidized because they lack benzylic hydrogens and therefore cannot be converted to the initial benzylic radicals.

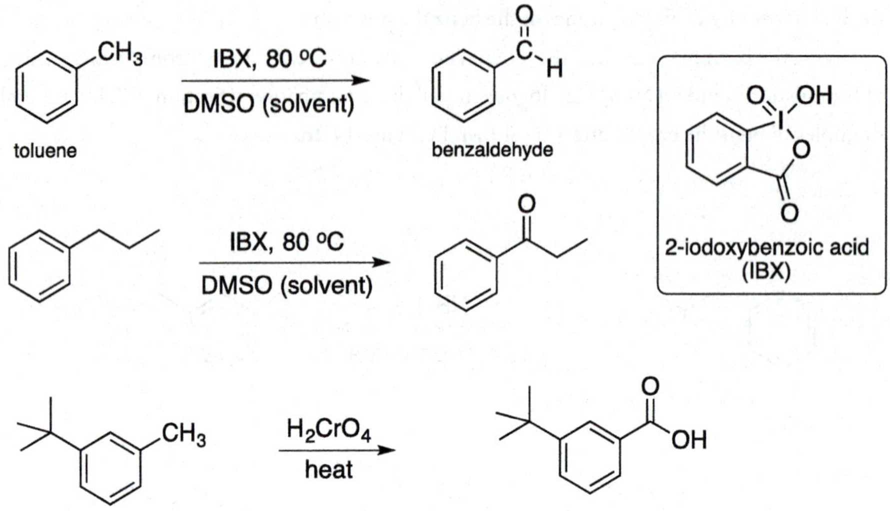

FIGURE 14.27. Oxidation of alkylbenzenes at the benzylic position.

Problems

14.14. Propose a detailed mechanism (i.e., initiation, propagation, and termination steps) for the radical chlorination of ethylbenzene.

14.15. What are the major products of the following transformations?

a)

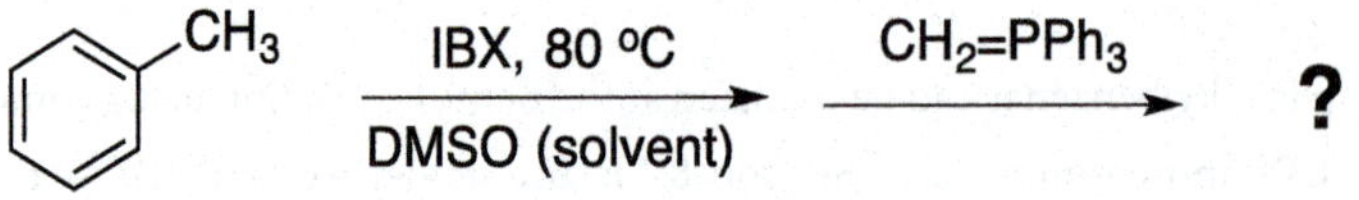

b)

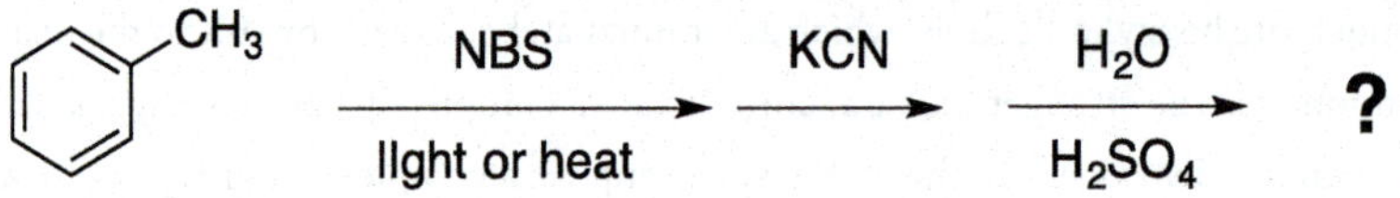

302   ORGANIC CHEMISTRY

14.16. Benzyl bromide is a primary halide. In nucleophilic substitution reactions, why does this compound react via S_N1 and S_N2 reaction mechanisms, dependent on the reaction conditions?

14.7. PHENOLS

Phenols are aromatic analogs of alcohols with a hydroxyl group directly attached to the benzene ring. Phenolic compounds are very common in nature and widely used as pharmaceuticals and industrial chemicals. The phenolic hydroxyl group has much higher acidity compared to alcohols, which is reflected in its old common name "carbolic acid" for the parent phenol (PhOH). Phenol (pK_a = 9.95) is a stronger acid than ethanol (pK_a = 15.9) but weaker acid than acetic acid (pK_a = 4.76). The greater acidity of phenols is due to the resonance stabilization of the phenoxide anion, which is not possible for alkoxide anions (Figure 14.28).

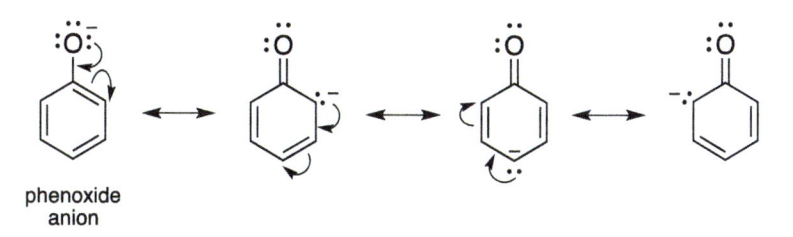

FIGURE 14.28. Resonance stabilization of phenoxide anion.

Phenolic compounds bearing electron-withdrawing substituents on the benzene ring have higher acidity compared to the parent phenol, while phenols with electron-donating substituents have slightly lower acidity. For example, p-nitrophenol (pK_a = 7.15) is a stronger acid than phenol (pK_a = 9.95) because the nitro group in para position can participate in the resonance stabilization of p-nitrophenoxide anions (Figure 14.29). The effect of the meta substituent is much lower since it is not involved in the resonance stabilization. Placement of two or three nitro groups in ortho and para positions of phenol further increases its acidity. In fact, 2,4,6-trinitrophenol (pK$_a$ = 0.38), which is also known as picric acid, has an acidity comparable to strong inorganic acids. Methyl substituted phenols, referred to as cresols, are slightly less acidic than phenol because the electron-donating inductive effect of the alkyl group increases the negative charge on the oxygen of the phenoxide anion, which destabilizes the anion (see Figure 2.6 and the related text in Section 2.3).

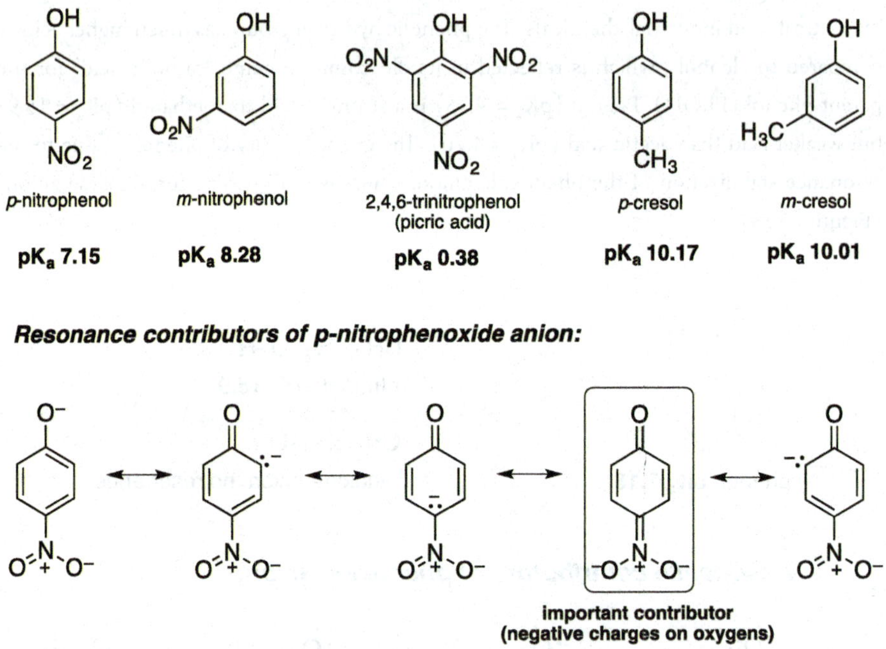

p-nitrophenol	*m*-nitrophenol	2,4,6-trinitrophenol (picric acid)	*p*-cresol	*m*-cresol
pK$_a$ 7.15	pK$_a$ 8.28	pK$_a$ 0.38	pK$_a$ 10.17	pK$_a$ 10.01

Resonance contributors of p-nitrophenoxide anion:

important contributor
(negative charges on oxygens)

FIGURE 14.29. Acidity of substituted phenols.

Nucleophilic reactivity of phenoxide anions is comparable to that of alkoxide anions and they can react with haloalkanes to form the corresponding aryl alkyl ethers (see Williamson ether synthesis in Section 8.3).

Phenols can be easily oxidized by common oxidants, forming unsaturated diketones commonly known as quinones (Figure 14.30). Because of the high reactivity toward oxidants, many natural phenolic compounds (e.g., flavonoids, polyphenols, tocopherols) are powerful biological antioxidants capable of protecting cells from oxidative damage.

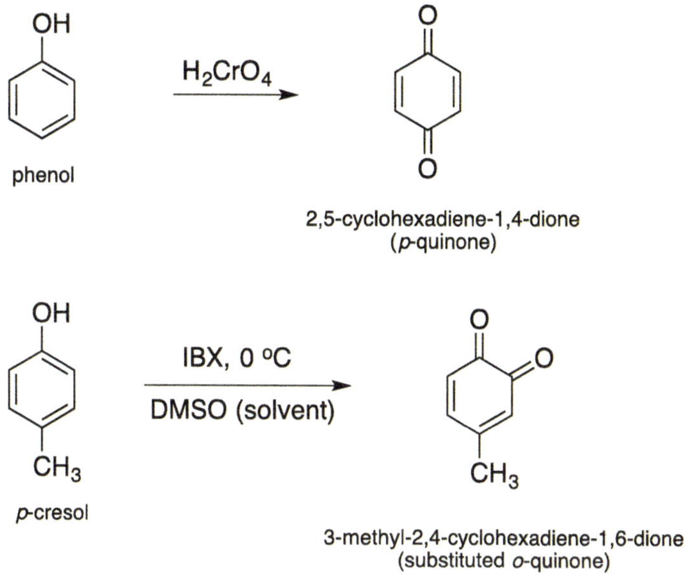

FIGURE 14.30. Oxidation of phenols to quinones.

Problems

14.17. What are the major products of the following transformations?

a)

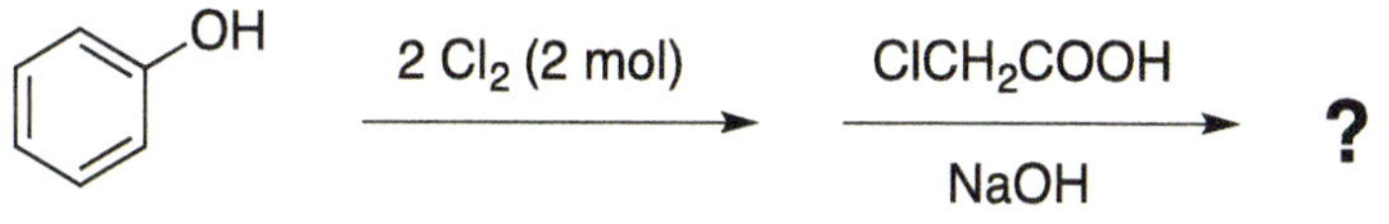

phenol —(2 Cl₂ (2 mol))→ —(ClCH₂COOH / NaOH)→ ?

b)

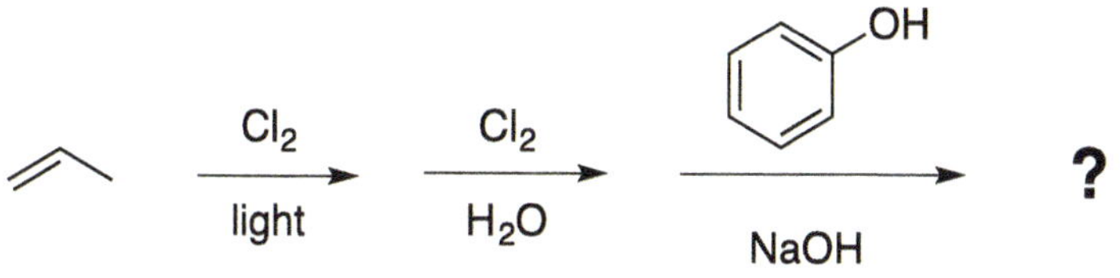

CH₂=CHCH₃ —(Cl₂, light)→ —(Cl₂, H₂O)→ —(phenol / NaOH)→ ?

CHAPTER 15

Amines

Amines are broadly defined as organic nitrogen compounds derived from ammonia (NH$_3$). Amines are very common in anture and play an exceptionally important role in biological systems. Proteins and nucleic acids (Chapter 16) include derivatives of amines (i.e., amino acids and nucleobases) as basic building blocks. Alkaloids are various naturally occurring amines produced by plants, bacteria, fungi, and other organisms. Because of their basic properties, alkaloids can be isolated from biological samples via acid-base extraction. Some common examples of alkaloids include morphine, cocaine, caffeine, and nicotine. Natural alkaloids, as well as many synthetic amines, have a broad range of biological functions and are widely used as pharmaceuticals and recreational drugs. Many amines are used in the chemical industry as starting materials for the manufacture of azo dyes and also in the production of polymers (epoxy resins). This chapter discusses the nomenclature, properties, and reactions of amines.

15.1. CLASSIFICATION AND NOMENCLATURE

Amines are generally classified as **primary**, **secondary**, and **tertiary amines** according to the number of organic groups R connected to the nitrogen (Figure 15.1). Ionic compounds containing positively charged nitrogen atom with four organic groups are called **quaternary ammonium salts**. **Alkylamines** (or aliphatic amines) have only alkyl group(s) R, while **arylamines** (or aromatic amines) have at least one aromatic phenyl ring directly connected to the nitrogen atom.

$$\begin{array}{cccc}
\text{H} & \text{R} & \text{R} & \text{R} \\
| & | & | & |+ \\
\text{R–N} & \text{R–N} & \text{R–N} & \text{R–N–R} \\
| & | & | & \\
\text{H} & \text{H} & \text{R} & \text{X}^- \quad \text{R}
\end{array}$$

primary amines   secondary amines   tertiary amines   quaternary
(1° amines)   (2° amines)   (3° amines)   ammonium salts

Examples of alkylamines:

Common name: *t*-butylamine   cyclopentylamine   N,N-dimethylcyclopentanamine
IUPAC name: 2-methylpropan-2-amine   (cyclopentanamine)
(1° amine)   (1° amine)   (3° amine)

Examples of arylamines:

aniline   N-methylaniline   N,N-dimethylaniline
(1° amine)   (2° amine)   (3° amine)

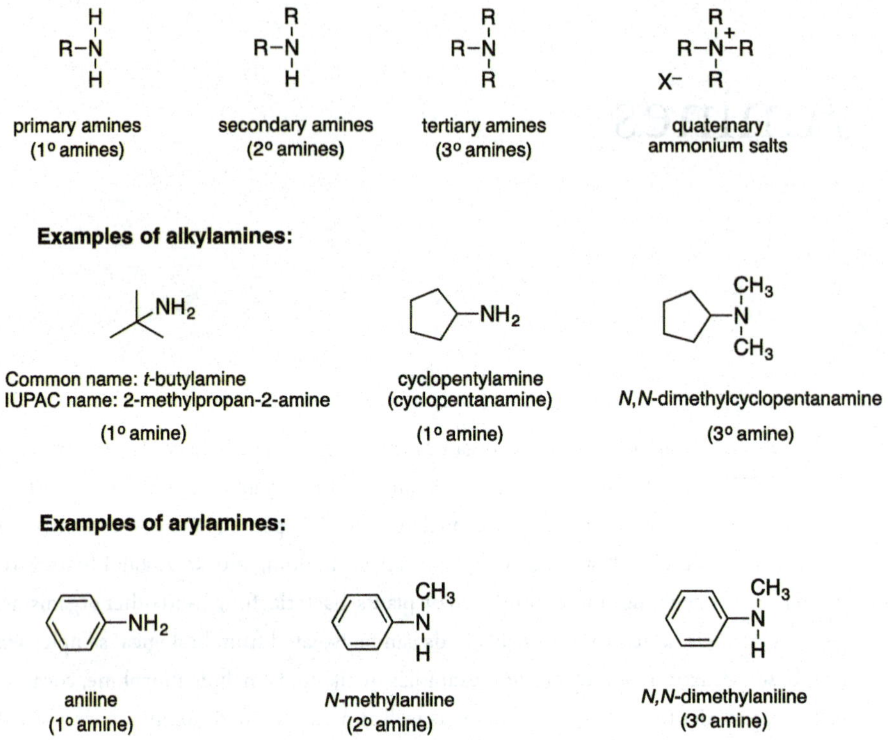

FIGURE 15.1. General classification and names of amines.

Common names of most alkylamines are derived by listing the alkyl groups bonded to the nitrogen in alphabetical order in a single word ending with the suffix -amine such as CH_3NH_2 methylamine, $(CH_3CH_2)_2NH$ diethylamine, $(CH_3CH_2)_2NCH_3$ diethyl(methyl)amine. Systematic IUPAC names are formed by dropping the suffix -e of the parent alkane and replacing it by the suffix -amine (Figure 15.1). Note that the prefix *N*- is used to indicate the presence of alkyl groups connected to the nitrogen atom in 2° or 3° amines analogous to the nomenclature of amides (see Figure 12.2 and the related text in Chapter 12). In compounds where a higher priority group is present (see Table 1.4 and the related text in Chapter 1), the prefix amino- is placed at the beginning of the name (Figure 15.2).

H₂N⁀⁀⁀OH　　　　H₂N⁀C(=O)OH　　　　HO-C₆H₃(CO₂H)(NH₂)

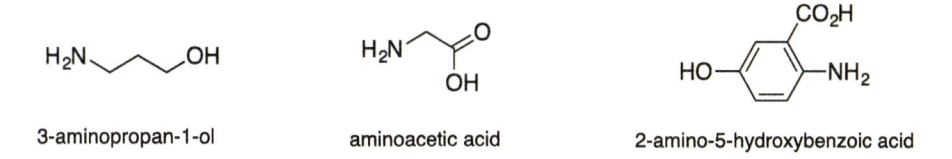

3-aminopropan-1-ol　　　aminoacetic acid　　　2-amino-5-hydroxybenzoic acid

FIGURE 15.2. Names of amines containing additional functional groups.

Amines which have nitrogen atom as part of a ring are classified as **heterocyclic amines**. Particularly important are aromatic heterocyclic amines such as pyridine and pyrrole (see Section 14.1). Structures and common names of several heterocyclic amines are shown in Figure 15.3.

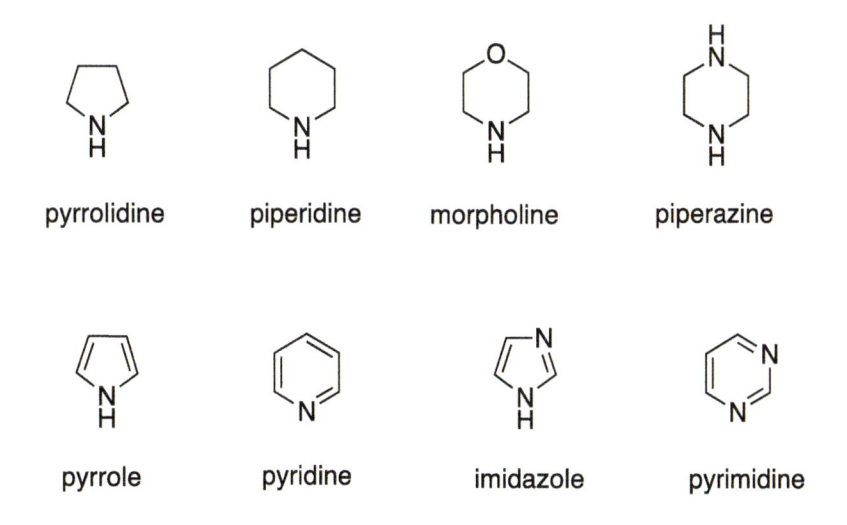

pyrrolidine   piperidine   morpholine   piperazine

pyrrole   pyridine   imidazole   pyrimidine

FIGURE 15.3. Important heterocyclic amines.

The nitrogen atom in amines has a non-bonded electron pair, making it both a base and a nucleophile. Therefore, amines can react with acids and electrophiles yielding the corresponding ammonium salts, ionic compounds with four bonds to nitrogen. If one of the four atoms bonded to the nitrogen atom is hydrogen, the ammonium salt is the conjugate acid of an amine. Quaternary ammonium salts have four carbon atoms connected to the nitrogen atom and are formed by S_N2 reaction of a 3° amine with an alkyl halide. The names of ammonium salts are formed by replacing the suffix -amine by the suffix -ammonium and adding the name of the anion (Figure 15.4). Likewise, protonated aniline is called anilinium, protonated pyridine is called pyridinium, and protonated imidazole is called imidazolium. An important oxidant pyridinium chlorochromate (PCC, see Section 8.7) is a salt formed by protonation of pyridine by chlorochromic acid (ClCrO$_3$H).

tetrabutylammonium bromide
(Bu$_4$N$^+$Br$^-$)

a quaternary ammonium salt

anilinium chloride

a conjugate acid of aniline, PhNH$_2$

pyridinium chlorochromate (PCC)

a conjugate acid of pyridine

FIGURE 15.4. Examples of ammonium salts.

Problems

15.1. Draw structures and provide names for all amines with the molecular formula C$_4$H$_{11}$N. Classify these amines as primary, secondary, and tertiary.

15.2. Provide the IUPAC name and the common name for the amines with the following ^{1}H NMR spectra.

a) The amine is a secondary amine.

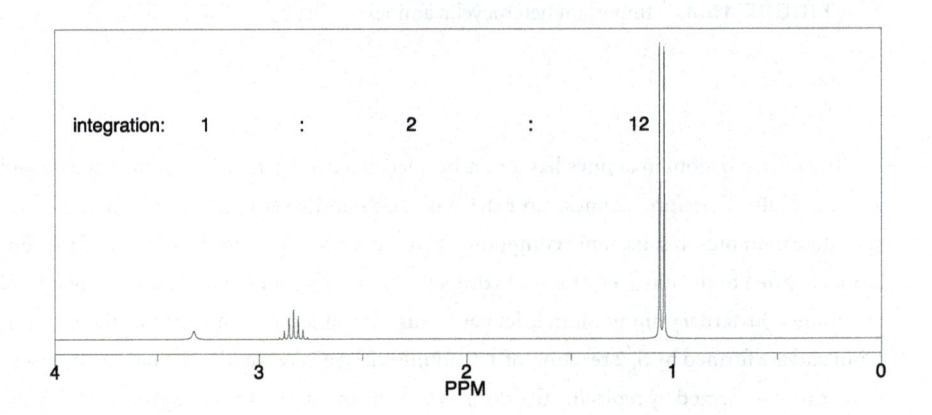

integration: 1   :   2   :   12

b) The amine is a derivative of aniline.

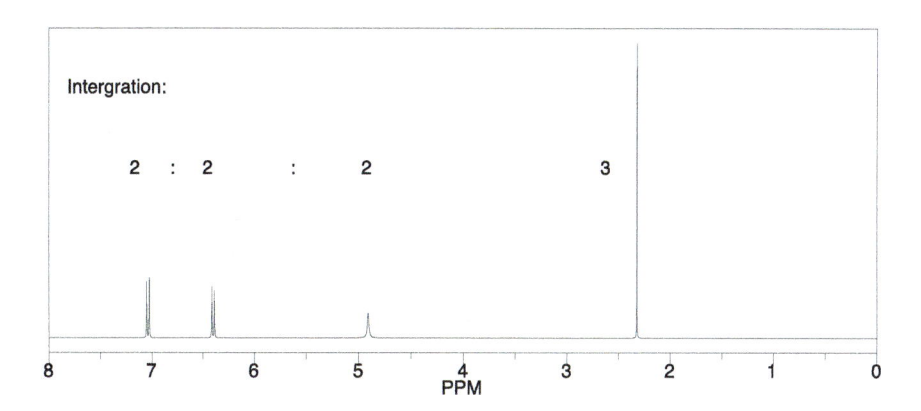

Intergration:

2 : 2 : 2   3

15.2. PHYSICAL PROPERTIES AND BASICITY OF AMINES

Amines are moderately polar compounds; they have boiling points that are higher than those of alkanes but generally lower than those of alcohols of comparable molecular weight. Methylamine and ethylamine are colorless gases with an ammonia-like odor and the heavier alkylamines are volatile liquids. Molecules of primary and secondary amines can form the N–H•••N hydrogen bonds with each other and the N–H•••O hydrogen bonds with water. Because of the N–H•••O hydrogen bonding, methylamine and other low molecular weight alkylamines are soluble in water. Tertiary amines cannot form hydrogen bonds with each other and generally have lower boiling point than primary and secondary amines of comparable molecular weight.

Aqueous solutions of amines are basic like aqueous solutions of ammonia. The position of acid-base equilibrium in the aqueous solution of an amine varies with the pK_a value of its conjugate acid (the protonated amine or ammonium ion). The equilibrium will shift to the right (forward direction) with increasing pK_a value of RNH_3^+ (or increasing basicity of RNH_2 and vice-versa (Figure 15.5). Therefore, the **amine with a larger pK_a value of its conjugate acid (ammonium ion) will be more basic** when comparing the relative basicity of two or more amines.

15.2.1. Basicity of alkylamines

Due to the electron-donating effect of the alkyl group(s), simple alkylamines such as ethylamine, diethylamine and triethylamine are significantly more basic than ammonia (Figure 15.6). Secondary alkylamines are usually slightly stronger bases than primary alkylamines

position of this equilibrium for the more basic amine is shifted to the right (higher concentration of HO⁻)

$$H_2O + RNH_2 \rightleftharpoons HO^- + R\overset{+}{N}H_3$$

an acid   a base   conjugate acid (ammonium ion)

$pK_a = 15.7$ pK_a between 9 and 12

larger pK_a value of $R\overset{+}{N}H_3$ shifts equilibrium to the right side

Example:

pK_a of $\overset{+}{N}H_4 = 9.24$ pK_a of $CH_3\overset{+}{N}H_3 = 10.64$

CH_3NH_2 is a stronger base than NH_3

FIGURE 15.5. Acid-base equilibrium in aqueous solution of an amine.

since the additional electron-donating alkyl group helps to stabilize the ammonium ion (similar to stabilization of carbocations, Section 5.2.2) and thus shifts the acid-base equilibrium (Figure 15.5) to the right. However, the difference in basicity of 1°, 2°, and 3° alkylamines is very small due to solvation effects in solution. In particular, tertiary alkylammonium cations (R_3NH^+) have reduced hydrogen bonding with water and therefore are slightly less stable compared to primary (RNH_3^+) and secondary ($R_2NH_2^+$) alkylammonium cations, and thus 3° alkylamines are slightly weaker bases.

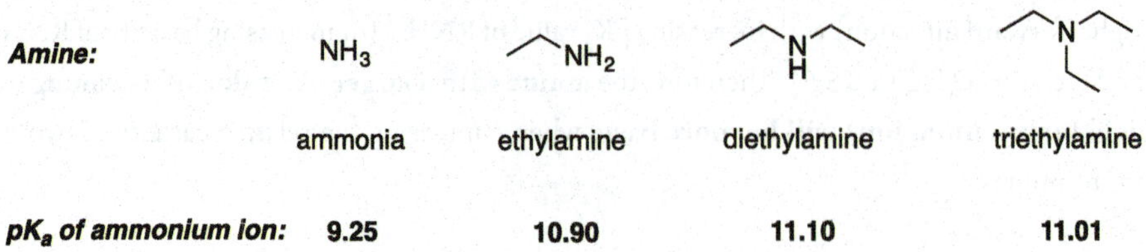

Amine:	NH_3	$CH_3CH_2NH_2$	$(CH_3CH_2)_2NH$	$(CH_3CH_2)_3N$
	ammonia	ethylamine	diethylamine	triethylamine
pK_a of ammonium ion:	9.25	10.90	11.10	11.01

FIGURE 15.6. Relative basicity of alkylamines.

ORGANIC CHEMISTRY

15.2.2. Basicity of arylamines

In general, aromatic amines such as derivatives of aniline are far less basic than alkylamines (Figure 15.7). Main reason for this reduced basicity is resonance conjugation of the amino group with aromatic ring which makes the unshared electrons less available for protonation. The presence of electron-donating substituents on the phenyl ring slightly increase basicity of arylamines, while electron-withdrawing substituents on the phenyl ring decrease basicity of arylamines.

Amine:	cyclohexylamine	aniline
pK_a of ammonium ion:	10.63	4.6

resonance conjugation makes unshared electron pair on nitrogen less available for protonation

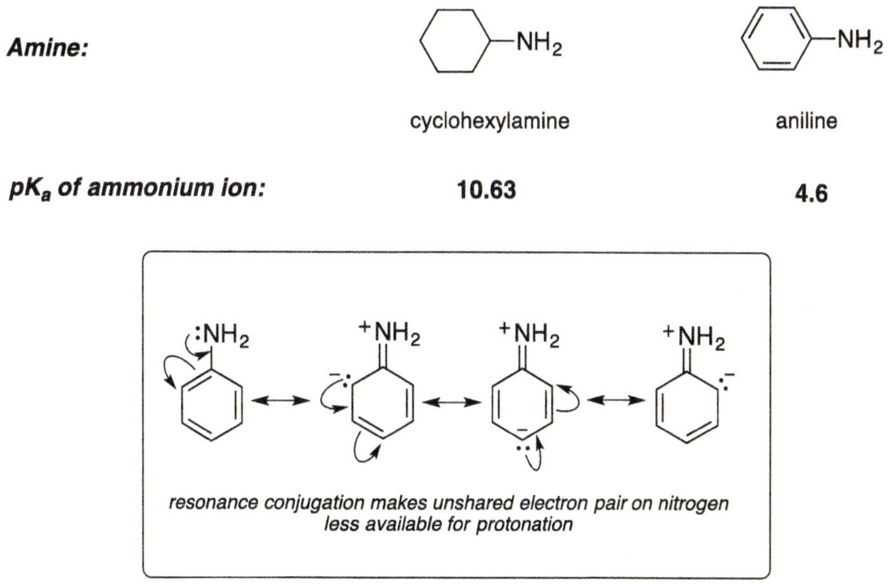

FIGURE 15.7. Basicity of aniline in comparison with an alkylamine.

15.2.3. Basicity of heterocyclic amines

The non-aromatic heterocyclic amines piperidine or pyrrolidine (Figure 15.8) have basicity comparable to that of acyclic alkylamines, such as diethylamine. Morpholine on the other hand is significantly less basic than piperidine due to the electron-withdrawing inductive effect of an oxygen atom in the ring. Piperazine contains two nitrogen atoms within the six-membered ring system. Either one or both of these nitrogens can be protonated.

Nonaromatic amine:	pyrrolidine	piperidine	morpholine	piperazine
pK_a of ammonium ion:	11.31	11.22	8.49	9.73 and 5.33

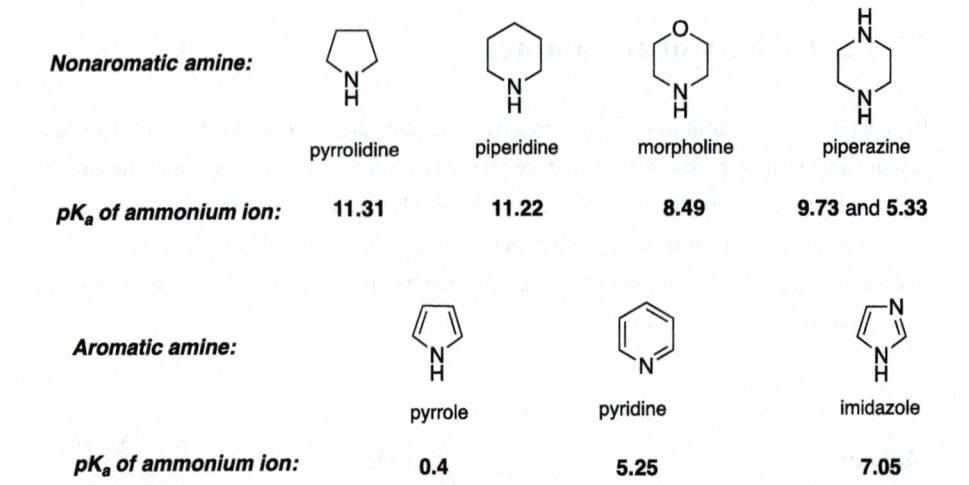

Aromatic amine:	pyrrole	pyridine	imidazole
pK_a of ammonium ion:	0.4	5.25	7.05

FIGURE 15.8. Basicity of heterocyclic amines.

Aromatic heterocycles such as pyridine and imidazole in general have much lower basicity. The protonated form of pyridine (the pyridinium cation, see Figure 15.4) is aromatic. The nitrogen atom within the aromatic ring system of pyridinium cation is sp^2 hybridized. The sp^2 molecular orbitals have higher s character than the sp^3 molecular orbitals in an alkylammonium cation and therefore are less stable (see Sections 1.7 and 2.3). Because of the lower stability of the corresponding conjugate acids, aromatic nitrogen heterocycles are less basic than structurally similar nonaromatic nitrogen heterocycles. Note that pyrrole, in contrast with pyridine, is not basic at all (its pK_a value is close to 0) because the lone pair in pyrrole is part of an aromatic system (Section 14.1) unavailable for protonation. For the same reason, only one nitrogen in imidazole can be protonated.

Problems

15.3. Within the aniline derivatives, as a rule of thumb electron-donating substituents such as $-CH_3$, $-NH_2$, and $-OCH_3$, which increase the reactivity towards electrophilic aromatic substitution, also increase the basicity. On the other hand, electron-withdrawing substituents such as $-Cl$, $-NO_2$, and $-CN$, which decrease reactivity towards electrophilic aromatic substitution, also decrease the basicity. Based on this information, rank the following sets of compounds in order of decreasing basicity:

a) O₂N—⌬—NH₂   H₃C—C(=O)—⌬—NH₂   Cl—⌬—NH₂

b) Br—⌬—NH₂   OHC—⌬—NH₂   H₃C—⌬—NH₂

c) F₃C—⌬—NH₂   H₃C—⌬—NH₂   ⌬—NH₂

d) Cl—⌬—NH₂   H₃CO—⌬—NH₂   O₂N—⌬—NH₂

15.4. Diphenylamine (Ph₂NH) is an extremely weak base (pK$_a$ = 0.8), while triphenylamine (Ph₃N) is by ordinary standards not basic at all. Explain.

15.5. Within each molecule identify the most basic nitrogen atom:

a) – b) – c) – d) – e) – f) – g) – h)

AMINES   315

15.3. SYNTHESIS OF AMINES

Because of the importance of amines in the chemical and pharmaceutical industries, numerous approaches to the synthesis of these compounds have been developed. In previous chapters we have studied several reactions that produce amines as final products. A summary of these reactions with additional examples is provided below.

1. A Nucleophilic ring opening of epoxides with amines (Section 6.5.2) forms a new amine with an additional N-C bond (Figure 15.9).

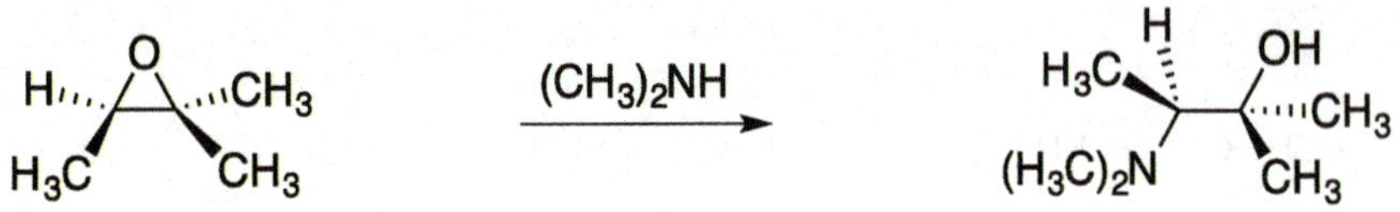

FIGURE 15.9. Nucleophilic ring opening of epoxides.

2. Reduction of imines (Section 11.6) with a source of hydride anion, such as sodium borohydride (NaBH$_4$) or sodium cyanoborohydride (NaBH$_3$CN), yields secondary amines. The reaction sequence converting a carbonyl compound to an amine is called **reductive amination** (Figure 15.10).

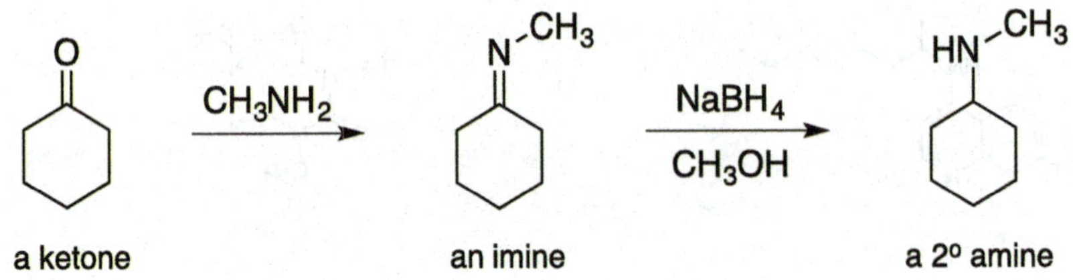

FIGURE 15.10. Reductive amination of a ketone.

3. Hydrolysis of amides under acidic conditions (Section 12.4.4) gives the corresponding carboxylic acid and ammonium salt. Ammonium salt can be easily converted to amine by treatment with a base, such as aqueous solution of sodium bicarbonate (Figure 15.11). Basic hydrolysis of amides produces the corresponding amines directly (Figure 12.4 in Section 12.4.4).

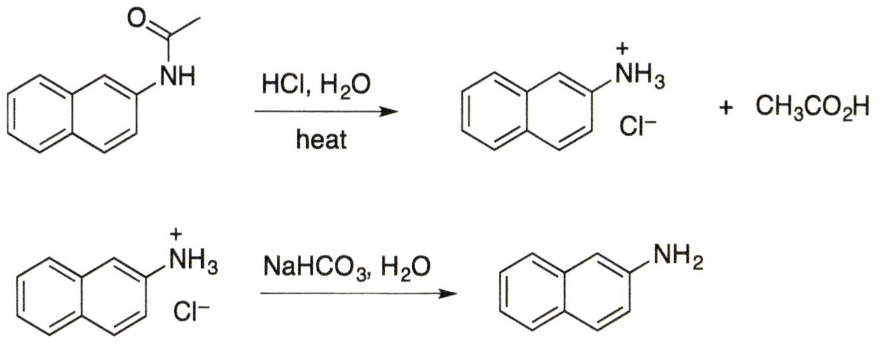

FIGURE 15.11. Synthesis of amines by hydrolysis of amides.

4. Reduction of amides with lithium aluminum hydride is used for the preparation of 1°, 2°, and 3° amines (Figure 12.17 in Section 12.5). Additional examples of this reaction are shown below in Figure 15.12.

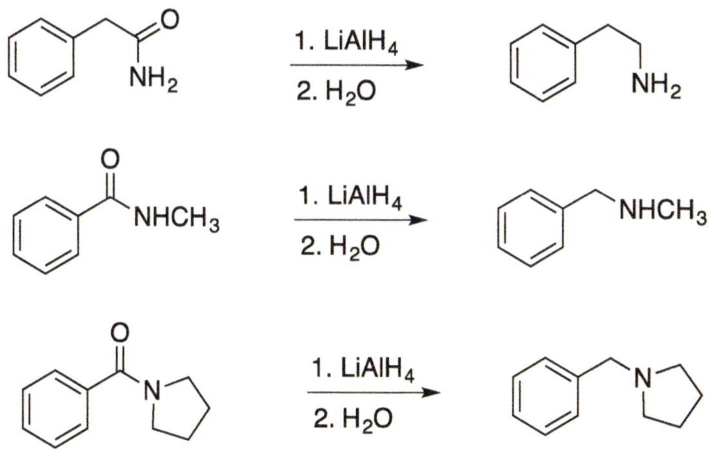

FIGURE 15.12. Synthesis of amines by reduction of amides.

5. Reduction of nitriles (Section 12.5) by LiAlH$_4$ produces the corresponding 1° amines (Figure 15.13).

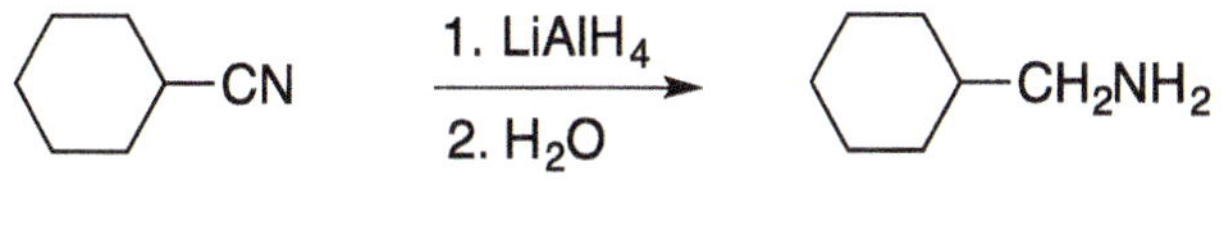

FIGURE 15.13. Synthesis of amines by reduction of nitriles.

6. Nucleophilic aromatic substitution reactions via an addition-elimination sequence, (Section 14.5.1) or a benzyne mechanism (Section 14.5.2), allow conversion of aryl halides to the corresponding aromatic amines (Figure 15.14).

Addition-Elimination Mechanism:

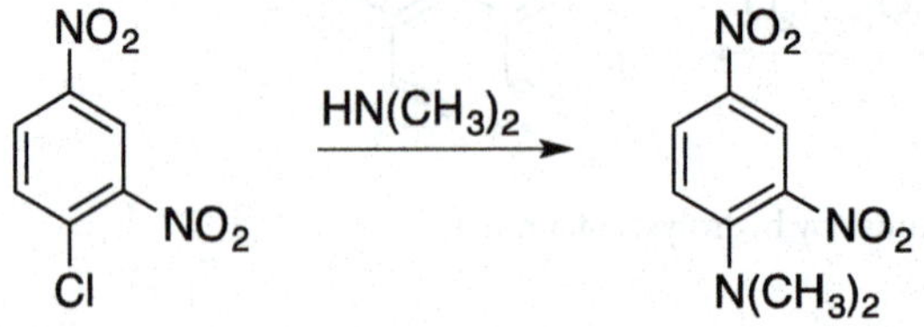

[2,4-dinitrochlorobenzene] + HN(CH$_3$)$_2$ → [2,4-dinitro-N,N-dimethylaniline]

Benzyne Mechanism:

H$_3$C—C$_6$H$_4$—Cl $\xrightarrow{\text{NaNH}_2,\ \text{liquid NH}_3}$ H$_3$C—C$_6$H$_4$—NH$_2$ + H$_3$C—C$_6$H$_4$—NH$_2$

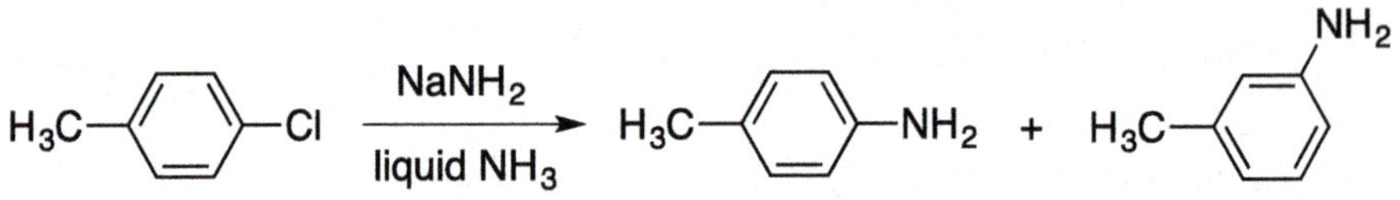

FIGURE 15.14. Synthesis of amines by nucleophilic aromatic substitution reactions.

A nucleophilic substitution reaction of ammonia (or alkylamines) with alkyl halides is generally not a good way to synthesize amines. Reactions of alkyl halides with ammonia usually produce a complex mixture of 1°, 2°, and 3° alkylamines (see problem 15.5). Only tertiary amines can be selectively alkylated with alkyl halides to render quaternary ammonium salts as single products (Figure 15.15).

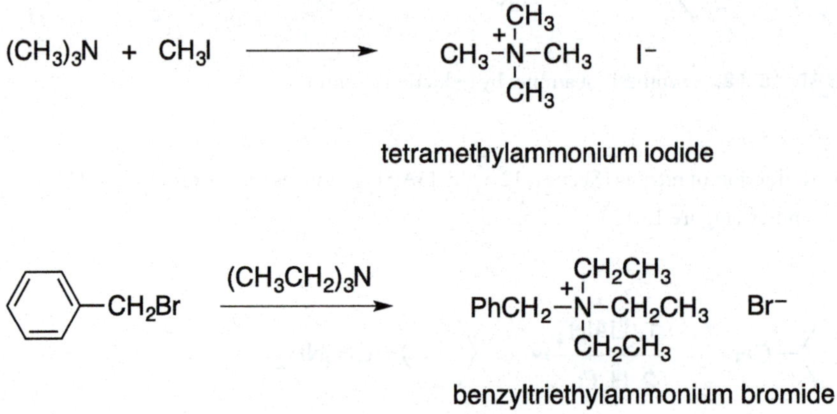

(CH$_3$)$_3$N + CH$_3$I ⟶ (CH$_3$)$_4$N$^+$ I$^-$

tetramethylammonium iodide

PhCH$_2$Br $\xrightarrow{(CH_3CH_2)_3N}$ PhCH$_2$–N$^+$(CH$_2$CH$_3$)$_3$ Br$^-$

benzyltriethylammonium bromide

FIGURE 15.15. Synthesis of quaternary ammonium salts from 3° amines.

318 ORGANIC CHEMISTRY

Instead of using ammonia as a nucleophile in reaction with alkyl halides, an alternative approach to the selective synthesis of a primary alkylamine involves the initial preparation of an alkyl azide followed by reduction of the azido group to an amino group. Azide anions are strong nucleophiles that readily react with alkyl halides by an S_N2 mechanism, producing the corresponding alkyl azides (Chapter 5). The azido group in alkyl azides can be reduced with $LiAlH_4$, and yields the 1° amine upon treatment with water (Figure 15.16).

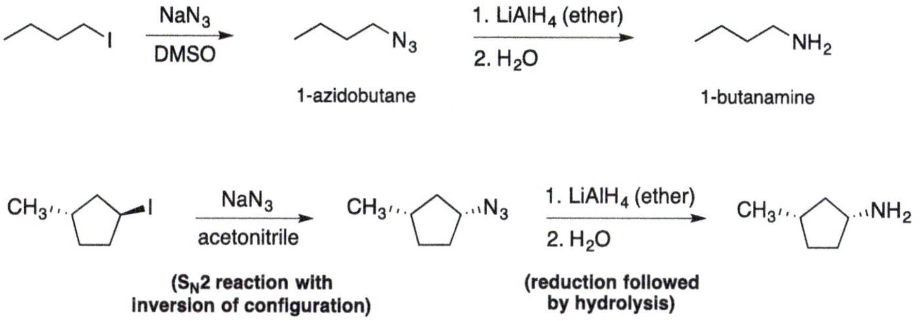

FIGURE 15.16. Synthesis of 1° amines by reduction of alkyl azides.

Another common approach to synthesize primary amines is the Gabriel synthesis, named after the German chemist Siegmund Gabriel. This reaction uses potassium phthalimide, a salt of the corresponding imide (Chapter 12), which is a relatively strong acid (pK_a = 8.3). The anion of phthalimide is a strong nucleophile that readily reacts with an alkyl halide via an S_N2 mechanism to produce N-alkylphthalimide. This intermediate product is further converted to the final 1° amine by acidic hydrolysis (Chapter 12) or by treatment with hydrazine (Figure 15.17).

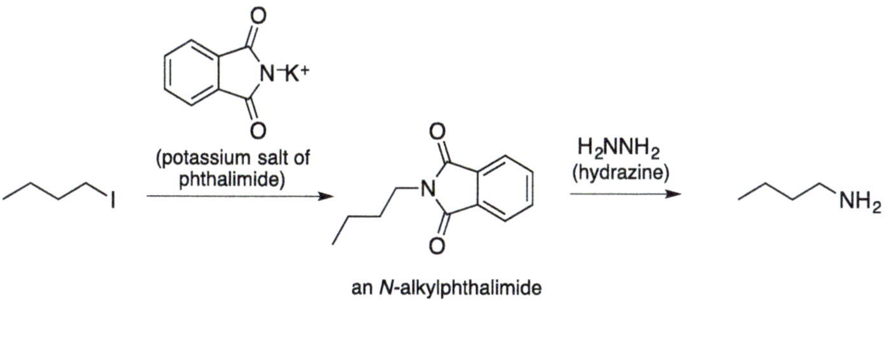

FIGURE 15.17. Synthesis of 1° amines by Gabriel synthesis.

An amino group cannot be introduced into an aromatic ring by electrophilic aromatic substitution. Therefore, chemists use a two-step procedure to synthesize aniline and its derivatives. The synthesis of aniline from benzene is shown in Figure 15.18. In the first step, benzene is nitrated to yield nitrobenzene. In the second step, the nitro group is reduced to an amino group by reaction with iron or tin in hydrochloric acid.

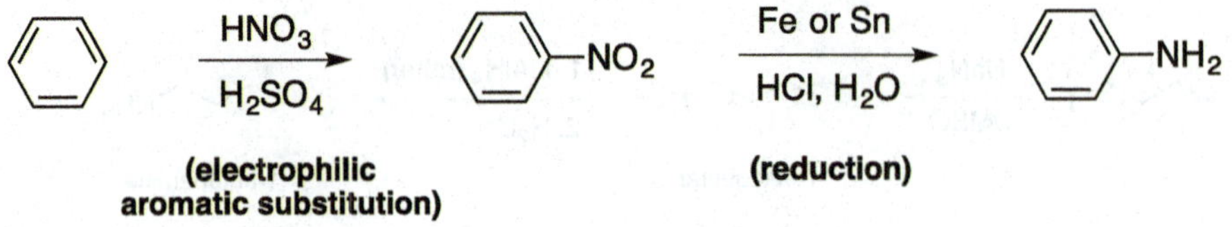

benzene →(HNO₃, H₂SO₄)→ nitrobenzene (–NO₂) →(Fe or Sn, HCl, H₂O)→ aniline (–NH₂)

(electrophilic aromatic substitution)   (reduction)

FIGURE 15.18. Synthesis of aniline from benzene.

Problems

15.6. Attempts to synthesize 1-ethylamine in high yield by treating 1 mol bromoethane with 1 mol ammonia in the presence of sodium bicarbonate as a base to neutralize the developing HBr produces a mixture of ethylamine, diethylamine, and triethylamine. Some unreacted ammonia is also recovered, but all of the bromoethane is consumed during the reaction. Use the relative basicity of the alkyl amines that form during this reaction and ammonia as measures of nucleophilicity to explain this finding.

15.7. Provide the intermediate and final products in the following reaction sequences:

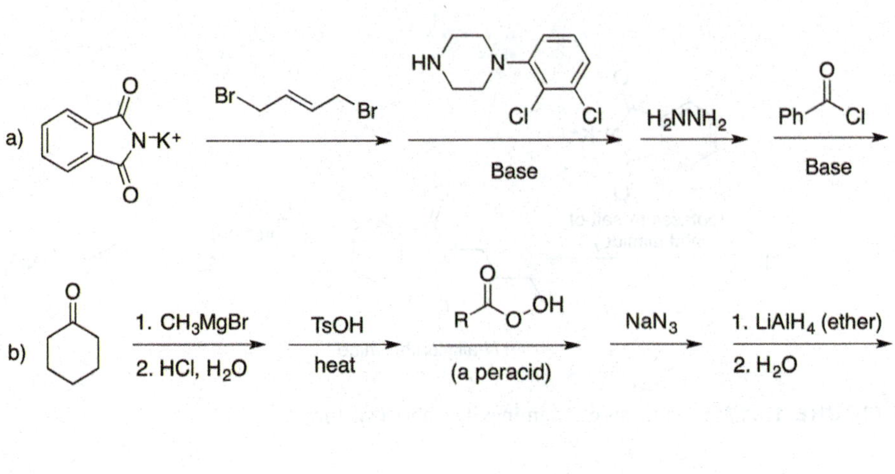

a) phthalimide N⁻K⁺ → Br-CH₂-CH=CH-CH₂-Br → HN-piperazine-N-(2,3-dichlorophenyl), Base → H₂NNH₂ → Ph(C=O)Cl, Base

b) cyclohexanone → 1. CH₃MgBr, 2. HCl, H₂O → TsOH, heat → R(C=O)O-OH (a peracid) → NaN₃ → 1. LiAlH₄ (ether), 2. H₂O

15.4. HOFMANN ELIMINATION

Hofmann elimination is the elimination of a tertiary amine from a quaternary ammonium salt, producing alkene. The Hofmann elimination, first described by August W. von Hofmann in 1851, is a three-step process. It starts with the treatment of an amine with an excess of iodomethane to convert the amine to the corresponding quaternary ammonium iodide salt. This is followed by anion exchange via the reaction of ammonium iodide with silver oxide in water, resulting in the formation of quaternary ammonium hydroxide. Finally, the quaternary ammonium hydroxide is heated to induce the elimination reaction to form an alkene, a tertiary amine, and water (Figure 15.9).

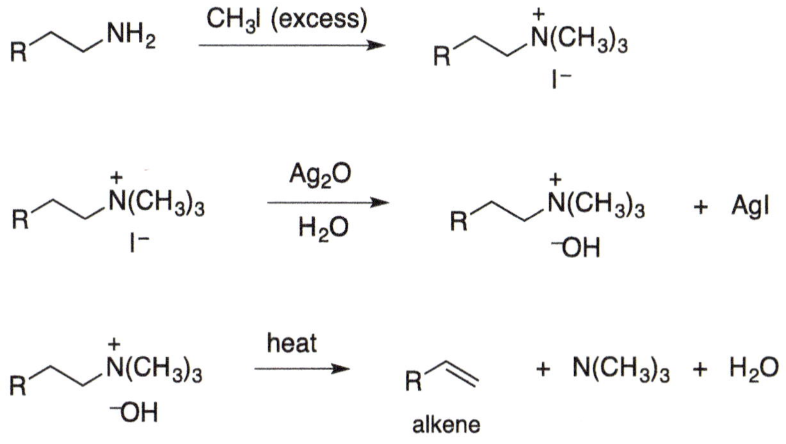

FIGURE 15.19. General scheme of Hofmann elimination.

Hofmann elimination is a β-elimination reaction that proceeds according to an E2 mechanism (Section 5.4). Similar to the β-elimination of alkyl halides, it is a stereoselective reaction that requires the anti-coplanar arrangement of the atoms involved (see Figure 5.12 and the related text in Section 5.4). For example, when (1R,2R)-1,2-diphenylpropanamine is subjected to Hofmann elimination, (Z)-1,2-diphenylpropene is produced (Figure 15.20). The diastereomeric (1S,2R)-1,2-diphenylpropanamine on the other hand, produces (E)-1,2-diphenylpropene.

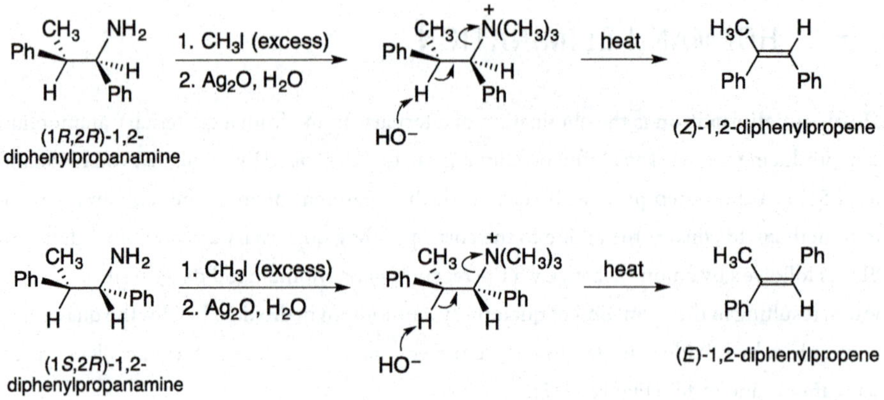

FIGURE 15.20. Stereochemistry of Hofmann elimination.

The regioselectivity of Hofmann elimination is different from the β-elimination reactions of alkyl halides. While the β-elimination reaction of alkyl halides proceeds according to Zaitsev's rule giving the more stable, more highly substituted alkenes as the major products (Section 5.4), the Hofmann elimination yields the less highly substituted alkenes as the major products. Figure 15.21 illustrates the opposite regioselectivity of these β-elimination reactions. When 2-iodo-3-methylbutane is treated with sodium methoxide, 2-methylbutene is formed as the major product. In contrast, the Hofmann elimination of a structurally related quaternary ammonium hydroxide yields 3-methyl-1-butene as the major product.

The regioselectivity observed in the elimination of quaternary ammonium salts is summarized by the Hofmann rule: the b-hydrogen atom is removed most readily if it is located on a CH_3 group, less readily from a CH_2 group, and least readily from a CH group. Therefore, Hofmann elimination produces the least substituted alkene as the major product. The Hofmann rule implies that steric effects have the greatest influence on the outcome of the elimination of the $N(CH_3)_3$ group. Compared to the iodine leaving group, the $N(CH_3)_3$ group is relatively bulky. Therefore, the hydroxide ion preferentially approaches and removes a proton from the least hindered carbon.

Problems

15.8. The exhaustive Hofmann degradation of 2-methyl-piperidine yields 1,5-hexadiene. Provide the structures of all intermediate products in this sequence.

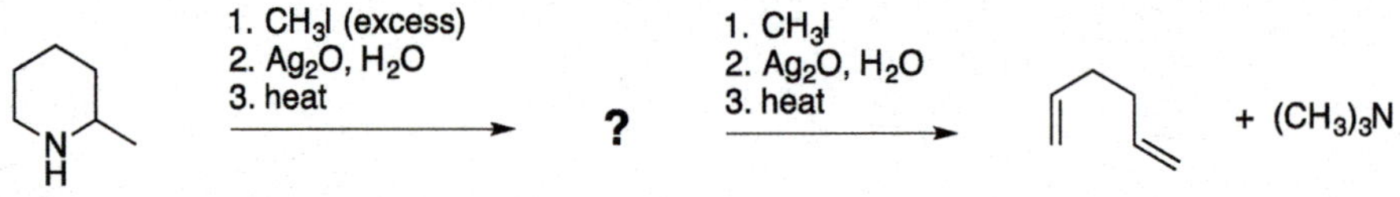

322   ORGANIC CHEMISTRY

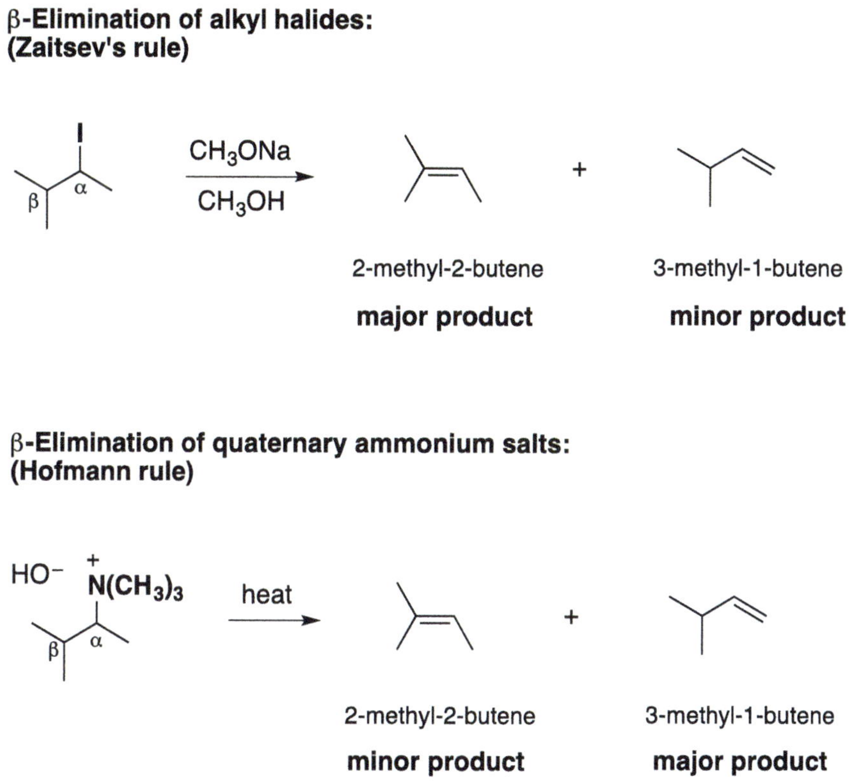

**β-Elimination of alkyl halides:
(Zaitsev's rule)**

2-methyl-2-butene — **major product**

3-methyl-1-butene — **minor product**

**β-Elimination of quaternary ammonium salts:
(Hofmann rule)**

2-methyl-2-butene — **minor product**

3-methyl-1-butene — **major product**

FIGURE 15.21.   Regioselectivity of β-elimination reactions.

15.5. ARYLDIAZONIUM SALTS

Aryldiazonium salts ($ArN_2^+X^-$) can be synthesized by the so called **diazotization reaction,** from primary arylamines, such as aniline, by treating it with sodium nitrite ($NaNO_2$) in the presence of a strong acid (Figure 15.22).

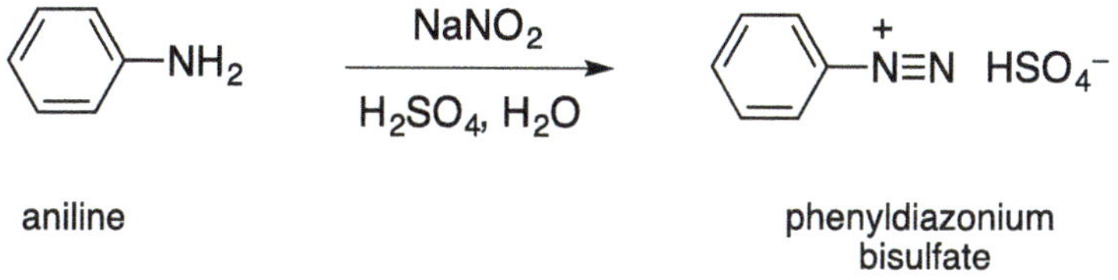

aniline

phenyldiazonium bisulfate

FIGURE 15.22.   Synthesis of aryldiazonium salts (i.e., the diazotization reaction).

The mechanism of this reaction involves the initial generation of nitrous acid (HNO$_2$, a weak acid) from sodium nitrite and a strong acid (HCl, HBr, or H$_2$SO$_4$). Nitrous acid is unstable and decomposes rapidly at room temperature. In aqueous solution, nitrous acid reacts with a strong acid to form nitrosonium ion (NO$^+$), a weak electrophile. The formation of a nitrosonium ion is analogous to the formation of nitronium ion (NO$_2^+$), which is a strong electrophilic reagent in the nitration reaction of aromatic compounds (Section 14.3.2). The nitrosonium ion further reacts with arylamine according to the multistep mechanism shown in Figure 15.23.

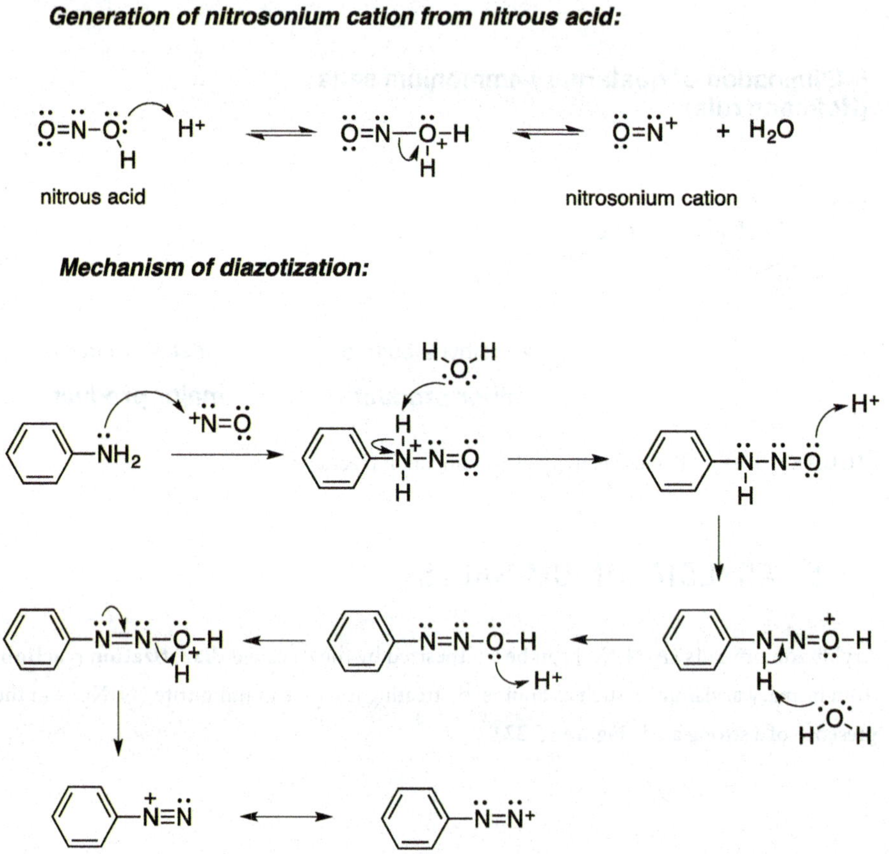

Generation of nitrosonium cation from nitrous acid:

Mechanism of diazotization:

FIGURE 15.23. Mechanism of diazotization reactions.

Aryldiazonium salts undergo numerous reactions in which the diazonium group is replaced by a nucleophile. Since the leaving group, a molecule of nitrogen, (N$_2$), is thermodynamically very stable, these reactions are energetically favored. Figure 15.24 summarizes some of the most common reactions of aryldiazonium salts. Aryldiazonium salts are versatile substrates for the synthesis of aryl halides, aryl nitriles, and phenols. The replacement of diazo group by hydrogen can also occur. The substitution reactions of aryldiazonium salts provide a valuable synthetic approach to various polysubstituted benzene derivatives.

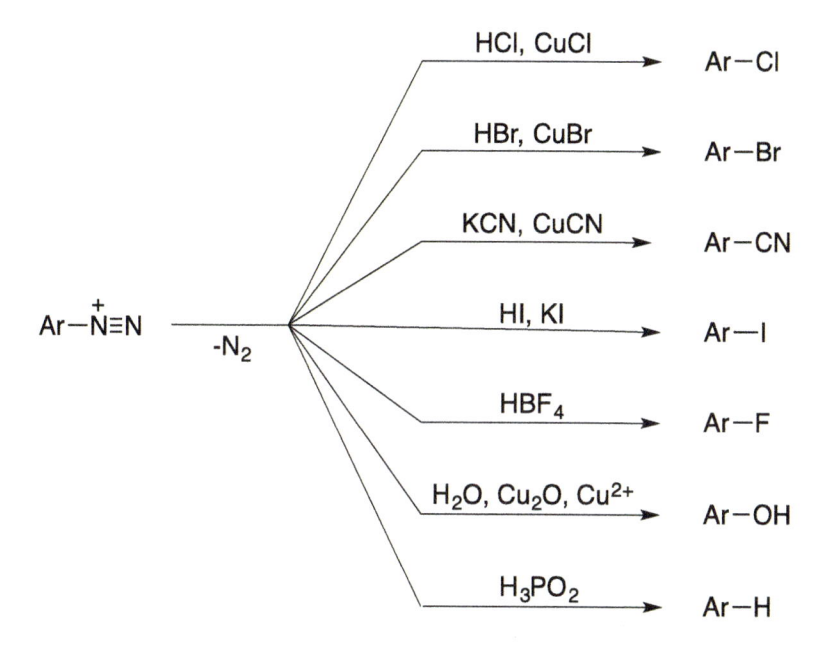

$$Ar-\overset{+}{N}\equiv N \xrightarrow{-N_2} \begin{cases} \xrightarrow{HCl, CuCl} Ar-Cl \\ \xrightarrow{HBr, CuBr} Ar-Br \\ \xrightarrow{KCN, CuCN} Ar-CN \\ \xrightarrow{HI, KI} Ar-I \\ \xrightarrow{HBF_4} Ar-F \\ \xrightarrow{H_2O, Cu_2O, Cu^{2+}} Ar-OH \\ \xrightarrow{H_3PO_2} Ar-H \end{cases}$$

FIGURE 15.24. Substitution reactions of diazonium ions.

15.5.1. Synthesis of aryl chlorides, aryl bromides, and aryl cyanides from diazonium salts (the Sandmeyer reaction)

Aryldiazonium salts undergo reactions with cuprous chloride (CuCl), cuprous bromide (CuBr), and cuprous cyanide (CuCN) to yield products in which the diazonium group is replaced by a chloro, bromo, or cyano group respectively. This reaction is called the **Sandmeyer reaction** in honor of the German chemist, Wilhelm A. Sandmeyer, who discovered the reaction. The mechanism of this reaction appears to proceed through a radical rather than a polar pathway, and the copper cation is required to enable the single electron transfer process. Examples of the Sandmeyer reaction are shown in Figure 15.25.

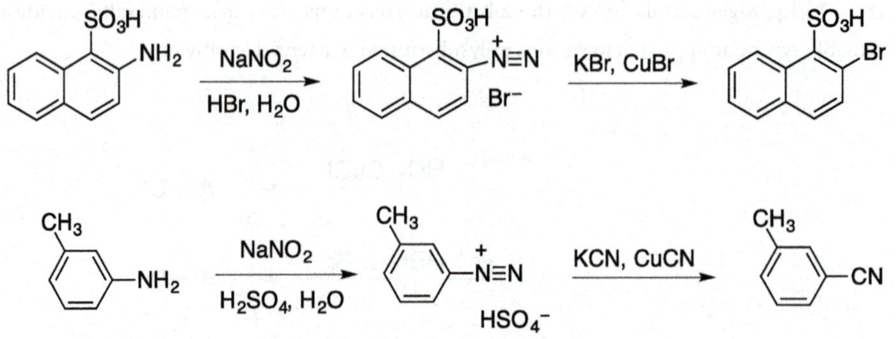

FIGURE 15.25. Examples of the Sandmeyer reaction.

15.5.2. Synthesis of aryl iodides from diazonium salts

Compared to the chlorination or the bromination of an aromatic system, the direct iodination of such compounds via electrophilic aromatic substitution is experimentally challenging. Therefore, the synthesis of an iodinated aromatic compound by means of a diazonium salt offers an alternative approach. Aryldiazonium salts react with potassium iodide (KI) to yield the corresponding aryl iodides (Figure 15.26).

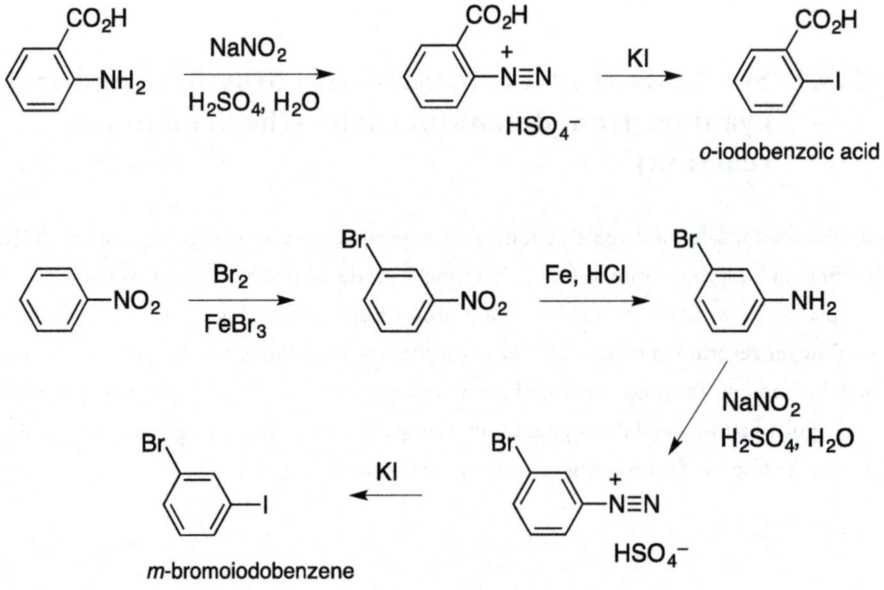

FIGURE 15.26. Synthesis of aryl iodides.

326 ORGANIC CHEMISTRY

15.5.3. Synthesis of aryl fluorides from diazonium salts (the Balz-Schiemann reaction)

The diazonium group can be replaced by fluorine by treating the diazonium salt with tetrafluoroboric acid (HBF$_4$) and heating the dried diazonium tetrafluoroborate salt (Figure 15.27). Mechanistic studies of the **Balz–Schiemann reaction** indicate that this reaction proceeds by an S$_N$1 mechanism via an aryl cation intermediate, which is quenched by fluoride transfer directly from the BF$_4^-$ ion.

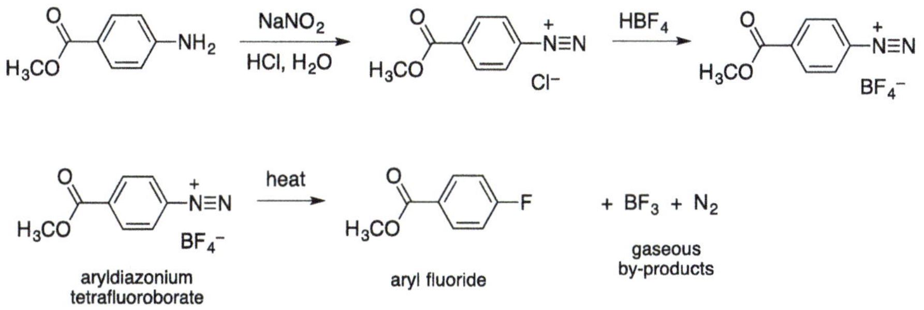

FIGURE 15.27. Synthesis of aryl fluorides by the Balz-Schiemann reaction.

15.5.4. Synthesis of phenols from diazonium salts

Phenols are formed by heating a solution of diazonium salt in diluted sulfuric acid. The presence of cuprous oxide (Cu$_2$O) and copper salts as a catalyst improves yield of phenols in this reaction (Figure 15.28).

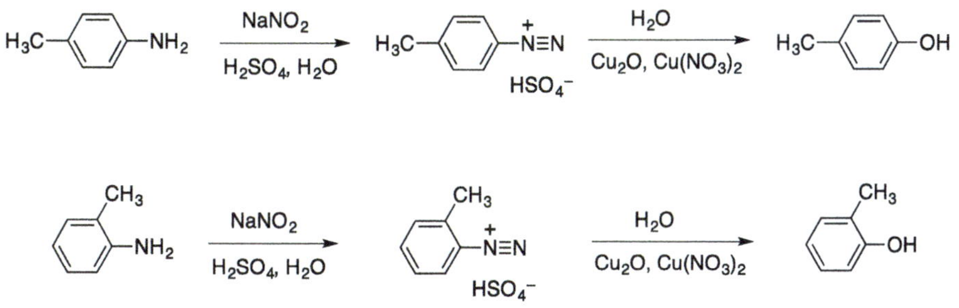

FIGURE 15.28. Synthesis of phenols from aryldiazonium salts.

15.5.5. Replacement of the diazo group with hydrogen

The diazonium group in aryldiazonium salts can be replaced with a hydrogen by treatment with hypophosphorus acid (H_3PO_2). Hypophosphorus acid is a strong reducing reagent due to the presence of P–H bonds, which serve as a source of nucleophilic hydride anion in the reaction. Figure 15.29 shows an example application of this reaction for the synthesis of *m*-bromotoluene from *p*-methylaniline. Note that *m*-bromotoluene cannot result from direct electrophilic bromination of toluene (see Figure 14.7 in Section 14.4).

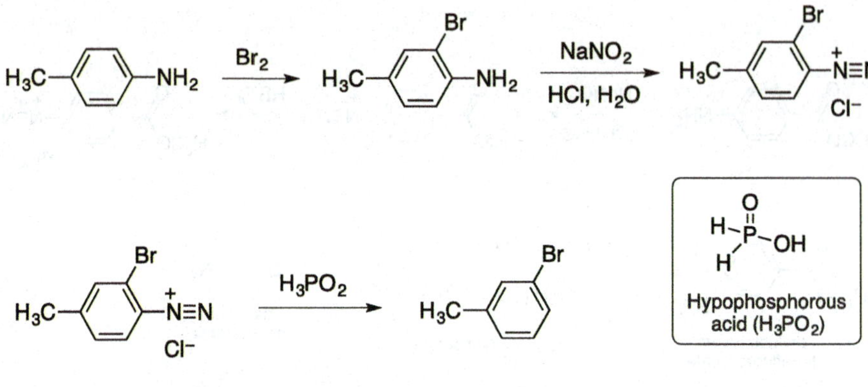

FIGURE 15.29. Synthesis of m-bromotoluene by reduction of aryldiazonium salt.

15.5.6. Diazonium salts as electrophiles in electrophilic aromatic substitution reactions

Aryldiazonium salts can react as electrophiles towards activated aromatic compounds such as phenols and arylamines. This type of reaction is known as azo-coupling and the products are azo compounds. An example of azo-coupling with aryldiazonium salt and phenol is shown in Figure 15.30.

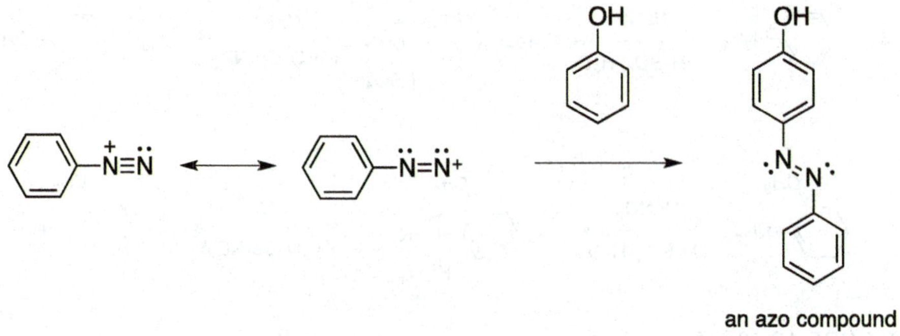

FIGURE 15.30. Azo-coupling reaction.

328 ORGANIC CHEMISTRY

Substitution usually occurs at the para position of phenols and arylamines, although ortho reaction can occur if the para position is blocked. Azo compounds are widely used as dyes because their extended conjugated π electron system causes them to absorb energy of light in the visible region of the electromagnetic spectrum. Some food colorings are azo compounds (Figure 15.31).

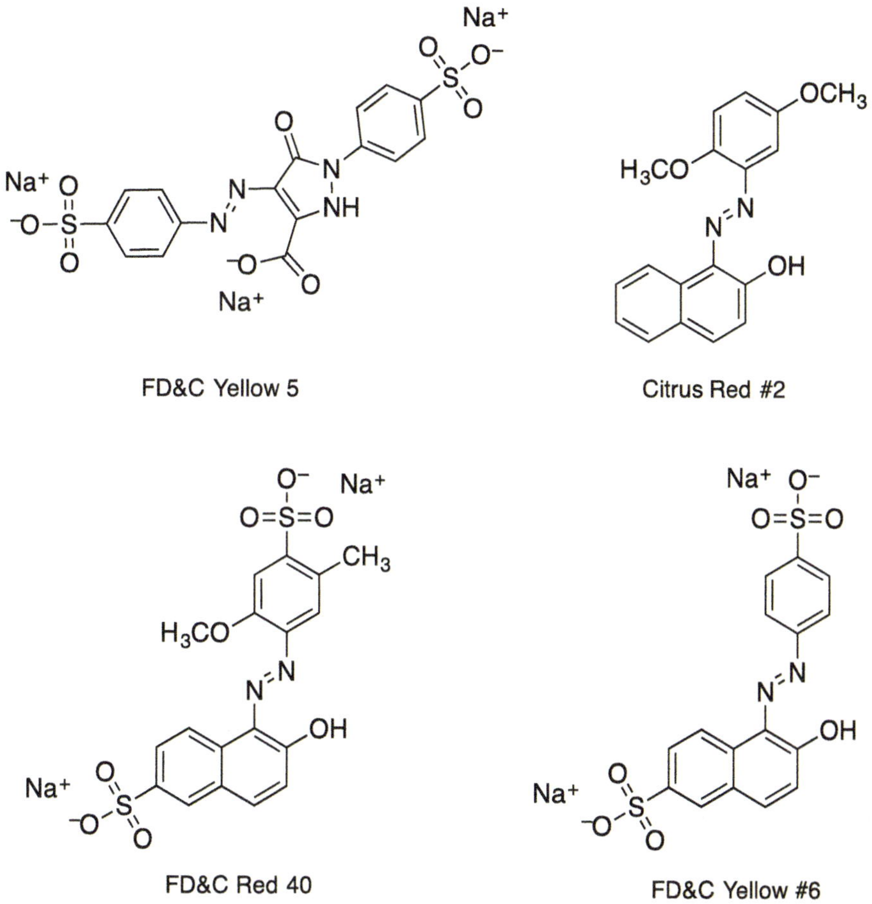

FIGURE 15.31. Azo compounds that are used as food colorings

Problems

15.9. Provide the intermediate products and the final product in the following reactions:

a) H₃CO—⟨⟩—C(CH₃)₃ $\xrightarrow{\text{HNO}_3, \text{H}_2\text{SO}_4}$ $\xrightarrow{\text{HCl, Fe}}$ $\xrightarrow{\text{NaNO}_2, \text{HCl, H}_2\text{O}}$ $\xrightarrow{\text{H}_2\text{O, Cu}_2\text{O, Cu(NO}_3)_2}$?

b) ⟨⟩—NO₂ $\xrightarrow{\text{Br}_2, \text{FeBr}_3}$ $\xrightarrow{\text{HCl, Fe}}$ $\xrightarrow{\text{NaNO}_2, \text{HBr, H}_2\text{O}}$ $\xrightarrow{\text{CuBr, KBr}}$?

15.10. Provide the reagents and intermediates for the synthesis of the following compounds, using benzene as the starting material:

a) 1,3,5-tribromobenzene   b) *m*-bromophenol

15.11. The key step in the synthesis of the food coloring Citrus Red #2 is the conversion of 2,5-methoxyaniline into its diazonium salt, followed by an azo-coupling using 2-naphthol. Propose a synthesis for 2,5-dimethoxyaniline from hydroquinone (benzene-1,4-diol).

2,5-dimethoxyaniline $\xrightarrow{\text{NaNO}_2, \text{H}_2\text{SO}_4, \text{H}_2\text{O}}$ diazonium intermediate $\xrightarrow{\text{2-naphthol}}$ Citrus Red #2

15.12. The key step in the synthesis of the pH indicator methyl red is the conversion of anthranilic acid into its diazonium salt, followed by an azo coupling using N,N-dimethylaniline. Propose a synthesis for anthranilic acid and N,N-dimethylaniline from benzene.

anthranilic acid (2-aminobenzoic acid) $\xrightarrow{\text{NaNO}_2, \text{HCl, H}_2\text{O}}$ diazonium intermediate $\xrightarrow{\text{N,N-dimethylaniline}}$ Methyl Red

CHAPTER 16

Introduction to Biomolecules

Biological molecules and chemical processes occurring in living organisms are studied in biochemistry courses. This chapter overviews important biomolecules, providing basic background for advanced biochemistry studies. Biomolecules consist of the same functional groups as other common classes of organic compounds and the mechanisms of chemical processes in living organisms are generally the same as those of the organic reactions discussed in previous chapters. In contrast to simple organic molecules, biomolecules are usually large in size and contain various functional groups and numerous stereocenters within the same molecule. This chapter provides a brief discussion of key structural features of the most important classes of biomolecules: lipids, amino acids, proteins, nucleosides, nucleotides, and nucleic acids.

16.1. LIPIDS

Lipids are broadly defined as biomolecules that have low solubility in water and can be isolated from biological samples by extraction with a nonpolar organic solvent. Biological roles of lipids include energy storage, functioning as structural components of cell membranes, and cell signaling, among others. Fats, oils, waxes, some vitamins and hormones, and most nonprotein cell-membrane components are all different types of lipids. Based on their chemical composition, lipids are classified as **triacylglycerols** (or triglycerides), **terpenoids**, **steroids**, or **prostaglandins** (Figure 16.1).

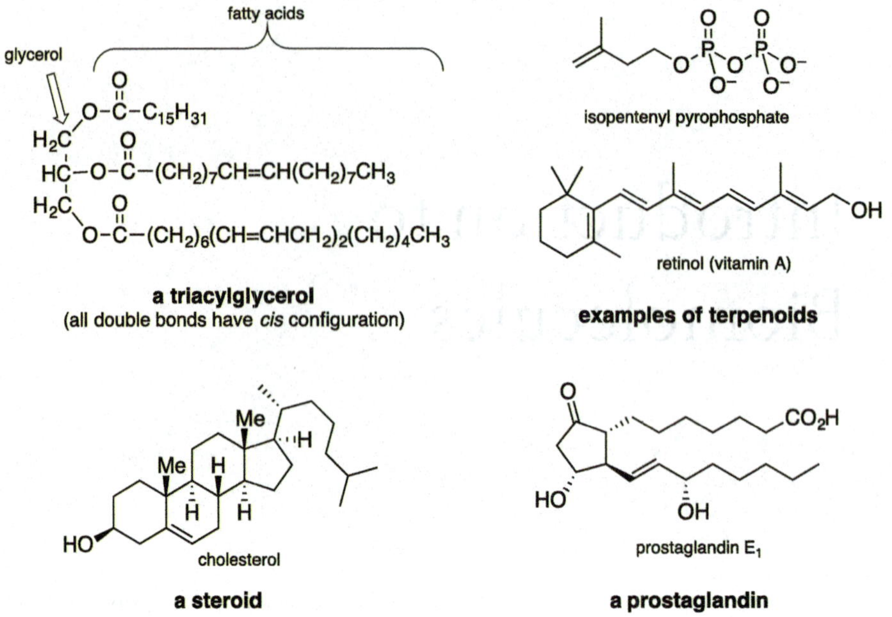

FIGURE 16.1. Main classes of lipids.

Triacylglycerols (or **triglycerides**) are triesters derived from glycerol (1,2,3-propanetriol) and three long-chain carboxylic acids (fatty acids). The fatty acids have an even number of carbons ranging from 12 to 20, and can have one or more double bonds with cis (Z) configuration. More than 500 different fatty acids are known, and about 40 occur widely. Vegetable oils have a higher proportion of unsaturated fatty acids than animal fats, which are mainly formed from saturated fatty acids. Basic hydrolysis of fats (**saponification**) is an industrial process that produces soap (salts of fatty acids) and glycerol (Section 12.4.3).

Terpenoids are lipids derived biosynthetically from the 5-carbon molecule of isopentenyl diphosphate. Several thousands structurally diverse terpenoids are known to exist.

Steroids are lipids based on a tetracyclic skeleton of three six-membered rings joined to a five-membered ring. Steroids have many structural variations and diverse biological functions. For example, cholesterol is a component of cell membranes that serves as a biological precursor to all steroidal hormones.

Prostaglandins are lipids derived biologically from fatty acids, containing twenty carbon atoms, including a cyclopentane ring. The several dozen known prostaglandins a have a broad range of physiological effects, including controlling inflammation, inducing labor, and lowering blood pressure.

Problems

16.1. Assign R or S configuration to each stereocenter present in cholesterol and prostaglandin E$_1$ (Figure 16.1).

16.2. CARBOHYDRATES

Carbohydrates (or **saccharides**) are generally defined as biological molecules with the empirical formula $C_m(H_2O)_n$ (hydrates of carbon). Carbohydrates are the most abundant organic compounds in plants, produced in green leaves by photosynthesis from CO_2 and water in the presence of chlorophyll (Figure 16.2). The function of chlorophyll is to absorb light energy and transfer it to other parts of the photosystem for conversion to chemical energy that accumulates in carbohydrates. Photosynthesis is a fundamental process, responsible for producing and maintaining the oxygen content of the Earth's atmosphere, and providing all of the organic compounds and most of the energy essential for life on Earth.

$$mCO_2 + nH_2O \xrightarrow[\text{chlorophyll}]{\text{sunlight}} C_m(H_2O)_n + mO_2$$

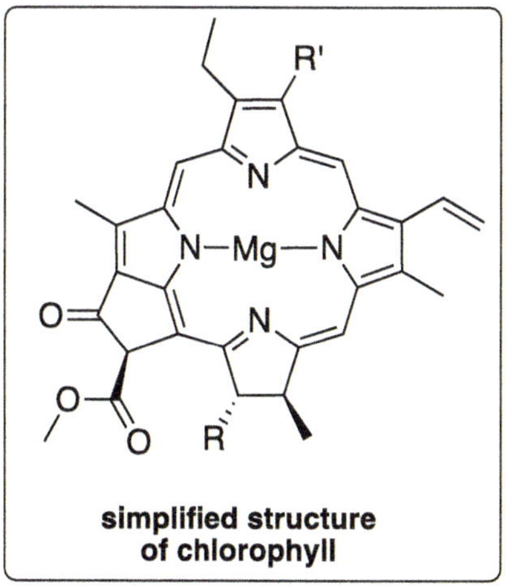

simplified structure of chlorophyll

FIGURE 16.2. Photosynthesis and the structure of chlorophyll.

Carbohydrates are broadly classified as simple sugars (**monosaccharides**) and complex carbohydrates (**oligosaccharides** and **polysaccharides**), which consist of several simple sugars joined together by an acetal link. Any complex carbohydrate, in principle, can be converted to simple sugars by hydrolysis.

Problems

16.2. Which functional groups and heterocyclic rings are present in the simplified structure of chlorophyll (Figure 16.2)? Why is this molecule aromatic?

16.2.1. General classification and nomenclature of monosaccharides

Monosaccharides have the chemical structure of polyhydroxy aldehydes (called **aldoses**) or polyhydroxy ketones (**ketoses**) with the general formula $C_n(H_2O)_n$, where n generally ranges from three to six. Monosaccharides with n = 3 are called **trioses**, n = 4 **tetroses**, n = 5 **pentoses**, and n = 6 **hexoses**. Carbohydrates usually have numerous chirality centers, and each molecule has many stereoisomers. The three-dimensional structures of monosaccharides are usually shown as Fischer projections (Section 4.3).

The simplest aldose (aldotriose) HOCH$_2$CH(OH)CHO has the IUPAC name 2-dihydroxypropanal and is known by the common name glyceraldehyde. This molecule has one stereocenter and exists in the form of R and S enantiomers. For a historic reason, instead of the IUPAC designations R and S, biomolecules, such as carbohydrates, generally use their common names with the stereochemical labels D and L (from the Latin words *dexter* for right side and *laevus* for left side). In the Fischer projection of an aldotriose, the carbon with the OH substituent on the right side of the chain is assigned a D configuration, the carbon with the OH substituent on the left side of the chain is assigned an L configuration (Figure 16.3). The simplest ketose HOCH$_2$C(O)CH$_2$OH (ketotriose, common name dihydroxyacetone) has no chirality centers.

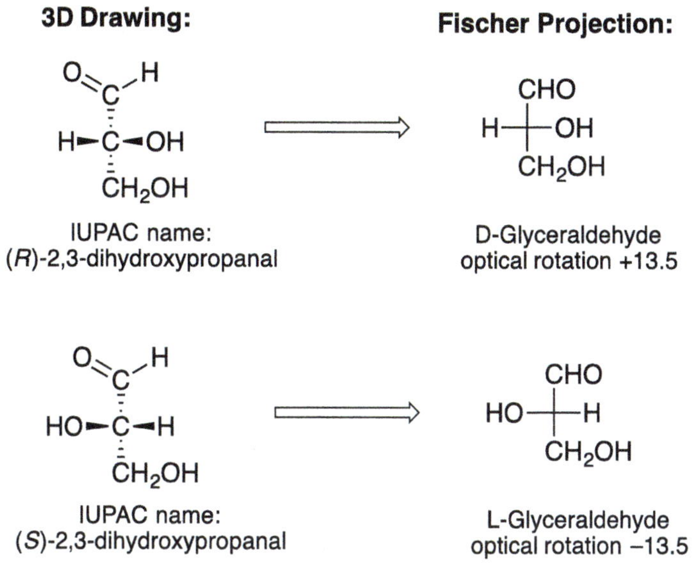

3D Drawing:

$$\underset{\underset{CH_2OH}{|}}{\overset{\overset{O\diagdown_C\diagup H}{|}}{H\text{—}C\text{—}OH}}$$

IUPAC name:
(R)-2,3-dihydroxypropanal

Fischer Projection:

```
      CHO
   H—┼—OH
      CH₂OH
```

D-Glyceraldehyde
optical rotation +13.5

$$\underset{\underset{CH_2OH}{|}}{\overset{\overset{O\diagdown_C\diagup H}{|}}{HO\text{—}C\text{—}H}}$$

IUPAC name:
(S)-2,3-dihydroxypropanal

```
      CHO
  HO—┼—H
      CH₂OH
```

L-Glyceraldehyde
optical rotation −13.5

FIGURE 16.3. Structures of enantiomeric aldotrioses.

Aldotetroses (HOCH$_2$CH(OH)CH(OH)CHO) have two stereocenters and four stereoisomers. In addition, two enantiomeric ketotetroses are known. It should be noted that in the Fischer projection of any aldose or ketose, the aldehyde or keto group is always placed in the upper part of the drawing. The D or L configuration of the lowest chiral center (the penultimate carbon) indicates the configuration of the enantiomeric monosaccharides. Each pair of enantiomeric monosaccharides has the same common name, but the D or L labels differ across pairs (Figure 16.4). Note that all naturally occurring carbohydrates have a D configuration at the penultimate carbon.

Two pairs of enantiomeric aldotetroses:

```
   CHO            CHO
H—┼—OH        HO—┼—H
H—┼—OH        HO—┼—H
   CH₂OH          CH₂OH

 D-Erythrose    L-Erythrose

    CHO           CHO
HO—┼—H         H—┼—OH
 H—┼—OH       HO—┼—H
    CH₂OH         CH₂OH

  D-Threose     L-Threose
```

Enantiomeric ketotetroses:

```
   CH₂OH          CH₂OH
    C=O            C=O
H—┼—OH        HO—┼—H
   CH₂OH          CH₂OH

 D-Erythrulose  L-Erythrulose
```

D or L configuration of the lowest chiral center is used to indicate the configuration of enantiomeric monosaccharides.
Only D enantiomers are naturally occurring monosaccharides.

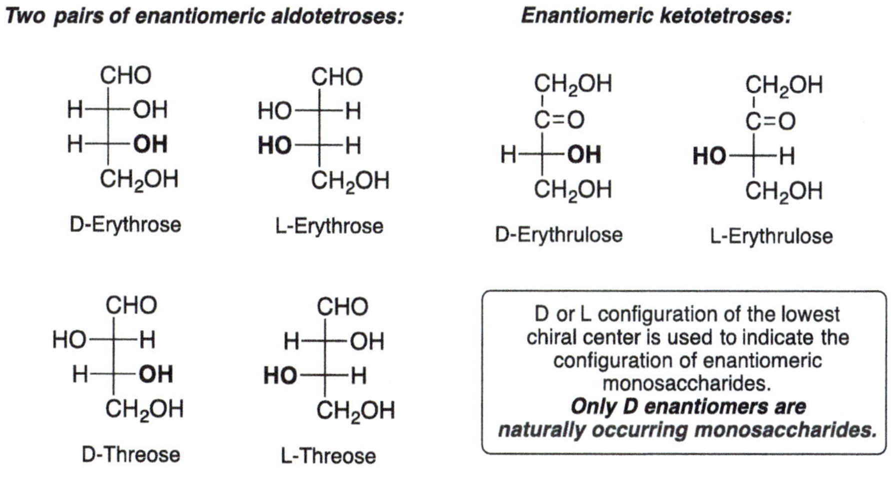

FIGURE 16.4. Fischer projections of tetroses.

An aldopentose has three stereocenters and 8 stereoisomers (i.e., 4 pairs of enantiomers) and an aldohexose has 4 chirality centers and 16 stereoisomers (i.e., 8 pairs of enantiomers). All D-enantiomers of aldopentose and aldohexose are naturally occurring sugars, while L-enantiomers are not found in nature. The structures of several naturally occurring D-pentoses and D-hexoses are shown in Figure 16.5. Particularly important are D-ribose (an essential building block of RNA), D-glucose (commonly known as grape sugar or blood sugar), and D-fructose (a ketohexose found in the disaccharide sucrose commonly known as table sugar; see Section 16.2.4).

Examples of aldopentoses: **Examples of aldohexoses:**   **Ketohexose:**

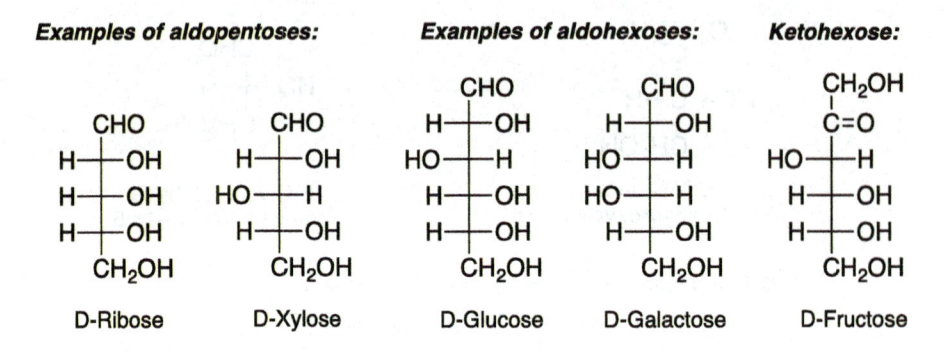

| D-Ribose | D-Xylose | D-Glucose | D-Galactose | D-Fructose |

FIGURE 16.5. Examples of naturally occurring D-pentoses and D-hexoses.

Problems

16.3. Draw the three-dimensional structure of D-xylose and assign the IUPAC name, including the R or S configurations of each stereocenter.

16.4. Draw Fischer projections of L-ribose, L-xylose, L-glucose, L-galactose and L-fructose.

16.5. The optical rotation of natural glucose in aqueous solution under standard conditions (the specific rotation) is +52.7. What is the optical rotation of unnatural L-glucose?

16.2.2. Cyclic structures of monosaccharides

Pentoses and hexoses exist predominantly as five- or six-membered cyclic hemiacetals (see Section 11.7). In most cases, the cyclic hemiacetal is formed via a reversible, acid-catalyzed, nucleophilic addition of the hydroxyl group at the penultimate carbon to the carbonyl carbon (C-1), as shown in Figure 16.6 for D-ribose. The five-membered cyclic hemiacetal formed by cyclization of an aldopentose has the general name **furanose** by analogy with a five-membered aromatic heterocycle furan (see Figure 14.5 in Chapter 14). As a result of this cyclization, a new chirality center is formed at the hemiacetal carbon C-1 (the **anomeric carbon**). Cyclic hemiacetals are usually shown as Haworth projections (Figure 16.6), which were originally invented by the English chemist Walter N. Haworth. In Haworth projections, groups below the plane of the ring correspond to the groups on the right side of the vertical carbon backbone in the acyclic Fischer projections. The two diastereomers with opposite configurations at the anomeric carbon are called **anomers**. The β-anomer (β-D-ribofuranose) has the OH at C-1 in cis configuration, relative to the CH$_2$OH group, while the α-anomer (α-D-ribofuranose) has the OH at C-1 and the CH$_2$OH group in trans configuration. In solution, both anomers exist in equilibrium with the open-chain form; however, individual anomers can be isolated in the solid state.

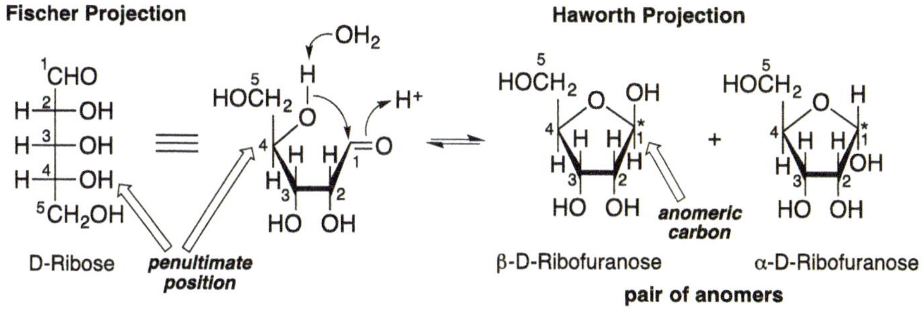

FIGURE 16.6. Formation of cyclic hemiacetal (D-ribofuranose) from D-ribose.

Aldohexoses usually exist as six-membered cyclic hemiacetals, generally named **pyranoses** by analogy with the six-membered heterocycle pyran. Figure 16.7 shows the formation of a cyclic hemiacetal as a Haworth projection, from the open-chain D-glucose. Similar to D-ribose, D-glucose exists in the form of two anomers (β-D-glucopyranose and α-D-glucopyranose) with a cis or trans configuration of the OH at C-1 and at the CH$_2$OH group. The individual anomers can be isolated in the solid state as pure crystals with different melting points. In aqueous solution at room temperature, both anomers exist in equilibrium with the open-chain form. The process of interconversion of β-D-glucopyranose and α-D-glucopyranose via intermediate formation of the open-chain D-glucose is called **mutarotation**. Mutarotation is a relatively slow process which can be monitored by the change in optical rotation of the solution. An aqueous solution of pure β-D-glucopyranose or α-D-glucopyranose slowly forms a mixture with a constant ratio of β:α anomers = 64:36 (optical rotation +52.7), with a trace amount of the open-chain D-glucose.

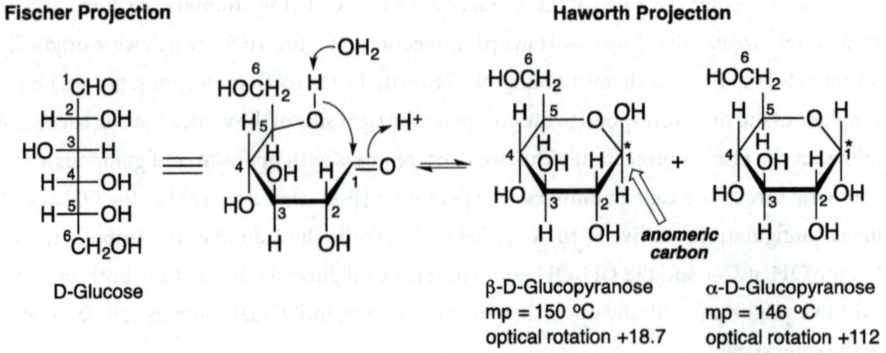

FIGURE 16.7. Formation of cyclic hemiacetal (D-glucopyranose) from D-glucose.

The Haworth projection of a pyranose gives a good representation of the configurations at each stereocenter in comparison with the Fischer projection of a monosaccharide. The actual three-dimensional structure of any pyranose is described by a chair conformation analogous to cyclohexane (Section 3.4). The chair representations of β-D-glucopyranose and α-D-glucopyranose are shown in Figure 16.8. Note that β-D-glucopyranose has all non-hydrogen substituents in the equatorial position, while the anomeric OH group occupies the axial position in α-D-glucopyranose.

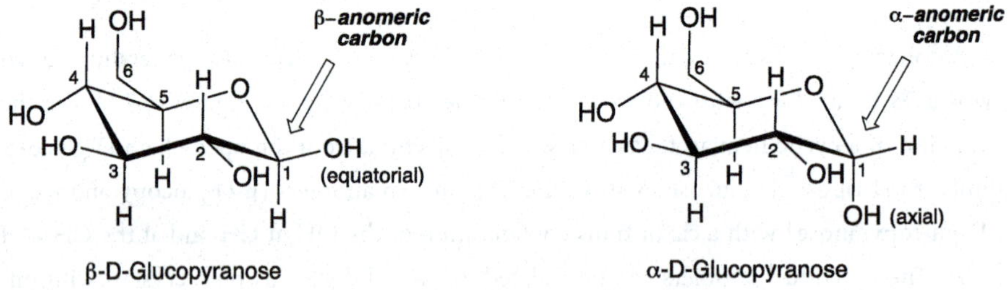

FIGURE 16.8. Chair conformation of D-glucopyranoses.

Problems

16.6. Draw the furanose form of the two anomers of D-xylose.

16.7. Draw β-D-galactopyranose in Haworth projection and as a chair conformation.

16.8. Draw the furanose form of the two anomers of D-fructose.

16.2.3. Chemical reactions of monosaccharides

Any chemical reaction typical of an alcohol (Chapter 8) or a carbonyl group (Chapter 11) can occur in carbohydrates. Any hydroxyl group in a monosaccharide can be oxidized or converted to an ester or an ether. The aldehyde group in an aldose molecule can be reduced to an alcohol, oxidized to an acid, or converted to hemiacetal, acetal, or aminoacetal. Several representative examples of such reactions are shown in Figure 16.9. Note that furanoses and

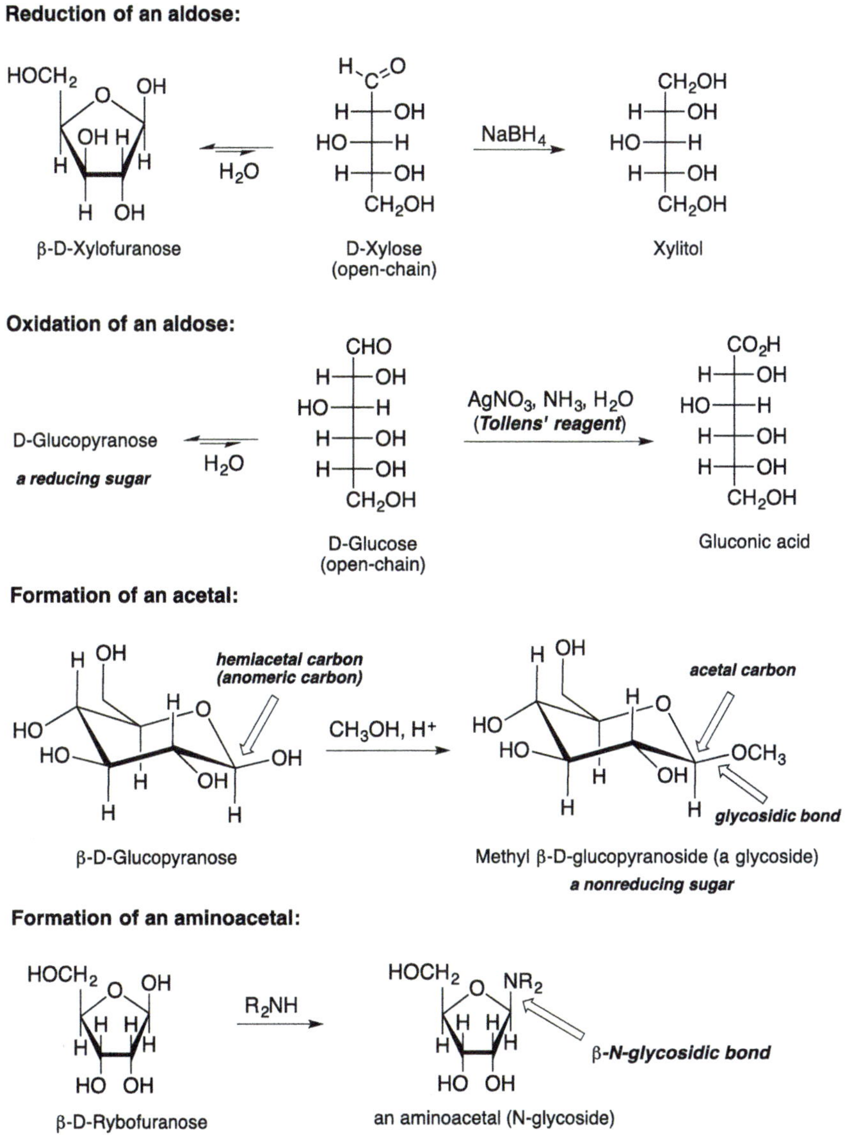

FIGURE 16.9. Reactions of aldoses.

INTRODUCTION TO BIOMOLECULES 339

pyranoses are cyclic hemiacetals which exist in solution in equilibrium with the open-chain aldehyde form. The open-chain aldehyde can be oxidized by Tollens' reagent, producing a characteristic "silver mirror" on the inner surface of the reaction flask. The carbohydrates that give a positive "silver mirror" test are called **reducing sugars** since they can reduce silver salt to silver metal. In contrast, acetals or **glycosides** (such as methyl β-D-glucopyranoside) are **nonreducing sugars,** that are stable in neutral solution and do not exist in equilibrium with an open-chain aldehyde form (see Section 11.7).

The reaction of cyclic hemiacetals (furanoses or pyranoses) with secondary amines (R_2NH) produces aminoacetals or N-glycosides. The N-glycosides formed from D-ribose or 2-deoxy-D-ribose and heterocyclic aromatic amines are key structural units of RNA and DNA (Section 16.4).

Problems

16.9. Is the molecule of xylitol (Figure 16.9) chiral?

16.10. What are the products of oxidation (with Tollens' reagent) and reduction (with $NaBH_4$) of D-ribose? Are these compounds chiral?

16.11. What is a likely product of the esterification reaction of D-ribose with excess acetic anhydride?

16.12. What is the product of the reaction of D-ribose with hydroxylamine (H_2NOH, see Figure 11.7)?

16.2.4. Disaccharides and polysaccharides

Disaccharides consist of two units of monosaccharides joined together by an acetal link (glycosidic bond). Formation of the disaccharide cellobiose from two molecules of β-D-glucopyranose is shown in Figure 16.10. The acetal link in cellobiose is created via interaction of the β-anomeric hemiacetal group at C-1 of one molecule with the OH group at C-4' of the second molecule, and is called β-1,4'-glycosidic bond. A molecule of cellobiose has one acetal group (at C-1) and one hemiacetal group (at C-1'). Because of the presence of the hemiacetal group, cellobiose in aqueous solution undergoes the process of mutarotation and gives a positive silver mirror test typical of reducing sugars.

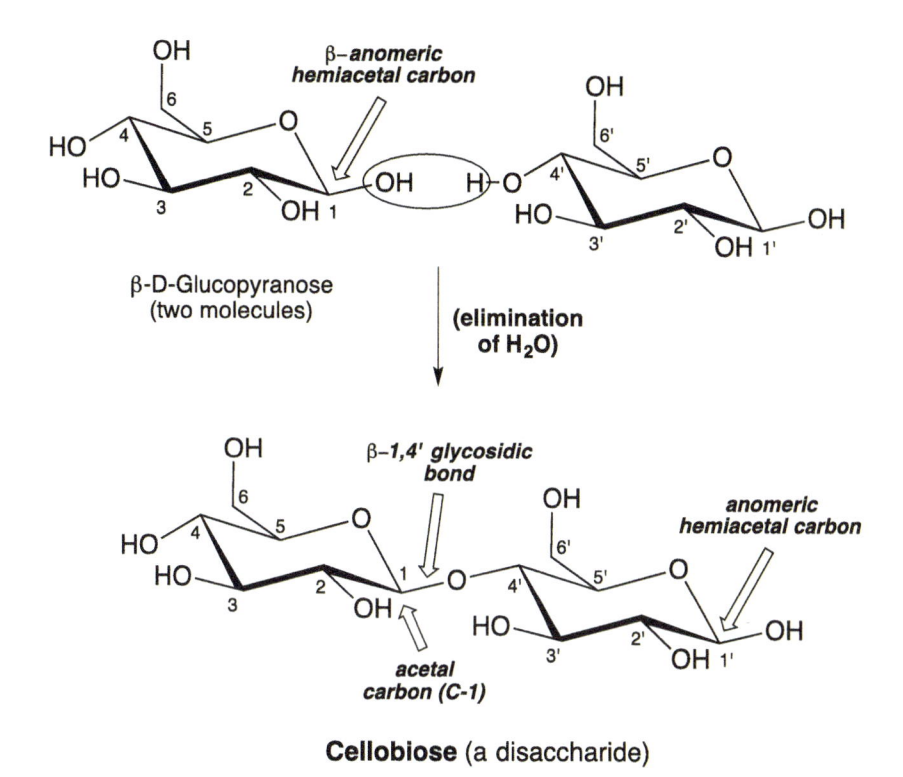

FIGURE 16.10. Formation of the disaccharide cellobiose from two molecules of β-D-glucopyranose.

Disaccharides can be formed from two different monosaccharides via the interaction of any hydroxyl group in one molecule with the hemiacetal group of the second molecule. One of the most common disaccharides, sucrose (table sugar), is formed by the interaction of the α-anomeric hemiacetal group at C-1 of glucose with the β-anomeric OH group at C-2' of the cyclic form of D-fructose (Figure 16.11). Sucrose does not have a hemiacetal group and therefore is not a reducing sugar.

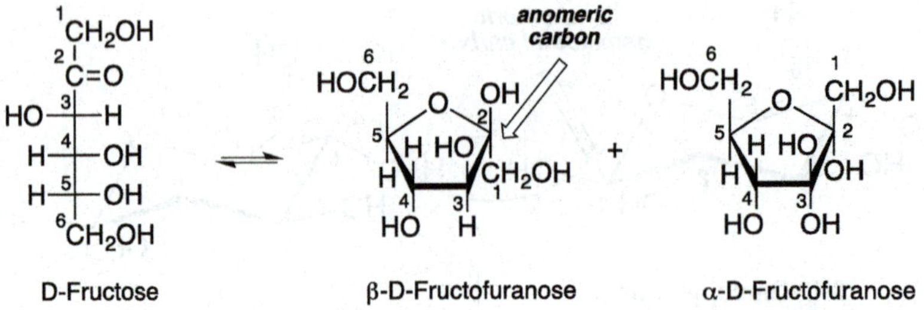

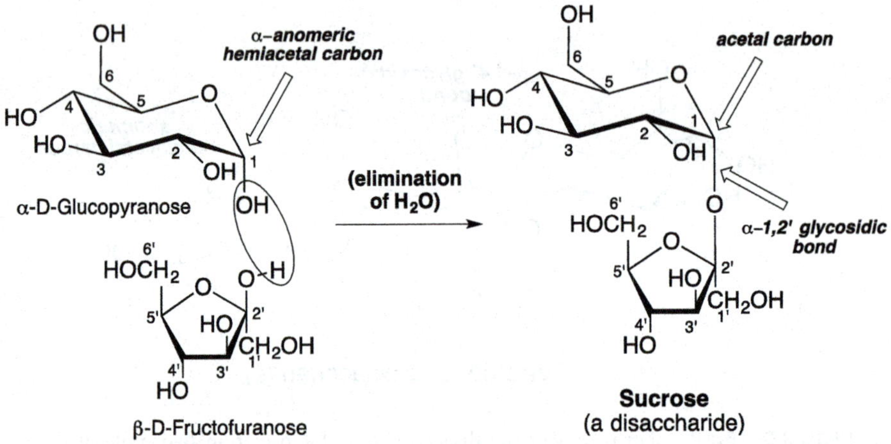

FIGURE 16.11. Formation of sucrose from α-D-glucopyranose and β-D-fructofuranose.

Complex carbohydrates formed by three or more units of monosaccharides are called **trisaccharides, tetrasaccharides**, and so on, and have the general name **oligosaccharides**. Polymeric carbohydrates consisting of numerous simple carbohydrate units linked by glycosidic bonds are called polysaccharides. Cellulose (found in wood or cotton) is a common polysaccharide, composed of β-D-glucopyranose units joined together by β-1,4>-glycosidic linkages (Figure 16.12). Starch is another important polysaccharide, containing α-D-glucopyranose units connected by α-1,4'-glycosidic bonds. Polymeric chains of cellulose or starch may contain several thousand D-glucopyranose units.

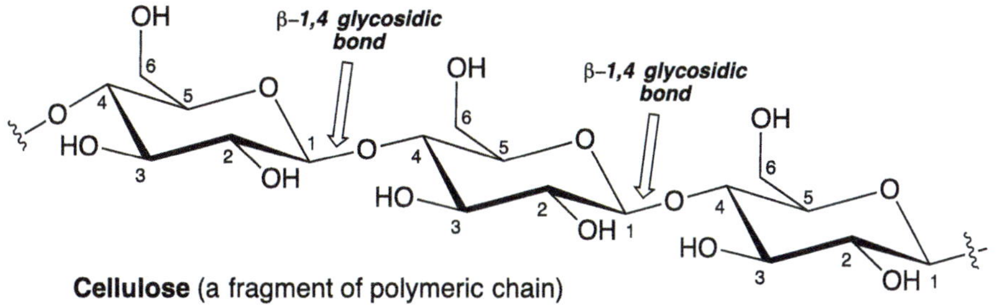

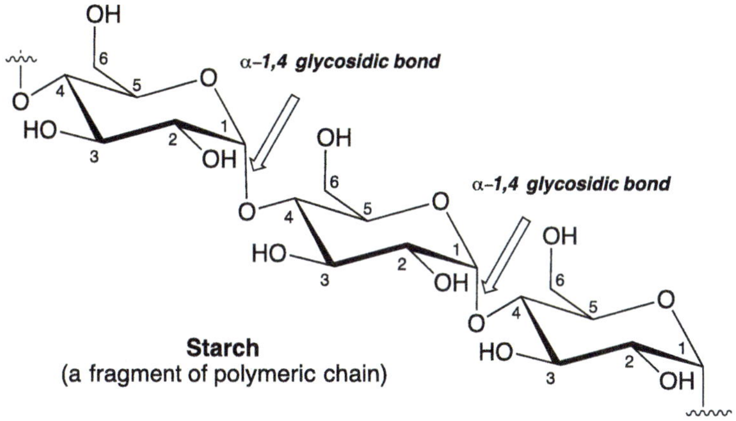

Cellulose (a fragment of polymeric chain)

Starch (a fragment of polymeric chain)

FIGURE 16.12. Important polysaccharides.

Problems

16.13 Draw the structure of disaccharide maltose, formed from two units of α-D-glucopyranose connected by α-1,4'-glycosidic bond. Is maltose a reducing sugar or a nonreducing sugar?

16.3. AMINO ACIDS, PEPTIDES, AND PROTEINS

Proteins are biological polymers formed from amino acid units. Many different types of proteins have various functions in living organisms, including structural functions, enzymatic catalysis of biochemical reactions, DNA replication, cell signaling, immune responses, etc. All proteins are formed from amino acid units connected by **amide bonds**, also known as **peptide bonds** (see Chapter 12).

Amino acids are simple biomolecules containing a basic amino group and an acidic carboxylic functional group. α-Amino acids are the building blocks of proteins. These contain an amino group and a carboxyl group attached to the same carbon (α carbon). In the presence of a strong acid (solution pH < 2), an amino acid exists in a protonated ammonium form, while in a basic solution (pH > 10) it exists as an amino carboxylate anion. In a neutral aqueous solution, amino acids undergo an intra-molecular proton transfer producing a dipolar form called a **zwitterion** (Figure 16.13). The exact pH value at which an amino acid exists in zwitterionic form and carries no net electrical charge is termed the **isoelectric point**. The value of the isoelectric point depends on the nature of the substituent R in the amino acid molecule.

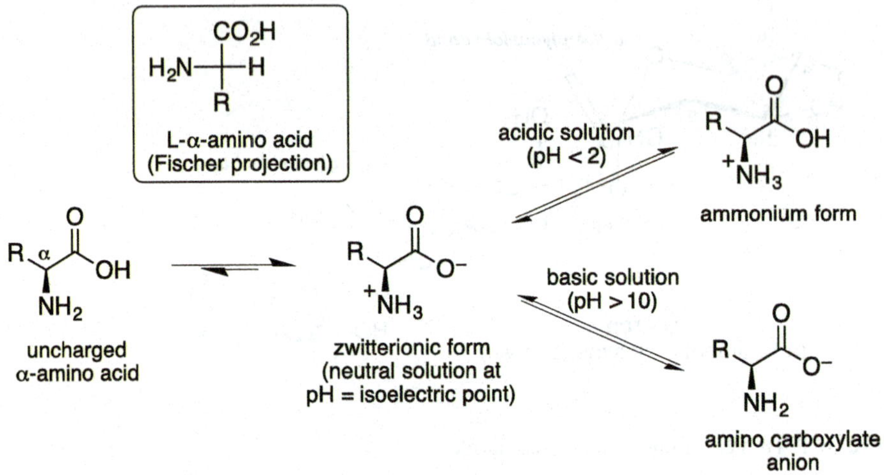

FIGURE 16.13. General structure of natural α-amino acids.

With the exception of non-chiral glycine ($H_2NCH_2CO_2H$), all natural amino acids are chiral with an L-configuration at the α-carbon. In the Fischer projection of a naturally occurring amino acid, the carboxylic group is placed at the top, the side chain is pointing down, and the amino group is placed on the left. Figure 16.14 shows the structures, names, and standard abbreviations (three letter and one letter abbreviations) of 20 common α-amino acids found in natural proteins. Note that the substituent R attached to the α-carbon of an amino acid can contain various functional groups, including alkyl, aryl, heteroaryl, phenol, alcohol, thiol (SH), thioether (CH_3S), carboxylic acid, amine, and amide.

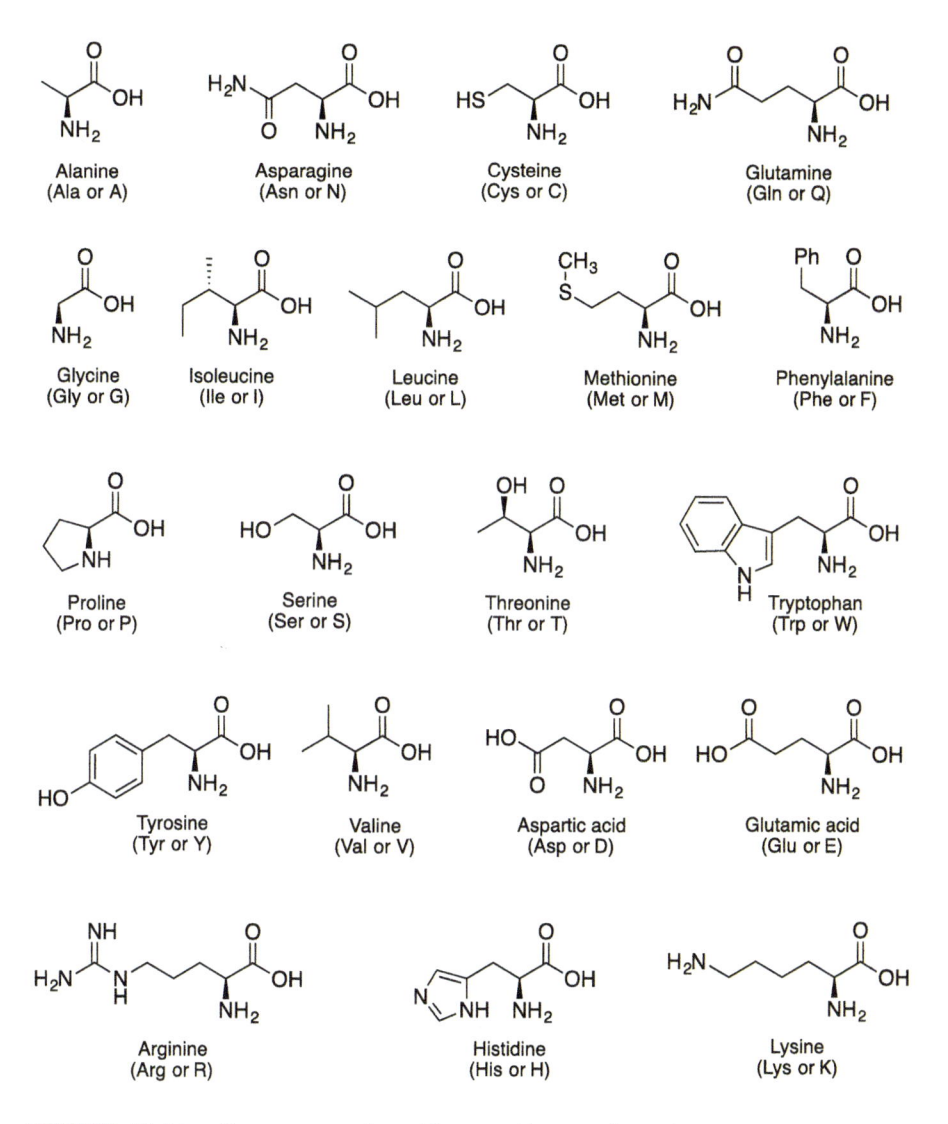

FIGURE 16.14. Common α-amino acids present in natural proteins.

Proteins consist of a polymeric chain of amino acid units connected by amide bonds (peptide bonds). The chain consisting of a repeating sequence of the (-N-CH-CO) fragments, including the α-carbon to which the R-groups are attached, is called a **protein backbone**, and the amino acid units are called **residues**. Smaller proteins with less than about 50 amino acid units are called **peptides**. Examples of small peptides (dipeptide, tripeptide, and tetrapeptide) are shown in Figure 16.15. By convention, the protein chain is written starting with the N-terminal amino acid (i.e., the amino acid unit with the free amino group) and ending with the C-terminal amino acid (the unit with free carboxyl group) on the right side. Note that even the simplest dipeptide consisting of two amino acids has two isomers with different N-terminal and C-terminal amino acid units (Figure 16.15). The peptide bond has a partial double-bond character and planar geometry, with restricted rotation due to resonance interaction (see Figure 12.7 in Section 12.3).

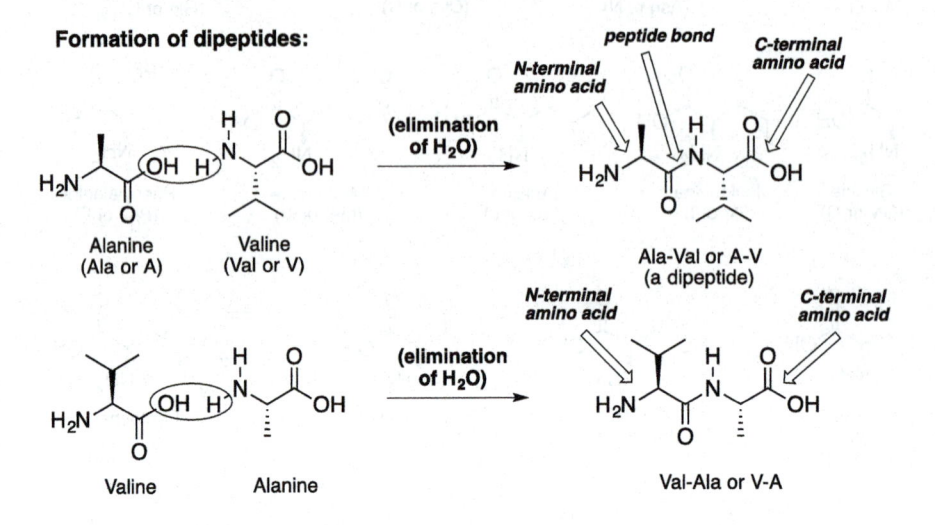

FIGURE 16.15. Examples of small peptides.

Large proteins can have up to 27,000 amino acid residues. An enormous number of different proteins exist in nature. Each protein has primary structure defined as the sequence of amino acid residues, and also secondary and tertiary structures describing the three-dimensional shape of protein molecules. Small proteins can be synthesized in a lab by common chemical methods (peptide synthesis), while larger proteins can be assembled from amino acids by biosynthesis using information encoded in genes.

Problems

16.14. There are three amino acids which are often referred to as basic amino acids. One of these basic amino acids is histidine, because it contains the basic imidazole moiety. Which two other amino acids are also considered to be basic? Furthermore there are also two amino acids that are labeled as acidic amino acids because they contain an additional carboxylic acid functional group. Which ones are these?

16.15. We commonly use the D/L-convention to distinguish between enantiomers of α-amino acids. The configuration of the α-carbon of all naturally occurring amino acids is L, and for most of the α-amino acids in Figure 16.14 the configuration of the α-carbon is S in the R/S convention. However, what is the configuration α-carbon in L-cysteine in the R/S convention?

16.16. The following monoamines serve important functions as neurotransmitters within the central nervous system. Which amino acids are these compounds derived from?

dopamine   histamine   serotonin

16.17. Draw the structure of tripeptide His-Tyr-Gln. Identify the N-terminal, C-terminal amino acids, the protein backbone, and the peptide bonds in your drawing. Which functional groups are present in this molecule?

16.4. NUCLEOSIDES, NUCLEOTIDES, AND NUCLEIC ACIDS

Deoxyribonucleic acid (DNA) and ribonucleic acid (RNA) are large biopolymeric molecules responsible for storage and processing of genetic information. **Nucleic acids** are made of nucleotide units connected together to form a long chain. A **nucleotide** unit is composed of a nucleoside and a phosphate group. A **nucleoside** consists of a cyclic aldopentose linked via a β-N-glycosidic bond to a heterocyclic amine (a nucleobase). The cyclic aldopentose unit in

DNA is β-D-2-deoxyribofuranose and RNA is β-D-ribofuranose. Nucleic acids contain five different nucleobases: adenine, guanine, cytosine, thymine, and uracil (Figure 16.16).

Cyclic aldopentose units of nucleic acids:

β-D-2-Deoxybofuranose (present in DNA)

β-D-Rybofuranose (present in RNA)

Nucleobases:

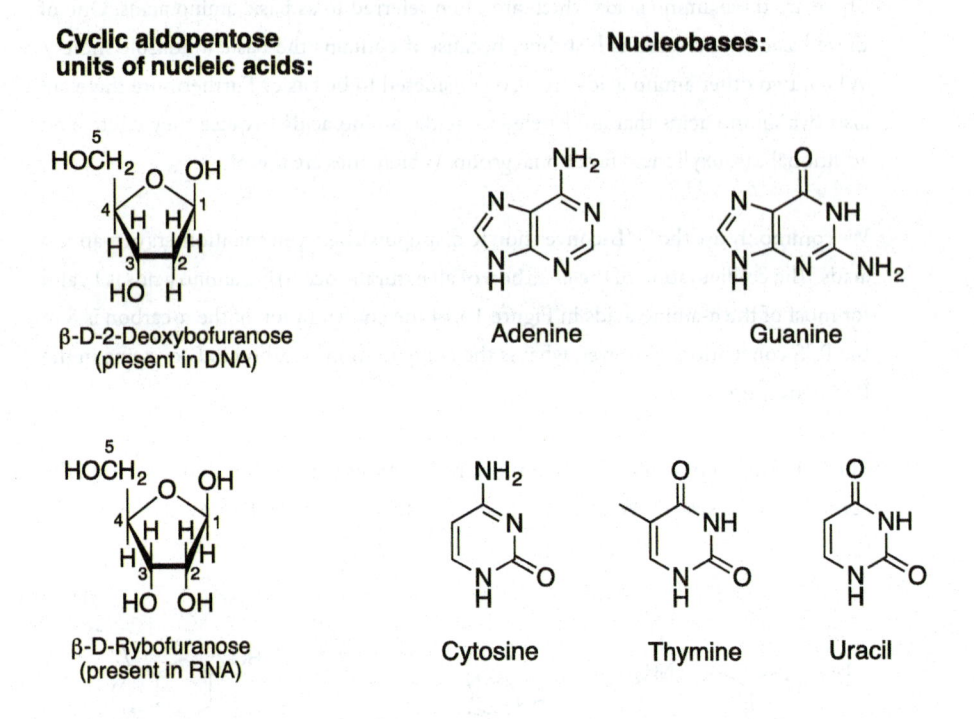

Adenine

Guanine

Cytosine

Thymine

Uracil

FIGURE 16.16. Small biomolecules present in nucleic acids.

Nucleosides are β-N-glycosides formed from β-D-2-deoxyribofuranose or β-D-ribofuranose and a nucleobase (Figure 16.17). A nucleotide is a phosphate ester of a nucleoside at carbon number 5. Nucleotides in a nucleic acid are connected by phosphate links as shown in Figure 16.18.

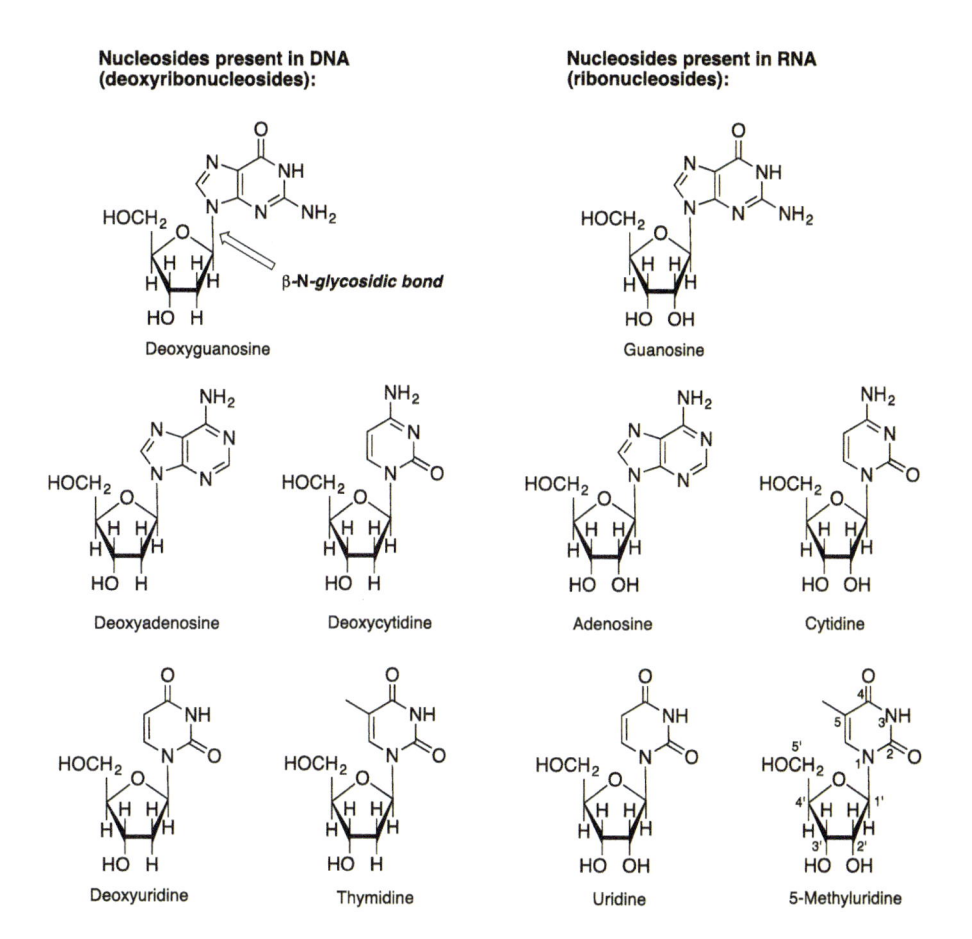

FIGURE 16.17. Nucleosides.

The sequence of nucleotides in a chain is indicated by nucleobases using the abbreviations A (adenine), G (guanine), C (cytosine), T (thymine), and U (uracil). An example of a small fragment of DNA, a trinucleotide (TGA), is shown in Figure 16.18. Actual DNA molecules consist of two polymer strands, bound together as a double helix by hydrogen bonds between complementary pairs of bases, A with T and C with G. DNA is the largest biopolymer containing up to 2.5 million nucleotide pairs. When uncoiled, a single DNA molecule can be up to 8.5 cm long. The human genome, responsible for storage of genetic information in human cells, contains 23 DNA molecules. Unlike DNA, RNA is a much smaller, single-stranded molecule. RNA molecules serve as transmitters and processors of genetic information in cells. For example, messenger RNA (mRNA) transfers genetic information from DNA and functions as the template for protein synthesis.

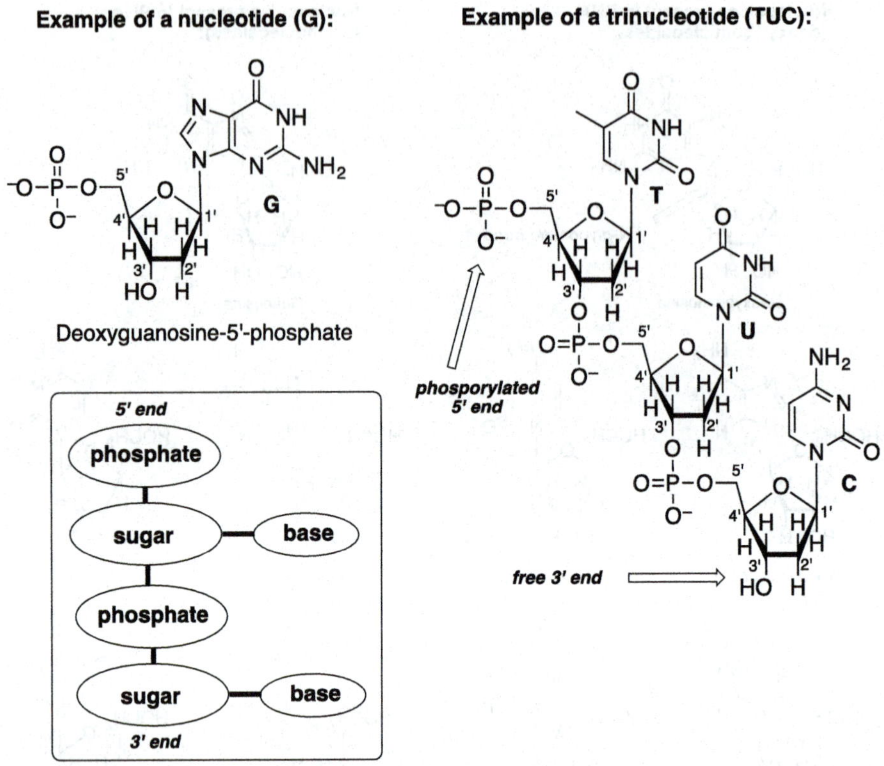

Example of a nucleotide (G):

Deoxyguanosine-5'-phosphate

Example of a trinucleotide (TUC):

phosporylated 5' end

free 3' end

5' end
phosphate — sugar — base
phosphate — sugar — base
3' end

FIGURE 16.18. Nucleotides and principal structure of a DNA strand.

Problems

16.18. Draw the structure of an RNA dinucleotide GC and identify its 5' and the 3' ends.

GLOSSARY

Chapter 1. Covalent Bonding and Structure of Molecules

Antibonding molecular orbitals (Section 1.7) The higher energy unoccupied molecular orbitals.

Atom (Section 1.1) A basic unit of a chemical element, and the smallest particle of a substance, that can exist by itself or be combined with other atoms to form a molecule.

Atomic mass (Section 1.1) The combined number of protons and neutrons in a nucleus.

Atomic number (Section 1.1) The number of protons in the nucleus of an atom.

Atomic orbitals (Section 1.1) Areas of space within electronic shells where electrons can be localized with the highest probability according to the laws of quantum mechanics.

Bond dipole moment (Section 1.1) A physical measure of bond polarity calculated as charge multiplied by bond length.

Bond strength (or **bond dissociation energy**, **BDE**) (Section 1.7) The amount of energy required to break a covalent bond.

Bonding molecular orbitals (Section 1.7) The lower energy molecular orbitals occupied by electron pairs.

Carbocations (Section 1.2) Molecular ions containing positively charged carbon atoms.

Condensed structure (Section 1.2) A simplified Lewis structure in which single bonds between elements are not shown.

Curved arrow (Section 1.6) An arrow used to show the movement of a pair of electrons from one position to another.

Dipole (Section 1.1) A pair of oppositely charged atoms connected by a polar covalent bond

Dipole-dipole attraction (Sections 1.1 and 1.5) Intermolecular attractive forces between polar molecules.

Dispersion forces (or **van der Waals forces**) Weak intermolecular attractive forces between nonpolar molecules due to temporary dipoles induced in adjacent atoms of molecules.

Electron shells (Section 1.1) Orbits occupied by electrons around an atom's nucleus.

Electronegativity (Section 1.1) A measure of the tendency of an atom to attract a bonding pair of electrons.

Formal charge (Section 1.2) The charge assigned to an atom in a molecule.

Functional groups (Section 1.3 and Table 1.3) Structural fragments within organic molecules responsible for their physical properties, chemical reactions, and general classification.

Hydrocarbons (Section 1.3) Compounds with molecular formulas that include only hydrogen and carbon atoms.

Hydrogen bonding (Section 1.5) Especially strong dipole-dipole attraction typical of molecules with a hydroxyl functional group (OH).

Ionic attractive forces (Section 1.5) Strong electrostatic attraction between opposite ions in ionic compounds.

Ionic compound (Section 1.1) Chemical compound consisting of an anion and a cation that are formed by full transfer of an electron from one atom to another.

Isotopes (Section 1.1) Atoms of the same element that contain equal numbers of protons but different numbers of neutrons in their nuclei.

IUPAC nomenclature (Section 1.3) Systematic naming of chemical compounds using names of functional groups in combination with the name of the parent carbon chain.

Lewis acids (Section 1.2) Uncharged molecules with six valence electrons on the central atoms that can accept an electron pair from a donor compound (Lewis base)

Lewis model of bonding (Section 1.1) When forming a molecule, the atoms of elements combine together either by sharing valence electrons, or by a full transfer of electrons from one to another, in order to reach the most stable electronic configuration, similar to that of a noble gas.

Lewis structures of atoms (Section 1.1) Symbols of elements in which dots (•) indicating the number of valence electrons.

Lewis structures of molecules (Section 1.2) Drawings that show the bonds between atoms of a molecule, the lone pairs of electrons, and positive or negative charges that may exist in the molecule.

Line-angle structure (Section 1.2) A simplified Lewis structure that shows only the chain of bonds between carbon atoms and not the actual symbols of attached C and H atoms.

Molecular dipole moment (Section 1.5) A physical measure of the polarity of a molecule, expressed in Debye units (μ).

Molecular formula (Section 1.2) The formula that shows the number and kinds of atoms in a molecule or chemical compound.

Molecular ions (Section 1.2) Charged chemical species composed of two or more covalently bonded atoms.

Molecular orbital (MO) theory (Section 1.7) A theory providing basic methodology for theoretical chemistry.

Molecule (Section 1.1) The smallest fundamental unit of a chemical compound responsible for its physical properties and chemical reactions.

Nonpolar covalent bond (Section 1.1) A bonding pair of electrons located exactly between two nuclei.

Nonpolar molecule (Section 1.5) A molecule with a dipole moment where $\mu = 0$.

Orbital hybridization (Section 1.7) A key concept of VB theory explaining the shape, reactivity, and electronic structure of organic compounds with single, double, and triple bonds.

Pi (π) bonds (Section 1.7) Bonds formed by the overlap of two lobes of one nonhybridized p orbital, side-by-side with two lobes of the p orbital on a second carbon.

Polar covalent bond (Section 1.1) A bonding pair of electrons shifted closer to the more electronegative atom.

Polar molecule (Section 1.5) A molecule with a dipole moment different from zero.

Resonance arrow (Section 1.6) A straight double-headed arrow showing the relationship of Lewis structures as the resonance contributing structures.

Resonance contributing structures (Section 1.6) Several Lewis structures with the nuclei in the same positions, but electrons in different location.

Resonance hybrid (Section 1.6) The real structure of a molecule created by mixing all contributing resonance structures.

Saturated hydrocarbons (Section 1.3) Hydrocarbons with the highest possible content of hydrogen and a general molecular formula C_nH_{2n+2}.

Sigma (σ) bonds (Section 1.7) Single bonds with a cylindrical shape along the axis between the two connected nuclei, formed by head-on overlapping of atomic orbitals.

sp hybridization (Section 1.7) A mixing of the 2s orbital and one 2p orbital of carbon, while two 2p orbitals stay unchanged, that explains the formation of a linear molecule.

sp² hybridization (Section 1.7) A mixing of the 2s orbital and two 2p orbitals of carbon, while the third 2p orbital stays unchanged, that explains the formation of a trigonal molecule.

sp³ hybridization (Section 1.7) A mixing of an electron from the 2s orbital with three electrons in the 2p orbitals of carbon that explains the formation of tetrahedral molecules.

Structural (or **constitutional**) **isomers** (Section 1.2) Compounds with the same molecular formula but differ in how atoms are connected.

Tetrahedral atom (Section 1.4) An atom with four bonds with bond angles of 109.5° and the overall shape of a tetrahedron.

Trigonal planar atom (Section 1.4) An atom with three bonds (single or double bonds) with bond angles of 120° and the overall shape of a triangle.

Units of unsaturation (or the **degree of unsaturation**, or **index of hydrogen deficiency**) (Section 1.3) The number of hydrogen molecules (H_2) that should be added to a molecule to produce a saturated hydrocarbon.
Unsaturated hydrocarbons (Section 1.3) Hydrocarbons with double or triple bonds.
Valence Bond (VB) theory (Section 1.7) A simplified version of MO theory that explains the structure and reactivity of organic compounds.
Valence electrons (Section 1.1) Electrons in the outer electron shell that can participate in bonding with other atoms.
Valence Shell Electron Pair Repulsion (VSEPR) theory (Section 1.4) Posits that repulsion between negatively charged bonds and valence shell electron pairs determine the three-dimensional geometry of molecules.

Chapter 2. Proton Transfer Reactions in Organic Chemistry

Acid dissociation constant (K_a) (Section 2.2) A quantitative measure of an acid's strength (acidity) in aqueous solution, defined as the proton transfer equilibrium constant K_{eq} multiplied by the molar concentration of water.
Basicity (Section 2.2) The tendency of a compound to act as proton acceptor in the acid-base equilibrium; quantitatively measured by the pK_a of the corresponding conjugate acid.
Brønsted–Lowry theory (Section 2.1) A general description of acid-base reactions defining acid (A–H) as a donor of protons and a base (:B⁻) as an acceptor of protons in a proton-transfer equilibrium.
Conjugate acid (Section 2.1) Species HB produced from a base B:⁻ as a result of a proton-transfer reaction.
Conjugate base (Section 2.1) Species A:⁻ produced from an acid A–H as a result of a proton-transfer reaction.
Electron-donating inductive effect (Section 2.3) A weak donation of electronic density away from an alkyl group due to the presence of less electronegative hydrogen atoms.
Electron-withdrawing inductive effect (Section 2.3) A shift of electronic density toward the electronegative substituent in the molecule.
Electrophile (Section 2.1) An acceptor of electrons.
Lewis definition of acids (Section 2.1) A Lewis acid is defined as any compound acting as an acceptor of a pair of electrons from a base.
Nucleophile (Section 2.1) An electron donor.
pK_a value (Section 2.2) A common measure of an acid's strength, defined as a logarithm of the acid dissociation constant with a negative sign.
Strong bases (Section 2.2) The most powerful acceptors of protons, which always act as a base in the acid-base equilibrium.

Strong inorganic acids (Section 2.2) The most powerful donors of protons, which always act as an acid in the acid-base equilibrium.

Chapter 3. Alkanes and Cycloalkanes

Alkanes (Section 3.1) Saturated hydrocarbons with a general molecular formula of C_nH_{2n+2}.

Alkyl group (Section 3.2) A fragment of alkane attached to the parent carbon chain, with a general molecular formula of C_nH_{2n+1}.

Angle strain (Section 3.4) Strain in cycloalkane due to deviation of the C-C-C angle in the ring from the normal tetrahedral angle (109.5°).

Axial bonds (Section 3.4) Bonds pointing straight up or straight down in chair conformation of cyclohexane.

Axial-axial (or diaxial) interaction (Section 3.4) Steric strain due to the spatial proximity of atoms or groups in axial positions of the chair conformation of cyclohexane.

Chain propagation (Section 3.6) A sequence of steps in which an initial radical reacts with a molecule to produce final products and regenerate the initial radical.

Chain termination (Section 3.6) A chemical reaction that ceases radical regeneration in the chain propagation process.

***cis* isomers** (Section 3.5) Cycloalkanes with two substituents on the same side of the ring.

Configuration of carbon atoms (Section 3.5) The spatial arrangement of substituents at carbon atoms.

Conformations (or **conformers**, or **rotamers**, or **conformational isomers**) (Section 3.3) Three-dimensional representations of the shapes of flexible molecules originating due to rotation about carbon-carbon single bonds.

Cycloalkanes (Section 3.4) Hydrocarbons with a ring of carbon atoms in their structure and a general molecular formula of C_nH_{2n}.

Equatorial bonds (Section 3.4) Bonds pointing away from the ring in chair conformation of cyclohexane.

Homolytic bond cleavage (Section 3.6) Even cleavage of a covalent bond to form two uncharged radicals.

Initiation step (Section 3.6) The initial formation of radical intermediates in a radical chain reaction.

Newman projection (Section 3.3) A view of a molecule along the C–C bond from front to back, with the front carbon represented by a dot and the back carbon as a circle.

Primary (1°) carbon atom (Section 3.1) A carbon atom bonded directly to one carbon and three hydrogen atoms.

Quaternary (4°) carbon atom (Section 3.1) A carbon atom bonded directly to four carbon atoms.

Radical chain reaction (Section 3.6) Reaction involving radical intermediates, consisting of three consecutive processes: 1) chain initiation; 2) chain propagation; and 3) chain termination.

Radicals (or **free radicals**) (Section 3.6) Chemical species that have unpaired valence electrons.

Reaction mechanism (Section 3.6) A detailed description of a chemical reaction showing the movement of electrons and the structures of intermediate products in each step.

Regioselective reactions (Section 3.6) Reactions that give predominantly one regioisomer (positional isomer) out of several possible isomers.

Secondary (2°) carbon atom (Section 3.1) A carbon atom bonded directly to two carbon and two hydrogen atoms.

Steric strain (Section 3.3) Strain due to the spatial proximity of atoms or groups in a molecule.

Strain (Section 3.3) In comparison with an unstrained molecule, the extra energy accumulated in a molecule due to structural distortions.

Substitution reactions (Section 3.6) Reactions resulting in the substitution of an atom or group with a different atom or group in a molecule.

Tertiary (3°) carbon atom (Section 3.1) A carbon atom bonded directly to three carbon and one hydrogen atom.

Torsional strain (Section 3.3) Strain in the eclipsed conformation due to the repulsion between the C–H (or C–C) bonds on adjacent carbons.

trans **isomers** (Section 3.5) Cycloalkanes with two substituents on opposite sides of the ring.

Chapter 4. Stereochemistry

Chiral center (or an **asymmetrical carbon atom**) (Section 4.2) A tetrahedral carbon atom with four different substituents.

Chiral molecules (Section 4.2) Molecules that cannot be superposed on their mirror image.

Configurational isomers (Section 4.1) Stereoisomers that cannot be interconverted by rotation around single bonds.

Diastereomers (Section 4.3) Stereoisomers that are not related as enantiomers.

Enantiomeric excess (ee) (Section 4.4) Numerical measure of the purity of an enantiomer.

Enantiomers (Section 4.2) Stereoisomers that are non-superposable mirror images.

Fischer projection (Section 4.3) A two-dimensional representation of a three-dimensional organic molecule by projection.

Meso compounds (Section 4.3) Achiral molecules with two or more chiral centers, characterized by the presence of internal symmetry.

R/S system for naming enantiomers (Section 4.2) The IUPAC nomenclature rules for assigning configuration of enantiomers based on the priority rules for substituents at the chiral center.

Racemate (racemic mixture) (Section 4.4) A mixture with an equal quantity of each enantiomer that does not exhibit optical activity.

Stereochemistry (Section 4.1) The study of molecules in three-dimensional space.

Stereoisomers (Section 4.1) Isomers with the same molecular formula and connectivity of atoms, but different orientation of atoms in space.

Chapter 5. Nucleophilic Substitution and ß-Elimination Reactions

E1 Elimination (Section 5.4) A two-step unimolecular mechanism of β-elimination involving initial formation of carbocations followed by proton elimination from the β-carbon of the carbocation.

E2 elimination (Section 5.4) A one-step bimolecular mechanism of β-elimination resulting in simultaneous removal of the proton at the β-carbon, formation of a double bond, and departure of a leaving group.

Good leaving groups (Section 5.1) The most stable anions of strong acids with the highest reactivity as a leaving group in nucleophilic substitution reactions.

Inversion of configuration (Section 5.3) A process resulting in the change of stereochemical configuration of an atom to the opposite configuration.

Nucleophilic substitution (Section 5.1) A reaction that replaces a suitable substituent (i.e., **leaving group**) at a carbon atom of an organic molecule (i.e., **substrate**) with an electron-donating reagent (i.e., **nucleophile**).

Polar aprotic solvents (Section 5.2.1) Solvents consisting of highly polar molecules incapable of forming hydrogen bonds to nucleophiles.

Poor leaving groups (Section 5.1) The less stable anions of weaker acids with low reactivity as a leaving group in nucleophilic substitution reactions.

Protic solvents (Section 5.2.1) Solvents (i.e., water and alcohols) that serve as sources of protons due to the presence of a relatively acidic hydroxyl function.

Racemization (Section 5.3) A process resulting in the formation of a pair of enantiomers (racemate) in the reaction of a chiral, nonracemic substrate.

Regioisomers (Section 5.4) Constitutional isomers that differ by the position of a substituent or a double bond in the parent chain.

S_N1 mechanism (Section 5.2) A two-step unimolecular mechanism of nucleophilic substitution in which the leaving group leaves first, forming a carbocationic intermediate, which then combines with a nucleophile in the second step of the reaction.

S_N2 mechanism (Section 5.2) Bimolecular mechanism of nucleophilic substitution involving two reactants (i.e., the nucleophile and the substrate) in the transition state.

Solvolysis reactions (Section 5.2.2) Nucleophilic substitution reactions in protic solvents acting as nucleophiles.

Stereoselective reactions (Section 5.3) Reactions that selectively produce a single stereoisomer as the product.

Strong nucleophiles (Section 5.1) Anionic species that are good donors of electrons in nucleophilic substitution reactions.

Weak nucleophiles and **moderate nucleophiles** (Section 5.1) Neutral molecules that are weaker donors of electrons in nucleophilic substitution reactions.

Zaitsev's rule (Section 5.4) Predominant formation of the more highly substituted alkenes in elimination reactions.

β-Elimination reaction (Section 5.1) A reaction of alkyl halide that forms alkene due to the elimination of halogen and hydrogen at adjacent carbon atoms.

Chapter 6. Alkenes

Alkenes (Section 6.1) Unsaturated hydrocarbons with a double bond in their structure and a general molecular formula of C_nH_{2n}.

Allylic halogenation (Section 6.6.1) Substitution of hydrogen with halogen at the allylic position via radical mechanism.

***anti*-addition** (Section 6.3) The stereoselective addition producing a *trans* product.

Catalytic hydrogenation (Section 6.5.4) Reduction of alkenes to the corresponding alkanes by addition of molecular hydrogen in the presence of a transition metal catalyst.

***cis*-Alkenes** (Section 6.1) Unbranched alkenes with two substituents on the same side of the double bond.

Dihydroxylation reaction (Section 6.5.1) Conversion of an alkene to a diol by introducing two hydroxyl groups (OH).

E/Z-Nomenclature (Section 6.1) IUPAC nomenclature rules assigning Z configuration to alkenes with high priority substituents on the same side of the double bond, and E configuration to alkenes with the high priority substituents on opposite sides of the double bond.

Electrophilic addition (Section 6.2) The addition reaction of an electrophilic reagent to a double or a triple bond.

Epoxidation reaction (Section 6.5.2) Conversion of an alkene to epoxide (oxirane).

Hydroboration (Section 6.4) A reaction involving the addition of borane, BH_3, to alkenes.

Markovnikov's rule (Section 6.2) Regioselective addition of an acid, HX, to an unsymmetrical alkene yields products with hydrogen added to the carbon with more hydrogen atoms, and the halogen atom, X, added to the carbon with fewer f hydrogen atoms.

Oxidation (Section 6.5) Adding an oxygen atom (or any element more electronegative than carbon) to a carbon atom, or removing a hydrogen atom from a carbon atom.

Ozonolysis reaction (Section 6.5.3) Oxidative cleavage of an alkene double bond into two carbonyl groups.

Polymerization (Section 6.6.3) The process of building a polymer from monomeric molecules.

Polymers (Section 6.6.3) Large molecules consisting of repeating units of small molecules called monomers.

Radical addition (Section 6.6.2) Addition of HBr (or other reagents) to alkenes by radical chain mechanism in the presence of peroxides as radical initiators.

Reduction (Section 6.5) The addition of a hydrogen atom (or any element that is more electronegative than carbon) or removal of an oxygen atom.

syn-**addition** (Section 6.4) Stereoselective additions producing *cis* products.

trans-**Alkenes** (Section 6.1) Unbranched alkenes with two substituents on opposite sides of the double bond.

Chapter 7. Alkynes

Acetylide (or **alkynyl anion**) (Section 7.2) A conjugate base of terminal alkyne.

Alkynes (Section 7.1) Unsaturated hydrocarbons with a triple bond and the general molecular formula of C_nH_{2n-2} with two units of unsaturation.

Internal alkynes (Section 7.1) Alkynes with a triple bond within the parent chain.

Organic synthesis (Section 7.5) A subdiscipline of organic chemistry dealing with the construction of complex organic molecules from simple organic and inorganic compounds.

Retrosynthetic analysis (Section 7.5) Development of the best synthetic route to a target molecule going backwards, step-by-step, starting from the target molecules and tracing back to the smaller building blocks.

Tautomeric equilibrium (or **tautomerization**) (Section 7.3.3) An equilibrium between tautomers.

Tautomers (Section 7.3.3) Constitutional isomers that differ by the position of a hydrogen atom and a double bond.

Terminal alkynes (Section 7.1) Alkynes that have a triple bond at the end of the chain.

Chapter 8. Alcohols

Acid-Catalyzed Dehydration (Section 8.5) Conversion of alcohols to alkenes by treatment with concentrated sulfuric acid.

Alcohols (Section 8.1) Organic compounds with a hydroxyl functional group, (-OH) connected to sp^3 hybridized carbon.

Alkoxides (Section 8.2) The conjugated bases of alcohols.

Alkyl tosylates (ROTs) (Section 8.6.2) Esters of tosic acid used as highly reactive substrates in nucleophilic substitution reactions, due to the excellent leaving group reactivity of the tosylate anion (TsO⁻).
Diols (or **glycols**) (Section 8.1) Compounds with two hydroxyl groups.
p-toluenesulfonic acid (tosic acid, TsOH) (Section 8.6.1) $CH_3C_6H_4SO_2OH$, an important representative of sulfonic acids with high acidity, comparable to the acidity of sulfuric acid.
Primary (1°) alcohols (Section 8.1) Alcohols with a carbon atom attached to the carbon bearing the hydroxyl group (-OH).
Secondary (2°) alcohols (Section 8.1) Alcohols with two carbon atoms attached to the carbon bearing the hydroxyl group.
Tertiary (3°) alcohols (Section 8.1) Alcohols with three carbon atoms attached to the carbon bearing the hydroxyl group.
Thiols (RSH) (Section 8.6.1) Sulfur analogs of alcohols.
Williamson ether synthesis (Section 8.3) Preparation of ethers by reacting 1° haloalkanes (RCH_2X) or halomethanes (CH_3X) with metal alkoxides.

Chapter 9. Spectroscopy of Organic Compounds

Chemical shifts (Section 9.2.2) The position of atomic signals in the NMR spectrum, measured in parts per million (ppm).
Coupling constant (Section 9.2.2) The distance between individual lines in a NMR signal.
Heteronuclear coupling Section 9.2.2) Magnetic interaction of ^{13}C carbon and ^1H hydrogen nuclei.
Infrared spectroscopy (IR spectroscopy) (Section 9.1) Spectroscopy dealing with the infrared region of the electromagnetic spectrum.
Integral (Section 9.2.2) NMR signal intensities that are directly proportional to the number of nuclei emitting the signal.
Mass spectrometry (Section 9.3) Method of measuring the mass of a molecule and its fragments using ionization of a sample in vacuum, and separation of ions in an external magnetic field according to their mass.
Nuclear Magnetic Resonance (NMR Spectroscopy) (Section 9.2) Spectroscopy dealing with the nuclei of ^1H or ^{13}C absorbing electromagnetic radiation in the presence of strong external magnetic field.
Spectroscopy (Section 9.1) Measuring and interpreting spectra arising from the interaction of electromagnetic radiation with matter in order to determine the structure of organic molecules.
Spin-spin splitting (Section 9.2.2) Splitting of NMR signals due to the magnetic interaction of nonequivalent nuclei.

Chapter 10. Organometallic Compounds and Transition Metal Catalysis

Alkene metathesis (Section 10.2.3) An organic reaction between two molecules of alkene resulting in exchange of the carbons by breaking and rebuilding the carbon-carbon double bonds.

Carbanions (Section 10.1) An anionic species bearing a formal negative charge on a carbon atom.

Carbenes (Section 10.2.2) Important intermediates in organic reactions with an uncharged carbon atom, two substituents, and an unshared electron pair.

Coupling reactions (Section 10.1) Reactions of organic compounds of transition metals with alkyl halides that result in formation of new carbon-carbon bonds.

Organometallic compounds (Section 10.1) Organic compounds with a carbon-metal bond in the molecule.

Oxidative addition (Section 10.1) Addition of a reactant to metal resulting in the formation of an oxidized metal species.

Chapter 11. Aldehydes and Ketones

Acetal (Section 11.7) The product of the reaction of aldehyde or ketone with two molecules of alcohol.

Aldehydes (Section 11.1) Compounds RHC=O with at least one hydrogen bonded to the carbon of the carbonyl group.

Aldol condensation (Section 11.10) Nucleophilic addition of enolate anions to a carbonyl group of aldehydes or ketones.

Carbonyl (Section 11.1) Functional group composed of a carbon atom double-bonded to an oxygen atom, C=O.

Deuterium exchange (Section 11.9) The exchange of hydrogen atoms of acidic C–H bonds with deuterium (D = ^{2}H isotope) by treatment with D_2O.

Enol (Section 11.9) The tautomer of a carbonyl compound formed by protonation of the oxygen atom of an enolate anion.

Enolate anion (Section 11.9) The anion formed by deprotonation of a carbonyl compound at the α-position.

Hemiacetal (Section 11.7) The product of nucleophilic addition of one molecule of alcohol to the carbonyl group of aldehyde or ketone.

Hydride anion (Section 11.4) A negatively charged hydrogen atom, H:$^-$.

Ketones (Section 11.1) Compounds R_2C=O with two alkyl substituents bonded to the carbon of the carbonyl group.

Nucleophilic addition (Section 11.2) A reaction resulting in the addition of a nucleophile to a carbon atom of an organic molecule.

Protective group (Section 11.7) An unreactive group, such as acetal, temporarily placed on a carbonyl group or other reactive functional group, and removed after a sequence of chemical reactions involving other functional groups in a multifunctional molecule.

Tollens' reagent (Section 11.8) An aqueous solution of silver nitrate and ammonia used for selective oxidation of aldehyde groups in the presence of other sensitive functional groups, such as alcohol.

Wittig reaction (Section 11.5) The reaction of phosphonium ylides with carbonyl compounds, yielding alkenes.

Wolff-Kishner reduction (Section 11.6) Conversion of carbonyl functionality (C=O) into a methylene group (CH_2) by heating aldehyde or ketone with hydrazine (H_2NNH_2) and potassium hydroxide.

Ylide (Section 11.5) An uncharged molecule containing a negatively charged carbon atom directly bonded to a positively charged atom of phosphorus or other element.

α-position in carbonyl compounds (Section 11.9) Carbon atoms (together with attached hydrogens) adjacent to the carbonyl group.

Chapter 12. Carboxylic Acids and their Derivatives

Acetoacetic ester synthesis (Section 12.7) A sequence of condensation-decarboxylation reactions based on acetoacetic ester.

Acyl group (Section 12.1) A common structural fragment, RCO, present in carboxylic acids and acid derivatives.

Carboxylic acids (Section 12.1) Compounds RCO_2H with a carboxyl functional group (CO_2H) in their structure.

Claisen condensation (Section 12.7) Nucleophilic acyl substitution reactions of enolate anions generated from ester with another molecule of ester producing b-keto esters.

Decarboxylation (Section 12.7) Elimination of carbon dioxide from carboxylic acid.

Fischer esterification (Section 12.4.3) Preparation of esters by reaction of carboxylic acids with alcohols in the presence of catalytic amounts of a strong acid.

Malonic ester synthesis (Section 12.7) A sequence of condensation-decarboxylation reactions based on malonic ester.

Nucleophilic acyl substitution (Section 12.3) Reactions of acids or acid derivatives resulting in the replacement of a substituent (i.e., the leaving group) at the carbon atom of an acyl group with a nucleophile.

Saponification (Section 12.4.3) Hydrolysis of esters by an aqueous base.

Chapter 13. Non-cyclic Conjugated Systems

Conjugate addition (Section 13.2) Addition of an electrophilic or nucleophilic reagent to atoms 1 and 4 of a conjugated system of four atoms.

Conjugated dienes and polyenes (Section 13.1) Molecules of hydrocarbons with alternating single and multiple bonds in their structure.

Conjugated system (Section 13.1) A system of connected sp^2 or sp hybridized atoms of carbon, nitrogen, or oxygen bearing p orbitals on each atom.

Cumulated dienes (Section 13.1) Hydrocarbons with adjacent double bonds sharing sp hybridized carbon atoms.

Diels-Alder reaction (Section 13.4) A cycloaddition reaction between a conjugated diene and a substituted alkene or alkyne, termed a dienophile, producing substituted cyclohexene or 1,4-cyclohexadiene.

Michael addition (Section 13.3) The addition of enolate anions to a β-carbon atom of an α,β-unsaturated carbonyl compound.

Chapter 14. Benzene and Aromatic Compounds

Aromatic compounds (Section 14.1) Conjugated cyclic systems of sp^2 hybridized atoms with completely filled bonding π-orbitals.

Aromaticity (Section 14.1) A theoretical concept providing an explanation for the increased stability and lower reactivity of aromatic compounds.

Benzylic carbon (benzylic position) (Section 14.6) The alkyl carbon adjacent to the aromatic ring.

Electrophilic aromatic substitution (Section 14.3) A reaction of benzene with an electrophilic reagent, resulting in the replacement of a hydrogen atom with the electrophile.

Friedel-Crafts reactions (Section 14.3.4) Electrophilic alkylation or acylation of benzene with alkyl chlorides or acyl chlorides in the presence of $AlCl_3$ as a catalyst.

Heterocyclic aromatic compounds (or **aromatic heterocycles**) (Section 14.1) A completely conjugated, aromatic cyclic system of sp^2 hybridized atoms that includes nitrogen, oxygen, or sulfur atoms, along with carbons.

Hückel's Rule (Section 14.1) A cyclic, planar, completely conjugated system of sp^2 hybridized atoms is aromatic if it is occupied by 2, 6, 10, 14, 18, ... (4n + 2 where *n* is zero or any positive integer) p electrons.

Meta-directing substituents (m-directors) (Section 14.5) Substituents in the phenyl ring that direct electrophilic aromatic substitution reactions to meta-positions.

Nucleophilic aromatic substitution (Section 14.5) Replacement of a leaving group (usually, atom of halogen) in an aromatic substrate with a nucleophile.

Ortho, para-directing substituents (o,p-directors) (Section 14.4) Substituents in the phenyl ring that direct an electrophilic aromatic substitution reaction to ortho- and para-positions.

Chapter 15. Amines

Alkylamines (Section 15.1) Alkylamines (or aliphatic amines) that have only alkyl groups connected to nitrogen.

Amines (Section 15.1) Organic nitrogen compounds derived from ammonia (NH_3) by replacing one, two, or three hydrogen atoms with organic substituents.

Arylamines (Section 15.1) Arylamines (or aromatic amines) have at least one phenyl ring directly connected to a nitrogen atom.

Balz-Schiemann reaction (Section 15.5.3) Treating a diazonium salt with tetrafluoroboric acid (HBF_4) and heating the dried diazonium tetrafluoroborate salt to synthesize aryl fluorides.

Diazotization reaction (Section 15.5) Synthesis of aryldiazonium salts ($ArN_2^+X^-$) from primary arylamines, such as aniline, by treatment with sodium nitrite ($NaNO_2$), in the presence of a strong acid.

Gabriel synthesis (Section 15.3) Synthesis of primary amines using a nucleophilic substitution reaction of alkyl halides with potassium phthalimide.

Heterocyclic amines (Section 15.1) Amines with a nitrogen atom as part of a cycle.

Hofmann elimination (Section 15.4) The elimination of tertiary amine from quaternary ammonium salt, producing alkene.

Primary (1°) amines (Section 15.1) Amines that have one carbon atom attached to the nitrogen atom, RNH_2.

Quaternary ammonium salts (Section 15.1) Ionic compounds containing positively charged nitrogen atoms with four organic groups, $R_4N^+X^-$.

Sandmeyer reaction (Section 15.5.1) Reaction of aryldiazonium salts with cuprous chloride (CuCl), cuprous bromide (CuBr), and cuprous cyanide (CuCN), yielding products in which the diazonium group is replaced by a chlorine, bromide or cyano group.

Secondary (2°) amines (Section 15.1) Amines that have two carbon atoms attached to the nitrogen atom, R_2NH.

Tertiary (3°) amines (Section 15.1) Amines that have three carbon atoms attached to the nitrogen atom, R_3N.

Chapter 16. Introduction to Biomolecules

Amino acid (Section 16.3) A simple biomolecule containing a basic amino group and an acidic carboxylic functional group.

Anomeric carbon (Section 16.2.2) The hemiacetal carbon in furanose or pyranose.

Anomers (Section 16.2.2) Diastereomeric furanoses, or pyranoses, with opposite configuration at the hemiacetal carbon (the anomeric carbon).

Carbohydrates (or **saccharides**) (Section 16.2) Biological molecules with the empirical formula $C_m(H_2O)_n$.

Disaccharides (Section 16.2.4) Two units of monosaccharides joined together by an acetal link (i.e., glycosidic bond).

Furanose (Section 16.2.2) A five-membered cyclic monosaccharide.

Glycoside (Section 16.2.3) A cyclic carbohydrate with an acetal functional group.

Isoelectric point (Section 16.3) The pH value at which an amino acid exists in a zwitterionic (i.e., diionic) form and carries no net electrical charge.

Lipids (Section 16.1) Biomolecules with low solubility in water that can be isolated from an organism by extraction with a nonpolar organic solvent.

Monosaccharides (or simple sugars) (Section 16.2) Carbohydrates that have the chemical structure of polyhydroxy aldehydes (called aldoses) or polyhydroxy ketones (ketoses) with general formula $C_n(H_2O)_n$ where n is most commonly 3 to 6.

Mutarotation (Section 16.2.2) The process of interconversion between β-D-glucopyranose and α-D-glucopyranose via intermediate formation of the open-chain form of D-glucose (monitored by the change of optical rotation of the solution).

Nonreducing sugar (Section 16.2.3) A carbohydrate that cannot reduce silver salt to silver metal.

Nucleic acids (DNA and RNA) (Section 16.4) Large biopolymeric molecules made of nucleotide units, responsible for storage and processing of genetic information.

Nucleoside (Section 16.4) A unit consisting of a cyclic aldopentose linked via b-N-glycosidic bond to a heterocyclic amine (i.e., a nucleobase).

Nucleotide (Section 16.4) A building unit of nucleic acids composed of a nucleoside and a phosphate group.

Peptides (Section 16.3) Proteins formed from less than about 50 amino acid units.

Polysaccharides (Section 16.2) Complex carbohydrates consisting of several simple sugars joined together by an acetal link (i.e., glycosidic bond).

Prostaglandins (Section 16.1) A group of lipids derived biologically from fatty acids, containing 20 carbon atoms, and including a 5-carbon ring.

Proteins (Section 16.3) Biological polymers formed from amino acid units.

Pyranose (Section 16.2.2) A six-membered cyclic monosaccharide.

Reducing sugar (Section 16.2.3) A carbohydrate that can act as a reducing agent.

Steroids (Section 16.1) A group of lipids based on a tetracyclic skeleton of three 6-membered rings joined to one 5-membered ring.

Terpenoids (Section 16.1) Various lipids derived biosynthetically from the 5-carbon molecule of isopentenyl diphosphate.

Triacylglycerols (or **triglycerides**) (Section 16.1) The triesters derived from glycerol (1,2,3-propanetriol) and three long-chain carboxylic acids (fatty acids).

CPSIA information can be obtained
at www.ICGtesting.com
Printed in the USA
LVOW02s2021300817
547001LV00003B/5/P

9 781634 878975